D0623041

DISCARDED

The Copernican Achievement

Published under the auspices of the
CENTER FOR MEDIEVAL AND RENAISSANCE STUDIES
University of California, Los Angeles

Contributions of the

UCLA CENTER FOR MEDIEVAL AND RENAISSANCE STUDIES

1: Medieval Secular Literature: Four Essays. Ed. William Matthews

2: Galileo Reappraised. Ed. Carlo L. Golino

3: The Transformation of the Roman World. Ed. Lynn White, jr.

4: Scientific Methods in Medieval Archaeology. Ed. Rainer Berger

5: Violence and Civil Disorder in Italian Cities, 1200–1500. Ed. Lauro Martines

6: The Darker Vision of the Renaissance: Beyond the Fields of Reason. Ed. Robert S. Kinsman

7: The Copernican Achievement. Ed. Robert S. Westman

UCLA CENTER FOR
MEDIEVAL AND RENAISSANCE STUDIES
CONTRIBUTIONS: VII

The Copernican Achievement

Edited by

ROBERT S. WESTMAN

UNIVERSITY OF CALIFORNIA PRESS

BERKELEY • LOS ANGELES • LONDON 1975

University of California Press
Berkeley and Los Angeles, California
University of California Press, Ltd.
London, England

ISBN: 0-520-02877-5
Library of Congress Catalog Card Number: 74-22974
Printed in the United States of America

For Joseph, Claire, Walter,

and Robin

Contents

Preface

On November 1-2, 1973, a symposium was held at the University of California at Los Angeles, entitled, "The Copernican Achievement," of which the papers in this volume represent the published results. The conference's four sessions were chaired, respectively, by: John G. Burke (U.C.L.A.), Edward Rosen (City University of New York), Julian Schwinger (U.C.L.A.) and Robert G. Frank, Jr. (U.C.L.A.). The first session was preceded by an introduction from David S. Saxon, Executive Vice-Chancellor of U.C.L.A.

Few who attended the last session will forget the animated exchange, sometimes heated, sometimes witty, between the late Imre Lakatos and Stephen Toulmin. It was with great sorrow that we learned of Lakatos' untimely death in February, 1974 only a short time after he had made a few further revisions in the paper which appears in this volume. The clarity, simplicity and brilliance with which he expressed his rarely uncontroversial views will be greatly missed by all who work in the history and philosophy of science.

In a year when Copernicus celebrations were held all over the world, it was not easy to convince the customary financial foundations that still another symposium—and especially one characterized by so many young contributors—was warranted. It is with special gratitude, therefore, that I wish to thank the University of California at Los Angeles for its recognition of an important event and its generous financial support of a serious scholarly gathering. The conference from which this book emanates was funded by the U.C.L.A. Department of History, College of Letters and Sciences, Office of the Chancellor and the Committee on Public Lectures.

For preparing the index, I am especially indebted to Christe Whitaker and also to Bruce and Barbara Moran.

Finally, I wish to express my deepest thanks and appreciation to John G. Burke, without whose wise advice and never-failing support in organizing the symposium, it would never have achieved the success that it did.

Los Angeles, California
November, 1974

Robert S. Westman

Preface

List of Contributors

ERNEST S. ABERS is Associate Professor of Physics at the University of California, Los Angeles, and, with Charles Kennel, joint author of *Physics and its Development* (1975); with Benjamin Lee, "Gauge Theories," *Physics Reports,* 1973; and other articles in elementary particle physics.

SALOMON BOCHNER is Edgar Odell Lovett Professor of Mathematics at Rice University as well as Henry Burchand Fine Professor Emeritus of Mathematics at Princeton University. He is author of *The Role of Mathematics in the Rise of Science* (1966), *Eclosion and Synthesis* (1969); and "The Rise of Functions," *Rice University Studies,* 1970; and numerous mathematical studies.

WILLIAM H. DONAHUE, Tutor at St. John's College, Santa Fe, is author of "The Dissolution of the Celestial Spheres, 1595-1650," Cambridge University Doctoral Dissertation, 1972; and "A Hitherto Unreported Pre-Keplerian Oval Orbit," *Journal for the History of Astronomy,* 1973.

MAURICE A. FINOCCHIARO is Associate Professor of Philosophy at the University of Nevada, Las Vegas, and author of *History of Science as Explanation* (1973); "Galileo's Space-Proportionality Argument: A Role for Logic in Historiography," *Physis,* 1973; and several articles in the philosophy of science.

AMOS FUNKENSTEIN, Professor of History at the University of California, Los Angeles, has written *Heilsplan und natürliche Entwicklung* (1965) and, among other articles, "Descartes on Eternal Truths and the Divine Omnipotence," *Studies in History and Philosophy of Science,* 1975; as well as the forthcoming *Science and Imagination; The Share of Hypothetical Reasoning in the Formation of Early Modern Modes of Reasoning.*

OWEN J. GINGERICH is Professor in the Departments of Astronomy and History of Science at Harvard University and holds a joint appointment at the Smithsonian Astrophysical Observatory. He is editor of *The Nature of Scientific Discovery* (1975), and author of "The Role of Erasmus Reinhold and the Prutenic Tables in the Dissemination of the Copernican Theory," *Studia Copernicana*, 1973; and many articles in the history of astronomy.

JOHN L. HEILBRON, Professor of History at the University of California, Berkeley, is author of *H.G.J. Mosley. The Life and Letters of an English Physicist, 1887-1915* (1974), and, forthcoming, *A History of Electricity in the 17th and 18th Centuries.* With P. Forman and S. Weart, he has co-authored "Physics circa 1900. Personnel, Funding and Productivity of the Academic Establishment," *Historical Studies in the Physical Sciences,* 1975; and numerous articles in the history of physics.

JANICE A. HENDERSON is Assistant Professor in the Department of History, Queens College, City University of New York, and is with A. Aaboe joint author of "The Babylonian Theory of Lunar Latitude and Eclipses according to System A," *Archives internationales d'histoire des sciences,* 1975. She has also prepared with S. Gibbs and D.J. de Solla Price *A Computerized Checklist of Astrolabes* (Yale University, private edition, 1973).

CHARLES F. KENNEL is Professor of Physics and a member of the Institute of Geophysics and Planetary Physics at the University of California, Los Angeles. In 1974-1975, he was Visiting Professor at the Ecole Polytechnique in Paris. He is author of the recent article "Cosmic Ray Acceleration by Pulsars," *Physical Review Letters,* 1974; and many articles in plasma and space physics.

PAUL W. KNOLL, Associate Professor of History at the University of Southern California, is author of *The Rise of the Polish Monarchy. Piast Poland in East Central Europe, 1320-1370* (1972), and "Poland as *Antemurale Christianitatis* in the Late Middle Ages," *The Catholic Historical Review,* 1974; and other articles.

IMRE LAKATOS was Professor of Logic with Special Reference to the Philosophy of Mathematics at the London School of Economics in the Department of Philosophy, Logic and Scientific Method until his untimely death in 1974. A posthumous edition of his writings will be forthcoming as *Proofs and Refutations and Other Essays in the Philosophy of Mathematics.*

PETER MACHAMER is Associate Professor in the Department of Philosophy at Ohio State University. With R. Turnbull, he has edited *Motion and Time, Space and Matter* (1975), and *Perception: Historical and Philosophical Studies* (1976). He has also written "Feyerabend and Galileo: The Interaction of Theories and the Reinterpretation of Experience," *Studies in History and Philosophy of Science*, 1973; and other studies in the history and philosophy of science.

NICHOLAS H. STENECK is Assistant Professor of History at the University of Michigan, and is editor of *Science and Society: Past, Present, and Future* (1975); author of *Science and Creation,* forthcoming; "Albert the Great on the Classification and Localization of the Internal Senses," *Isis*, 1974.

NOEL M. SWERDLOW is Associate Professor of History at The University of Chicago, and author of "The Derivation and First Draft of Copernicus' Planetary Theory. A Translation of the Commentariolus with Commentary," *Proceedings of the American Philosophical Society*, 1973. He is currently working on a translation of and commentary on Book III of Copernicus' *De revolutionibus*; Books III and VII of Regiomontanus' *Epitome of the Almagest*; and Erasmus Reinhold's commentary on Book III of *De revolutionibus.*

STEPHEN E. TOULMIN is Professor of Social Thought and Philosophy in the Committee on Social Thought at The University of Chicago. His books include *Philosophy of Science* (1953); *The Uses of Argument* (1958); *Foresight and Understanding* (1961). With June Goodfield he has written *The Fabric of the Heavens* (1961); *The Architecture of Matter* (1963); *The Discovery of Time* (1965). He is also author of *Human Understanding, Vol. 1* (1972) and has written with Allan Janik, *Wittgenstein's Vienna* (1973). His recent articles include "The Alexandrian Trap," *Encounter*, 1974.

ROBERT S. WESTMAN, Associate Professor of History at the University of California, Los Angeles, is author of "Kepler's Theory of Hypothesis and the 'Realist Dilemma'," *Studies in History and Philosophy of Science*, 1972; "The Comet and the Cosmos: Kepler, Mästlin and the Copernican Hypothesis," *Studia Copernicana*, 1972; and "The Melanchthon Circle, Rheticus and the Wittenberg Interpretation of the

Copernican Theory," *Isis*, 1975. He is currently completing a book entitled *Scientific Generations and Innovation: The Diffusion of the Copernican Theory, 1540-1650,* and, with Owen Gingerich, a study of the group of copies of *De revolutionibus* annotated by Tycho Brahe, to appear in *Centaurus*, 1976.

CURTIS A. WILSON is Dean of St. John's College, and author of *William Heytesbury: Medieval Logic and the Rise of Mathematical Physics* (1956); and, recently, "Newton and Some Philosophers on Kepler's 'Laws'," *Journal of the History of Ideas,* 1974; "The Inner Planets and the Keplerian Revolution," *Centaurus,* 1973; and numerous studies in the history of 16th-17th century astronomy.

BRUCE WRIGHTSMAN, Associate Professor of Religion and the History and Philosophy of Science at Luther College, is author of "Man: Manager or Manipulator of the Earth," *Dialog*, 1971, and, forthcoming, *Conflict and Consensus in Religion and Science.*

ELIE ZAHAR is Lecturer in Philosophy in the Department of Philosophy, Logic and Scientific Method at the London School of Economics. He is author of "The Development of Relativity Theory: A Case Study in the Methodology of Scientific Research Programmes," University of London Doctoral Dissertation, 1973; "Why did Einstein's Programme supersede Lorentz's? (I)," *The British Journal for the Philosophy of Science,* 1973.

I

INTRODUCTION: THE COPERNICAN ACHIEVEMENT

Robert S. Westman

University of California, Los Angeles

1. *Different Senses of "Scientific Achievement"*

The concept of a scientific achievement is at once familiar and disconcertingly rich in meaning. It may connote, first of all, the actual construction of a major scientific theory. More broadly, it may refer to the subsequent evaluation of the meaning of that innovation by individuals and groups whose intellectual and social recognition is what initially elevates the theory to the status of a significant achievement. And yet, while recognition and use of a theory may lead to the belief that what has been discovered is *better than* what came before, such a belief is not sufficient to establish the rationality of a particular achievement. While all scientists assume certain standards of evaluation in their assessment of a new theory, the conscious study of what conditions are necessary for a theory to be considered rational has come to lie within the specialized province of philosophers. In short, the notion of a scientific achievement may have both descriptive and normative meanings and it may refer to an act of individual discovery as well as to the tradition of research founded upon that innovation.

The essays gathered in this volume to honor the 500th anniversary of the birth of Nicholas Copernicus well illustrate the diversity of issues that may be assembled under the rubric "Copernican Achievement." Finely honed, technical analyses of Copernicus' astronomy; studies of his early formative years at the university; the intriguing circumstances surrounding the publication of *De revolutionibus* and the virtually unknown career of Copernicus' "editor," Andreas Osiander; the complex and difficult question of how the new theory was received by academic natural philosophers and astronomers; the equally demanding philosophical problem of establishing universal conditions according to which the Copernican theory might be compared to the Ptolemaic as a *scientific* theory; the historical and conceptual preconditions for a scientific revolution of the sort inaugurated by Copernicus – such is the wide range of topics that are considered in this volume.

Our authors do not take their mandates lightly. Their immediate concerns are generally specialized; they are not given to bombastic eulogies and birthday rhetoric. Yet underlying the subtle complexities of their topics, a measure of the maturity of Copernican historiography, there reside a few, fundamental questions: How did Copernicus make his great discoveries? In the construction of a scientific theory, what relative importance should be assigned to philosophical preconceptions, empirical information, the deficiencies of earlier theories, accidental blunders or oversights, and the wider social context? Under what conditions was the new system actually accepted and according to what standards of rationality should it have been accepted?

The purpose of this introduction is to summarize some general themes in the papers and tentatively to suggest some of the relationships that may be found among them. The commentaries, many of them written by philosophers or practising scientists, extend or criticize the contentions of the principal papers frequently leading them into new and fruitful directions. While occasional references will be made to these contributions, no attempt will be made to provide an overall synopsis of them; they speak eloquently for themselves.

2. *The Theories of Copernicus*

Curtis Wilson's paper opens our volume with an attack on one of the most interesting questions in Copernicus scholarship: the derivation of the heliocentric theory. The reconstruction of the route to any scientific discovery is, at best, a matter of learned ignorance. Significant early clues and evidence of critical turning points do not always survive; the later recollections of the scientific creator are not always completely trustworthy, any more than are those of his friends. Then, there are questions of the initial problem area and the temporal order of subsequent conceptual moves. In the case of Copernicus, what concern was it that provided the initial context for his main discovery? Was it his dissatisfaction with Ptolemy's lunar theory?[1] his study of Regiomontanus' eccentric model of the second anomaly?[2] his perception of an annual element in the epicycles of the superior planets and the deferents of Venus and Mercury?[3] his reading of early Greek authors, such as Philolaus and Aristarchus? Or was the seminal problem area that suggested by Jerome Ravetz, in one of the most ingenious conjectures to appear in recent years: Copernicus' attempt to reform the calendar through a new model of the equinoctial precession?

Wilson takes Ravetz' claims as his point of departure. He shows first that Copernicus' interest in the calendar occurred nearly two decades after the time asserted by Ravetz. By itself, this dating would place Copernicus' work on precession *after* the period when most historians agree that he had already discovered his theory.[4] Wilson next exposes a basic conceptual flaw in Ravetz' argument. Ravetz had claimed that Copernicus was *logically* forced to accept the motion of the earth, in order to explain the precession of the equinoxes, because the precessional effect could only be produced by the motion of the celestial equator which, in turn, implied the earth's motion. This contention of Ravetz seems to be confirmed by Rheticus, no unzealous adherent of the new theory, who went so far as to maintain that the Copernican account

---

[1]See Aleksander Birkenmajer, "Comment Copernic a-t-il conçu et réalise son oeuvre?," *Organon,* 1936, *1*: 111-134; reprinted in *Etudes d'histoire des sciences en Pologne, Studia Copernicana* IV, (Warsaw: Ossolineum, 1972), pp. 589-611; and Edward Rosen, "Biography of Copernicus," in *Three Copernican Treatises,* 3rd ed., (New York: Octagon, 1971), p. 324.

[2]See Noel M. Swerdlow, "The Derivation and First Draft of Copernicus' Planetary Theory: A Translation of the *Commentariolus* with Commentary," *Proceedings of the American Philosophical Society,* 1973, *117*: 471 ff.

[3]J.L.E. Dreyer, *A History of Astronomy from Thales to Kepler,* 2nd ed., (New York: Dover, 1953), pp. 310-316.

[4]Edward Rosen believes that Copernicus may have had "the earliest glimmerings of the new astronomy" by 1509-10 with 1514 as the *terminus ante quem.* Noel Swerdlow accepts the latter date but is sceptical about the former: "I think there is insufficient evidence to determine how long before 1514 Copernicus developed his new planetary theory." (Swerdlow, *op. cit.,* p. 431.)

of precession could not be transferred to a geostatic reference frame. Wilson contests both Rheticus and Ravetz by showing that an observationally-equivalent geostatic precessional model can be constructed through the addition of a ninth and tenth spheres to the universe.

Wilson's reconstruction of Copernicus' heliocentric derivation begins with Copernicus' near-obsessive desire to replace Ptolemy's equants with circles moving about their own, proper centers. The result of Copernicus' adherence to this principle was a series of multi-epicyclic planetary models. With the addition of the annual epicycle to the superior planets (See Fig. 5), the stage was set for what Wilson conjectures to be Copernicus' critical move: the transformation of the annual epicycle into the orbit of the earth, a move which gave to this element a place and center of its own. For Copernicus, then, the principle of uniform circular motion is not a mere hypothesis. It is the primary component in our understanding of the world as a system of interrelated motions, a *machina mundi,* in contrast with Ptolemy's more positivistic view that *any* mechanisms are acceptable which save the phenomena. "The fact remains," concludes Wilson on a fascinating note, "that it was the man with the pedantic principle who instituted the revolutionary cosmology."

Copernicus the great, creative thinker, the conservative revolutionary who overthrew the old order, is a natural and compelling subject of investigation. Yet, had Copernicus merely presented an unarticulated Aristarchan heliocentric scheme, his impact on astronomical thought probably would have been no greater than that of Aristarchus or the French medieval natural philosopher, Nicole Oresme, or the sixteenth-century Italian writer, Celio Calcagnini—all of whom speculated on the motion of the earth. Copernicus made it possible for his insight to become an achievement, i.e. a basis for further research, precisely because he chose to defend it with arguments from technical mathematical astronomy. And, of the many problems faced by him, surely the irregular precession of the equinoxes was one of the most taxing and the most urgently demanding of resolution. For the motions of the moon and planets are all related to the apparent motion of the sun. A correct solar theory, however, involves determinations of the sidereal year, i.e. the period of the suns' revolution with respect to the stars, and the tropical year, i.e. its period of revolution with respect to the vernal equinox. Since Copernicus could not measure the sidereal year directly, he had to determine, first of all, the difference between the non-uniform tropical year and the uniform sidereal year. In short, he needed a model to account for the irregular precession of the equinoctial points.

Noel Swerdlow's richly detailed study of Copernicus' precession theory will allow those who can follow its technical pathways to obtain a fuller appreciation of Copernicus as a model builder. Swerdlow shows us Copernicus struggling to defend a theory of the precession from which *all* previous ancient observations might be deduced. Yet here lay the crux of Copernicus' problem: not all of the observations which he had from antiquity were of equal reliability and still he was forced to operate with them as though they were. In the face of these difficulties, Copernicus' strategies are both fascinating and instructive to watch. When necessary he will falsify observations to fit the predictions that he derives from his model. At times, his derivations are concealed from the printed text and only Swerdlow's reconstructions bring them into focus; moreover, at one point, when dealing with the period of the anomaly of the precession (1717 Egyptian years), Swerdlow finds that he cannot uncover Copernicus' derivation. These efforts at reconstruction are of considerable value not only for their intrinsic interest, but also because they will help to illuminate the comments of later sixteenth-century astronomers on the frequently read Book III of *De revolutionibus*.

Swerdlow' study also invites deeper consideration of the role of observations in theory construction. Einstein' remark, quoted by Owen Gingerich in his interesting commentary, is pertinent: ". . . it may be heuristically useful to keep in mind what one has actually observed. But on principle, it is quite wrong to try to found a theory on observable magnitudes alone." Observations are necessary but not sufficient to construct a new theory. The case of Copernicus would appear to suggest the importance of a mediating link between the final theory and the observations in the form of intermediate-level assumptions which provide the scaffolding on which the theory is built. In case of conflict between the model and observations, Copernicus will defend his hypothesis by altering or suppressing observations in order to preserve what he assumes to be true.

In view of what we have now seen of Copernicus' method, it is perhaps not surprising that those astronomers who read *De revolutionibus* carefully in the sixteenth century and who attempted to re-work Copernicus' computations, would have experienced some of the dismay and frustration that modern historians of science now encounter. As Janice Henderson shows in her study of Erasmus Reinhold's unpublished commentary on *De revolutionibus,* Reinhold was aware that the observations which Copernicus used to derive his parameters and tables could not be reproduced from the tables in his text. He therefore set about the long and tedious task of re-deriving all the parameters and re-computing all the tables. Henderson's paper is greatly concerned with Reinhold's treatment of the solar distance, the basic dimension of the solar system. Copernicus, not wishing to deviate too far from Ptolemy, modifies Ptolemy's value of 1210 terrestrial radii to obtain a figure of 1179^{tr}. But Reinhold, who normally follows Copernicus quite closely in other matter, rejects Copernicus' value in favor of 1208^{tr}, a solar distance more approximately equal to Ptolemy's. Here we find a revealing complementarity of scientific styles: Copernicus the brilliant innovator and careless calculator; Reinhold, with his obsessive concern for exact details. Yet both astronomers are anxious to stay close to Ptolemy's value. As Charles Kennel and Ernest Abers argue in their commentary, this situation was inevitable because a figure exceeding 1200^{tr} would have been inconsistent with error and known observations.

One is tempted to see in Reinhold the prototype of Thomas Kuhn's Normal Scientist, that is to say, a scientist who does not aim to invent new theories, but who accepts the theories and facts supplied by the paradigm and then attempts to "mop up" its errors and inconsistencies. While there is a certain truth in this characterization, Henderson is quite correct to point out that the construction of tables is entirely independent of the issue of heliocentricity. Indeed, in spite of Copernicus' revolutionary cosmological claims, his methodological procedures were still closely dependent upon ancient data and techniques—a fact which enabled astronomers like Reinhold to articulate and refine Copernicus' computations without uttering a word about the *true* place of the sun.

3. Preparatory Factors

Up to this point, the focus of our papers on Copernicus as a creative innovator has been decidedly technical. How could it be otherwise? To understand the creativity of any scientist requires an intimate acquaintance with the principles and methods of the science in which he worked and a deep sensitivity to critical points at which shifts in conceptualization occur. With Paul Knoll's useful study of the Arts Faculty at Cracow, however, we begin to change our focus to the preparatory conditions for Copernicus' innovation.

Knoll's purpose is a modest one. As he candidly acknowledge, it is perilously difficult to establish the way in which general cultural influences may affect the process of theory construction. His paper, which draws liberally upon generally inaccessible Polish scholarship, traces the evolution of the *studium Cracoviensis* from its shaky beginnings on the Bologna model in the early fourteenth century, through its re-founding on the Parisian model under Queen Jadwiga in the late fourteenth century, the radical fifteenth century increase in student population, the influence of humanism and, most importantly, the development of the arts curriculum. One significant theme in Knoll's account is the early trend at Cracow toward specialization in mathematics and astronomy. A special chair was founded for these subjects in 1405 by Nicholas Stobner and, in 1459, Marcin Krol established a chair in astronomy and astrology. But it is doubtful that one can take the latter as *conclusive* evidence of Aleksander Birkenmajer's claim that Cracow "emerged as the international center of astronomical education at the end of the middle ages." Even if the assertion has not been demonstrated, however, it is extremely valuable to have it as a working hypothesis for further investigation. Indeed, had Copernicus been trained initially at Oxford, Cambridge, Paris or Salamanca in the late fifteenth century, one wonders whether, in spite of his natural gifts, he would have been introduced to the field of astronomy at a level sufficiently deep to provoke in him the critical questions that eventually led to his conceptual breakthrough. And, as Nicholas Steneck persuasively argues in his commentary, the extent to which Copernicus' intellectual environment was discontinuous with earlier medieval patterns will depend ultimately upon our assessment of medieval science itself.

In contrast with Knoll's study, an entirely different kind of attempt to analyze the preparatory conditions for Copernicus' (and also, Galileo's) central discoveries, one based upon a novel evaluation of the role of hypothetical reasoning in Greek and Medieval natural philosophy, is provided by Amos Funkenstein's contribution.

Funkenstein's main thesis may be stated simply as follows: well-articulated theories, in the course of their own formulation, will frequently take pains to define and then to reject certain assumptions which threaten their own existence. Yet, by the very act of formulating what from the viewpoint of the original theory must be seen as an absurdity, the developing theory frequently lays the groundwork for its own destruction.

A particularly important example in this stage of theory construction, Funkenstein argues, was Aristotle's anticipation of the principle of inertia by his analysis of motion in a void. To conceive of a motion which is uniform, rectilinear and eternal required a great leap of imagination, a leap from observed, factual cases to an unobserved, imagined case which, while never occurring in nature, yet represented logically the *limiting* case of all factual cases. This act of inductive imagination, as the seventeenth century realized, required first an act of isolation, i.e. motion had to be abstracted from the particular cases in which it was observed. Aristotle, in the course of his refutation of the atomists' theory of motion, uses a technique which anticipates, but stops short of, Galileo's later method of "resolution." He imagines an "indefinite motion" by considering a series of cases where one variable (resistance) diminishes as a continuous function. Unlike Galileo, however, Aristotle distinguishes sharply between individual factual cases and the counterfactual case because he is determined to assure himself that what is physically absurd must also be logically and conceptually absurd. The revaluation of the status of the counterfactual condition will later form one important component of the Copernican Achievement with Galileo's recognition that mathematical models hold absolutely *only* under counterfactual conditions—with the consequent revolutionary abandonment of our expectation that mathematics can account for all phenomenal instances.

A second set of preparatory conditions for the Copernican Achievement were the medieval discussions of the axial rotation of the earth, the possibility of a plurality of worlds and the status of biblical statements about science. Funkenstein shows that after the fourteenth century, "the classical age of hypothetical reasoning," the impossibility of an axial rotation is raised to the status of a mere *improbability*. While the heliocentric hypothesis itself is barely mentioned in the Middle Ages, the even more adventurous issue of a plurality of worlds is no longer considered a logical impossibility. Since the uniqueness of our own universe is no longer a logical necessity, Ockham will argue that it cannot be a physical necessity either. A final way in which medieval thinkers prepared the way for a positive consideration of the earth's motion was in their exegetical arguments. Read literally, the Bible treats the earth's motion as an impossibility. But medieval thinkers were well aware of the principle of divine accommodation according to which Holy Scriptures "speak the language of men"; yet, when read allegorically, other meanings could be deciphered. Indeed, some will maintain that the Scriptures neither include nor imply the whole body of scientific knowledge. Even if Copernicus claimed that his theory was the assertion of an absurdity, then, it was, in medieval perspective, the assertion of an improbability.

4. *The Reception of the Copernican Theory*

The reception of any scientific theory involves two kinds of historical variables: internal factors, such as clarity and depth of articulation in the theory's initial statement, relationship between the new theory and other sciences; and external factors, such as the state of scientific communications, individual personalities, age, local political factors, national loyalties and resources. One of the most important points to keep in mind, however, is the following: since the full implications of a new theory are never manifest from the beginning, attitudes toward it will always be conditioned by particular interpretations of its significance. Indeed, even later "mature" readings of the theory are seen as "standard" (i.e. generally accepted) interpretations. Initially, however, these kinds of questions are frequently raised: What parts of the theory should be completely trusted? What statements in the new theoretical corpus are a good risk on which to base future research? What methodological status should be assigned to the theory's basic claims? The answers to such implicit questions as these determine whether the theory will be totally accepted, totally rejected, or used with a variety of amendments and provisos.

Copernicus' major work, *De revolutionibus orbium coelestium,* began its career with conflicting interpretations about how it should be read. Affixed to the preliminary part of the book is an anonymous "Letter to the Reader" which informs him that the hypotheses expounded within, "need not be true nor even probable; if they provide a calculus consistent with observations, that alone is sufficient. . . ." This methodological advice was contradicted by the main text itself where it was claimed that the earth actually moved with a three-fold motion. Most persons today with any knowledge of Copernicus and his work know that the true author of this letter was Andreas Osiander. His identity was first publicly divulged in 1609 by Kepler but extant copies of *De revolutionibus* show that many sixteenth century readers were already well aware of Osiander's editorial identity. What is not very well known today by most historians of science are the details of Osiander's life, his scholarly activities apart from his role as editor of Copernicus' opus and his motivations in prefixing the famous letter.

Bruce Wrightsman's study, "Andreas Osiander's Contribution to the Copernican Achievement," goes some way in throwing light on this hitherto obscure area. Osiander was a Lutheran reformer and preacher of influence, particularly in the city of Nuremberg, where he helped to shape the city's future along Reformation lines. As a zealous reformer and a strong anti-Papist, he was able to win the political support of certain important German princes although his combative personality provoked hostile attitudes toward himself among Catholics and Protestants alike. At the time of his death in 1552 he had been declared a heretic by Lutheran orthodoxy. It is Wrightsman's contention that Osiander's role in publishing Copernicus' work has been as much maligned by subsequent historiography as his religious positions were by his contemporaries.

Osiander's intentions, according to Wrightsman, were quite sincere: he believed that by adding *Ad Lectorem* he would protect Copernicus' work from possible adverse criticism. In his carefully worded letter, he nowhere says that Copernicus' theory is wrong but only that there is insufficient evidence to establish its truth. From a political standpoint, Wrightsman believes, this was a sound position for a controversial

Lutheran whose name on the new work would only have hindered acceptance of the theory. While this seems indisputable, it is unclear that Copernicus knew of and shared Osiander's fears. From a methodological standpoint, Wrightsman contends that Osiander makes a contribution to the philosophy of science by his assertion that all scientific knowledge, whether astronomical or physical, can only be *probable*. (Here, Funkenstein's analysis of medieval theories of evidence should be kept in mind.) Wrightsman believes, on the other hand, that Osiander was *strategically* wrong (from a scientific point of view) in not advocating a realist interpretation of the Copernican theory. The Osiander who emerges from this study is a far more interesting figure than the one to whom we are generally accustomed: a political-religious activist of sorts who helped the early acceptance of a new scientific hypothesis whilst shearing it of its profounder — and more controversial — implications.

If Osiander's role in the publication of *De revolutionibus* did not encourage a realist interpretation of the new hypotheses, by the 1570s this situation had begun to change. In the first place, serious attention was now paid to the realist claims of Book I in *De revolutionibus*. By itself, the methodological assumption that hypotheses in astronomy must also meet physical requirements was insufficient to cause astronomers or natural philosophers to convert to the Copernican theory. Both Tycho Brahe and Christopher Clavius, for example, were realists although neither was a Copernican. And yet, such a methodological position would at least force one to think carefully about the physical conditions under which an hypothesis in astronomy would be acceptable. For a variety of reasons—among which the Copernican theory can only be considered one contributing factor—the last quarter of the sixteenth century witnessed a complex, but definite, trend among natural philosophers away from the radical distinction between the heavens and earth. This interesting development is the subject of William H. Donahue's study, "The Solid Planetary Spheres in Post-Copernican Natural Philosophy."

Although we are accustomed to think of the destruction of the terrestrial-celestial dichotomy as the revolutionary work of such well-known figures as Copernicus, Tycho Brahe, Kepler and Galileo, it is Donahue's contention that the distinction was already suffering erosion in the writings of natural philosophers, both academic and non-academic. Thus, he will see Tycho Brahe's examination of the parallax of the Comet of 1577 as "inspired by a philosophical viewpoint which was not in the Aristotelian mainstream," namely, Paracelsus' concept of the heavens filled with fire. By the beginning of the seventeenth century, philosophers, whatever the cosmological arrangement to which they adhered, found it difficult to preserve the distinction between astronomy and physics. Few continued to believe in solid spheres; some watered down their claims for celestial solidity; others proclaimed their support for the property of celestial fluidity although this was a position that was either opposed or ignored by academics from 1595 to 1610. In general, Donahue finds a new and considerable optimism in this period about what could be known in the heavens.

The early seventeenth century also brought, according to Donahue, an intensified interaction within the universities between astronomers sympathetic to the claims of physics and natural philosophers hitherto neglectful of the work of astronomers. This cross-fertilization was particularly evident among the Jesuits. The old, but influential Clavius had already laid down a paradigm for the future union of astronomy and physics in the last edition of his *Commentary on the "Sphere"* where he called upon Peripetetics to explain Galileo's telescopic observations. Donahue characterizes the Jesuit response to newly discovered celestial phenomena as "renovative," i.e. reluctant to abandon old theories yet more willing than many of their contemporaries to accept the fact that new evidence should be seriously regarded. For the most part, the Jesuits will continue to maintain the incorruptibility of the heavens whilst relinquishing the property of solidity in favor of a fluid aether.

Non-Jesuits, such as Tycho Brahe's important student Longomontanus, were prepared to go further. While Aquinas had regarded the heavenly bodies as condensations of the aether, Longomontanus was ready to admit the alterability of the planets themselves. Gradually, even the Jesuit Christopher Scheiner abandoned the idea of incorruptibility. And by 1620, only a small minority of academics were willing to maintain the existence of solid spheres; those who did so between 1630 and 1650 had clearly left the main bandwagon. The heavens were no longer quintessential. They were either filled with a subtle, airy fluid or they were empty.

It appears from Donahue's work, therefore, that, due to a number of factors outside of astronomy itself, such as the revival of Stoic and Neoplatonic doctrines, late sixteenth century natural philosophers were beginning to conjecture more freely about the nature of celestial matter. At the same time, by the seventeenth century, the relationship between natural philosophers and mathematical astronomers was becoming mutually closer: to some extent, astronomers were stimulated to develop new hypotheses about the nature of their planetary devices; and, gradually, philosophers felt the necessity to explain the new celestial discoveries of those who studied the heavens empirically.

Yet initial reaction to the innovation of Copernicus among philosophers was either silent, critical or overly rhapsodic (Bruno)—in part, we may suggest, because for whatever reason, few had studied the technical sections of *De revolutionibus.* On the other hand, mid-sixteenth century mathematical astronomers—particularly within the German universities—were immediately drawn to Copernicus' models and tables while neglecting the cosmological claims of the new theory. Copernicus had predicted that the *mathematici* alone would appreciate what he was trying to do; yet, of these early *mathematici* only Copernicus' first disciple, Georg Joachim Rheticus, saw the theory as more than a series of calculating devices. It was Rheticus alone who recognized one of Copernicus' chief discoveries: a unique interconnexity of the distances and periods of the planets.

Robert S. Westman's paper, "Three Responses to the Copernican Theory: Johannes Praetorius, Tycho Brahe and Michael Maestlin," argues that the generation of astronomers born roughly in the 1540s came to appreciate the Copernican theory far more profoundly than the previous generation (born in the period ca. 1497-1525). The new generation noticed especially the theory's systemic properties, its equant-less models, and its realist claims about the earth's manifold motions, although, of those considered in the paper, only Maestlin proclaimed his belief in the new ordering of the planets. If deeper understanding of Copernicus' arguments did not produce immediate and total conversions to the novel theory, however, it did cause this generational group to think more seriously about the ordering of the planets and the sun's prominent participation in their motions. With Praetorius, this produces some interesting shuffling of the spheres, resulting at one point in a geoheliocentric model and a subsequent retreat to the Ptolemaic arrangement. With Tycho, it results in a commitment to a realist geoheliocentric scheme where Mars' orbit intersects the Sun's and causes Tycho to reject the existence of solid orbs. With Maestlin, the harmony of the spheres is sufficient to make him a Copernican, although a conservative one who was not especially sympathetic to Kepler's physics. It is interesting to note, in fact, that, unlike Tycho, Maestlin and Praetorius seem to have been far less affected by trends toward an immaterialization of the heavens. Only with Maestlin's student Johannes Kepler, a product of the strong German tradition of mathematical astronomy, do we find a serious and sustained effort to lay new methodological foundations for planetary theory in Renaissance theories of force and celestial matter.

Our volume ends with a related, but different, kind of problem: the normative appraisal of scientific change. If we are now closer to an explanation of why Copernicus' new hypotheses were actually accepted or rejected, can universal conditions be established under which that theory may be judged superior to, or more rational than, the one which it replaced? Such is the important question raised in the paper by Imre Lakatos and Elic Zahar: "Why did Copernicus' Research Program Supersede Ptolemy's?".

The analysis takes place within the framework of Lakatos' conception of scientific change. The basic unit of appraisal is not an isolated hypothesis but a developing group of theories which are linked together into what Lakatos calls a "research program." The structure of the research program is cell-like: it has a tenacious "hard core" composed of statements against which it is forbidden to seek refutations; this nucleic core is surrounded by a "protective belt" of auxilliary hypotheses against which the arrows of the *modus tollens* argument may be directed. The belt also includes a "heuristic" or set of techniques for solving problems. While the protective belt is constantly modified and increased, the hard core remains intact unless, of course, the entire research program is superseded by a new one. The rules for appraising research programs lead us to one of two possible judgments: the program is either progressive or degenerating. It is theoretically progressive if each modification leads to new and unexpected predictions; it is empirically progressive if at least some of these novel predictions are corroborated. Research programs, like Kuhnian paradigms, never solve all of their anomalies; refutations abound so that the research program exists in what Lakatos is fond of calling, "an ocean of anomalies." What matters, therefore, is that we find evidence of a few dramatic signs of empirical progress. In evaluating rival research programs in the "game of science," one program supersedes another if it has excess truth content over its rival in the sense that it predicts all that its rival truly predicts—and some more besides.

In applying his theory of rationality to the Copernican and Ptolemaic research programs, Lakatos first formulates and then rejects a number of alternative accounts of scientific change. These may be summarized briefly as follows: 1) "strict inductivism" requires that the superior theory be deduced from the facts while its rival cannot be; 2) "probabalistic inductivism" demands that the superior theory have a higher probability relative to the total evidence available at the time; 3) "falsificationism" sees refutability as the prime virtue of scientific theories and "crucial experiments" (one theory is refuted at the same time that its rival is corroborated) as the ultimate test of the winner; 4) "simplicism" states that one theory is better than another if it is simpler, more coherent or economical (Lakatos sees the "simplicity-balance" between the Copernican and Ptolemaic theories as roughly even); 5) a "scientific jury," which decides in each case what is or is not scientific through "tacit knowledge" (Polanyi) or *Fingerspitzengefühl* (Holton); and 6) "cultural relativism" (Feyerabend), a position which Lakatos states in the following way: "The Ptolemaists did their thing and the Copernicans did their's and at the end the Copernicans scored a propaganda victory." While Lakatos' summary of positions opposed to his own may not be acceptable to all, it has at least the virtue of clarifying the place he sees for his own demarcationist account, which places heavy emphasis on the criterion of *prediction.* For ultimately, Lakatos' and Zahar's paper amounts to a "predictivist" interpretation of the Copernican Achievement.

The authors argue that Copernicus' research program superseded Ptolemy's for two principal reasons. First, Copernicus explained in a straightforward manner, as natural consequences of his models, several important facts of planetary motion that Ptolemy had dealt with in an *ad hoc* manner by parameter adjustment. Secondly, from the Copernican theory, several "novel, though well-known facts" could be predicted *prior to any observation*, e.g. the relationship between a planet's elongation from the sun and the interval between two successive retrograde motions.

In his lively commentary, Stephen Toulmin disputes Lakatos' interpretation of Copernicus' achievement as an essentially predictive enterprise. Instead, he characterizes Copernicus' main intellectual strategy as *explanatory,* in the sense that well known facts were now explained by novel relations.

In sum, the papers in this volume show, both in their diversity and in their attention to common themes, the value of focusing different methodological beams on a great scientific achievement. Out of these varied perspectives there emerges a more sharply defined, and hence, a deeper view of scientific discovery and change.

II

RHETICUS, RAVETZ, AND THE "NECESSITY" OF COPERNICUS' INNOVATION

Curtis A. Wilson

St. John's College, Annapolis

In the preliminary account of Copernican astronomy that George Joachim Rheticus gave to the world in his *Narratio Prima* of 1540, there is more than one statement suggesting a certain "necessity" in Copernicus' innovation.

> . . . [my teacher] is far from thinking that he should rashly depart, in a lust for novelty, from the sound opinions of the ancient philosophers, except for good reasons and when the facts themselves coerce him.[1]
>
> . . . my teacher saw that only on this theory could all the circles in the universe be satisfactorily made to revolve uniformly and regularly about their own centers, and not about other centers — an essential property of circular motion.[2]

[1]Edward Rosen, *Three Copernican Treatises* (3d ed., New York: Octagon Books, 1971), p. 187. The Latin of the statement may be found in Johannes Kepler, *Gesammelte Werke*, I, ed. Max Caspar (Munich: C.H. Beck, 1938), p. 126, lines 35-37: ". . . sibi a Veterum philosophantium sententijs nisi magnis de causis, ac rebus ipsis efflagitantibus, studio quodam novitatis, temere discedendum putarit."

[2]*Ibid.*, p. 137. The Latin reads: ". . . hac unice ratione commode fieri posse D. Praeceptor videbat, ut quod maxime proprium circularis motus est, omnes revolutiones circulorum in mundo aequaliter, et regulariter super suis centris, et non alienis moverentur" (found in Kepler, *Gesammelte Werke*, I, p. 100, lines 9-11).

In the matter of the precession of the equinoxes, having described the apparent non-uniformity of the precession and the related inequality in the tropical year as accepted by Copernicus, Rheticus concludes:

> Hence we must say that the equinoctial points, like the nodes of the moon, move in precedence, and not that the stars move in consequence.[3]

It is on the "necessity" implied in the last-quoted statement that the argument of J.E. Ravetz in his *Astronomy and Cosmology in the Achievement of Nicolaus Copernicus,*[4] as well as in an article in *Scientific American,* "The Origins of the Copernican Revolution,"[5] turns. It is Ravetz's belief that Copernicus while still a young man in Krakow, before leaving for Italy in 1496, had undertaken an analysis of the foundations of astronomy, and had arrived at the conclusion that, if these foundations were to be logically coherent, the earth must be put into motion. "Thus," Ravetz concludes, "the rotation of the earth on a slowly tilting axis was necessary and sufficient to fix the frames of reference for a coherent science of astronomy. Because of this the rotation would not be merely hypothetical or speculative for Copernicus; it would be a necessary fact of the physical world."[6]

A number of features of Ravetz's reconstruction of Copernicus' journey of discovery prove untenable or highly questionable. According to Ravetz, we can be fairly certain that Copernicus' earliest account of his system, an untitled manuscript which circulated anonymously and which long afterward came to be called the *Commentariolus,* was written before the young Copernicus left Krakow; for (according to Ravetz) the *Commentariolus* shows no trace of acquaintance with the *Epitome in Almagestum* prepared by Georg Peurbach and Regiomontanus, and first printed in 1496. But to the *Epitome,* had it been available, Copernicus would naturally have turned (Ptolemy's *Syntaxis* or so-called *Almagest* itself being neither in print nor easily available in

[3]*Ibid.,* p. 115. The Latin reads: "Proinde statuendum puncta aequinoctialia moveri in praecedentia, quemadmodum in Luna nodos, et nequaquam stellas secundum signorum consequentiam progredi" (in Kepler, *Gesammelte Werke,* I, p. 90, lines 18-20).

[4]Warsaw, 1965.

[5]*Scientific American,* October, 1966, pp. 88-98.

[6]*Ibid.,* p. 93f.

manuscript) in order to obtain the values for the length of the tropical year given by Hipparchus and Ptolemy—the calendrical problem of establishing the length of the year being, according to Ravetz, the starting-point of Copernicus' train of thought.[7] In the *Commentariolus* Copernicus gives Hipparchus' and Ptolemy's length for the tropical year simply as 365¼ days. The *Epitome* gives the Hipparchan value as 365¼ days, but gives Ptolemy's value correctly as 365¼ days minus 1/300 day. In *The Revolutions of the Heavenly Spheres* Copernicus will correctly state that the value obtained by both Hipparchus and Ptolemy was 365¼ days minus 1/300 day.

Ravetz's dating of the *Commentariolus* to Copernicus' pre-Italian period is contradicted by findings of Birkenmajer and Rosen: in the *Commentariolus* there is cited a value for the length of the tropical year of $365^{d}\ 5^{h}\ 40^{m}$ due to "Hispalensis," and this Hispalensis proves to be Alfonso de Cordoba Hispalensis, whose corrected edition of Zacuto's *Almanach perpetuum,* with the just-mentioned length of the tropical year appearing on fol.a IV, was published on 15 July 1502.[8] The date 15 July 1502 thus becomes a *terminus post quem* for the *Commentariolus.* As *terminus ante quem* Rosen is able to give us 1 May 1514, when Matthew of Miechow, a professor of the University of Krakow, catalogued his books, including a "manuscript of six leaves expounding the theory of an author who asserts that the earth moves while the sun stands still." Despite absence of a title and author's name, Rosen points out, the description can only refer to Copernicus' *Commentariolus.*[9] It is Rosen's conclusion (the specific grounds will not be cited here) that "we may tentatively assign the earliest glimmerings of the new astronomy to 1509-1510 and the actual writing of the short treatise to 1511-1513."[10]

Copernicus' failure at such a late date as 1511-1513 to cite the Hipparchan and Ptolemaic value for the length of the tropical year correctly is explicable if he was writing at Frombork and still did not have a copy of Ptolemy's *Syntaxis* to consult (he will later make use of the 1515 edition). That he had available a copy of the *Epitome* of Peurbach and

[7]Ravetz, "The Origins . . . ," p. 92.

[8]Rosen, *op. cit.,* pp. 66-67, n. 21.

[9]*Ibid.,* p. 343.

[10]*Ibid.,* p. 345.

Regiomontanus is likely; thus it was probably from the *Epitome* that he obtained his information about Albategnius' estimate of the length of the tropical year. If so, and if in addition what he says about the Hipparchan and Ptolemaic values also derives from the *Epitome,* it would appear that he consulted the *Epitome* somewhat carelessly, noting merely the Hipparchan value as given by the *Epitome* to be 365¼ days, and assuming the Ptolemaic value to be identical with the Hipparchan. This would suggest a less than avid interest in the calendrical problem of the changing length of the tropical year. And indeed, as Rosen points out, Copernicus' own statement in the *Revolutions* is that he began his "closer study" of the length of the year only after being urged to do so by Paul of Middelburg during the years 1514-1516, when Paul was in charge of the investigation set in motion by Pope Leo X on the problem of calendar reform.[11] As a consequence we must reject Ravetz's suggestion that Copernicus' reconstruction of astronomy sprang from the problem of calendar reform.

But what about Ravetz's assertion that the Copernican introduction of the rotation of the earth on a slowly tilting axis is logically necessary, in order "to fix the frames of reference for a coherent science of astronomy?" In fact, Ravetz's assertion is deducible, if we will first accept the data assumed by Copernicus and his contemporaries, implying a changing rate of precession as well as a changing obliquity of the ecliptic, and if in addition we accept a premiss enunciated by Ravetz.

The data implying the change in the obliquity of the ecliptic were qualitatively correct, but quantitatively exaggerated; Ptolemy's value for the obliquity was nearly 11′ too large. The data implying a changing rate of precession, we can assert today, were spurious; they were, to identify them specifically, the observations of equinoxes described by Ptolemy for September 26, A.D. 139, and for March 22, A.D. 140. The first of these was too late by 33.0 hours, the second by 20.4 hours. Ptolemy's accuracy here first came to be seriously and persistently questioned by Tycho Brahe and by Kepler; since their time various attempts

[11]*Ibid.*, p. 360. The Copernican statement occurs at the end of the dedicatory epistle to Pope Paul III; see Nikolaus Kopernikus, *Gesamtausgabe,* II (Munich: R. Oldenbourg, 1949), p. 7.

have been made to explain the Ptolemaic errors. With a declination-measuring instrument such as Hipparchus used for determining the times of the equinoxes, the systematic errors in spring and fall will be in opposite directions; if the error for the fall equinox is negative, so that by measurement it is determined to be later than it was in fact, then the error for the spring equinox will be positive, and *vice versa*. The errors in Hipparchus' determinations as cited by Ptolemy are of this kind, and average about 7′ in declination, or about a quarter of a day in time. Ptolemy's fall and spring equinoxes show a systematic error not in declination but in time; they are both late. Another unsettling fact is that they agree precisely with Hipparchus' value for the length of the tropical year, and when used in conjunction with Ptolemy's reported observation of the solstice on 25 June 140 (also late by 35.4 hours) yield a solar theory agreeing exactly with the Hipparchan observations and with the Ptolemaic rate of precession, while independent observations would be expected to show a difference. John Phillips Britton in a recent careful study *On the Quality of Solar and Lunar Observations and Parameters In Ptolemy's Almagest*, has concluded:

> It is evident that Ptolemy's equinox observations cannot be understood as independent observations affected by an inadvertent systematic error or even as consistent observations designed to verify Hipparchus' solar parameters. . . . The conclusion that Ptolemy's equinox observations can have been scarcely more than the results of computations is unsatisfying, but I can find no other explanation of the errors in his reported times and their agreement with Hipparchus' observations and length of the year.[12]

That Ptolemy's reported equinox observations had been "made up" to fit a theory would never have been suspected in Copernicus' day; and on the assumption that they were in fact observations, no one at the time could have imagined that they would be as much in error as they are. Even if a reasonable observational error is allowed for, the Ptolemaic "observations" in conjunction with those of Hipparchus and

---

[12]John Phillips Britton, *On the Quality of Solar and Lunar Observations and Parameters in Ptolemy's Almagest* (Yale University Ph.D. Dissertation, 1967), pp. 42, 44.

those of Albategnius imply a changing rate of precession. Since Ptolemy allows for only 1° of precession per century since Hipparchus' time, while the rate was actually 1°24′ per century, Albategnius toward the end of the ninth century necessarily found from his (real) equinox observations that the precession since Ptolemy's time had been nearly 1°34′ per century, although it had in fact remained at 1°24′ per century all the while. It is apparently Tycho Brahe toward the end of the sixteenth century who first recognizes that precession can be assumed to be uniform if Ptolemy's equinox observations are ignored; not till the middle of the seventeenth century does the recognition begin to be general that the theory of the inequality of precession is "a needless multiplication of uncertainties."[13]

For Copernicus the way to such a conclusion had not been prepared by the accumulation of extensive and accurate observations; he could only assume that the ancient observations, including Ptolemy's, had been carefully made. He expresses his trust in the reliability of ancient observations, and his firm conviction that without those observations the enterprise of astronomy would be impossible, in the following passage of his *Letter against Werner* of 3 June 1524:

> The science of the stars is one of those subjects which we learn in the order opposite to the natural order. . . . First we learn that the apparent motions of the planets are unequal, and subsequently we conclude that there are epicycles, eccentrics, or other circles by which the planets are carried unequally . . . The situation is the same with respect to the motion of the eighth sphere. However, by reason of the extreme slowness of this motion, the ancient mathematicians were unable to pass on to us a complete account of it. But if we desire to examine it, we must follow in their footsteps and hold fast to their observations, bequeathed to us like an inheritance. And if anyone on the contrary thinks that the ancients are untrustworthy in this regard, surely the gates of this art are closed to

[13]Jeremy Shakerley, *The Anatomy of Urania Practica, or, A Short Mathematicall Discourse; Laying Open the Errors and Impertinencies delivered in a Treatise lately published by Mr. Vincent Wing and Mr. William Leybourne, under the title Urania Practica* (London: Thomas Brudenell, 1649), p. 8.

> him. Lying before the entrance, he will dream the dreams of the disordered about the motion of the eighth sphere and will receive his deserts for supposing that he must support his own hallucinations by defaming the ancients.[14]

Copernicus therefore accepts the inequality of the precession of the equinoxes.

Accepting the inequality of the precession of the equinoxes, Copernicus would be logically forced to accept the motion of the earth if he accepted in addition a premiss enunciated by Ravetz. The length of the tropical year, according to the data bequeathed to Copernicus, had varied, the discrepancy always corresponding to the motion of the equinoxes. As Copernicus argues in the *Commentariolus,*

> . . . when the cardinal points moved 1° in 100 years, as they were found to be moving in the age of Ptolemy, the length of year was then what Ptolemy stated it to be. When however in the following centuries they moved with greater rapidity, being opposed to lesser motions, the year became shorter; and this decrease corresponded to the increase in precession. For the annual motion was completed in a shorter time on account of the more rapid recurrence of the equinoxes. Therefore the derivation of the equal length of the year from the fixed stars is more accurate.[15]

The entwined "facts" that Copernicus cites are but twin artifacts produced by Ptolemy's erroneous equinoxes, but accepting them as Copernicus must, he is led to view the sidereal year, or annual return of the sun to a given star, as the only constant year. "But," Ravetz asks, "if one chooses this as the basic measure, what is one to make of the irregular changes in stellar longitudes due to the inequality in the precession? If one keeps the system of an irregularly moving stellar sphere carrying with it the sun's orbit, then one's fundamental measure of time is a motion along an irregularly moving orbit. This is strictly incoherent; as Copernicus later said of time: 'The measure and the thing measured are

[14]Rosen, *op. cit.*, pp. 98-99.

[15]*Ibid.*, p. 67.

interchangeable.' Thus the stellar sphere can have only its daily rotation; the changes in stellar longitudes must be explained by a motion of the equinoctial points. . . ."[16]

If indeed the stellar sphere can have at most its daily rotation, then the precession of the equinoxes must be due to the motion of the circle of the ecliptic, or the circle of the equator, or both, independently of the sphere of the stars. Now the circle of the ecliptic cannot move independently of the sphere of the stars, because in the course of the precession the latitudes of the stars, or their angular distances north or south of the ecliptic, are found to remain constant. Therefore the precession must be due to the motion of the celestial equator. But the celestial equator and its poles remain constantly aligned with particular locations on the earth's surface, and therefore as the celestial equator moves amidst the stars, the earth must be moving with it. Q.E.D.

At this point we may direct to Ravetz the following question: Faced with the pre-Copernican arrangement whereby the sun in its annual motion completes a sidereal year of constant length with respect to a stellar sphere that is itself moving with a long-term periodic motion, must we—and Copernicus—conclude that this arrangement is "strictly incoherent?" The answer would appear to be that it is just as coherent or incoherent as defining time in terms of a clock resting on a platform that is being very slowly rotated at a rate that is itself subject to a slow, periodic change. Assuming a diamond-point mechanism of the celestial variety, why should we suppose that the clock's accuracy would be affected?

Although Rheticus in the *Narratio Prima* writes that he does not see "how the Copernican explanation of precession is to be transferred to the sphere of the stars,"[17] it is not impossible to devise an arrangement of three spheres that will lead to all the phenomena implied by Copernicus' assumption of a rotating earth with slowly tilting axis.* Thus in

---

**Editor's Note.* The possibility of such an arrangement was suggested in 1568 by Caspar Peucer (1525-1602) in his *Hypotyposes orbium coelestium* . . . (Argentorati: Theodosius Rihelius), pp. 516-517.

[16]Ravetz, "The Origins . . . ," p. 92.

[17]Rosen, *op. cit.,* p. 164.

Figure 1, the east-to-west rotation of the tenth sphere about poles N and S in $23^h\ 56^m$ accounts for the diurnal rotation of the stars; the slow west-to-east rotation of the ninth sphere about poles E and F with a period of 25,816 Egyptian years accounts for the mean precession; the poles of the ecliptic are librated on the diameter of a small circle from A to B and back again in a period of 3434 Egyptian years, to account for the changing obliquity of the ecliptic; and the eighth sphere which actually carries the stars is librated through a small angle of about 70′ around the axis through poles J and K with a period of 1717 Egyptian years to account for the variation in the rate of precession.[18] The phenomena, such as Copernicus believed them to be, can thus be saved.

When the annual motion of the sun is introduced into the discussion, the situation assuredly becomes more complicated. Owing to the inequality in the precessional motion of the stars, the annual returns of the sun into alignment with a given star as observed from the earth will not be accomplished in exactly equal periods. The maximum discrepancy between the periods of the observed returns will be about 2.04 seconds of time, as between years that are 1717/2 years apart. But the same thing happens in the Copernican theory, where the precessional motions are attributed to the earth: the returns of the sun into alignment with a given star as observed from a given spot of the earth's surface will not be strictly periodic, but will vary by a maximum of 2.04 seconds. Ravetz to the contrary notwithstanding, the two theories are observationally equivalent; under both the apparent sidereal year varies slightly, the discrepancies being indetectible observationally but computable from theory.

That the Copernican solution, in which all the supposed motions are accomplished by the earth and its tilting poles, is superior in economy and elegance, cannot be denied. Did Copernicus regard the geostatic alternative, with the sun accomplishing its constant annual returns with respect to an eighth sphere that was itself moving with a motion of periodically varying rate, as "strictly incoherent?" He does not say so; his language is more modest. In the *Commentariolus* he says:

---

[18]Copernicus' derivation of these periods is examined by Kristian Peder Moesgaard, "The 1717 Egyptian years and the Copernican theory of precession," *Centaurus,* 1968, *13*: 120-138.

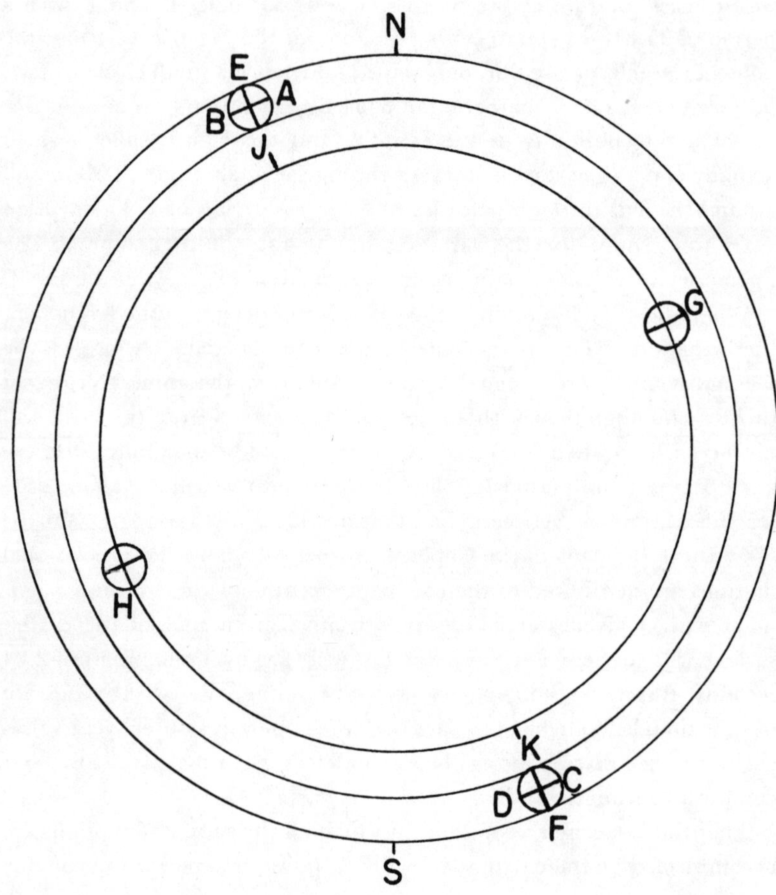

Figure 1. The Copernican Explanation of Precession Transferred to the Sphere of the Stars.

> . . . it is the common opinion that the firmament has several motions in conformity with a law not yet sufficiently understood. But the motion of the earth can explain all these changes in a less surprising way.[19]

In the *Revolutions* he says:

> No one perhaps will bring forward a better cause of this [the unequal precession] than that there is a certain deflexion of the axis of the earth.[20]

Copernicus' strongest statement in opposition to the motion of the stellar sphere does not bear particularly on the matter of the unequal precession but is rather the more general assertion that

> . . . it seems very absurd (*satis absurdum*) to ascribe movement to the container or to that which provides the place and not rather to that which is contained and has a place, that is, the earth.[21]

Unquestionably Ravetz is correct in stressing the importance to Copernicus of immobilizing the sphere of stars, so as to provide a stationary background against which all the celestial motions could be measured. Conclusive evidence is lacking, however, for Ravetz's claim that it was the aim of accomplishing this that triggered Copernicus' entire endeavor and made it inevitable. The fact is that Copernicus' statements in the *Commentariolus* assert a different starting-point—and in no uncertain or indefinite terms.

The *Commentariolus* begins as follows:

> Our ancestors assumed, I observe, a large number of celestial spheres for this reason especially, to explain the apparent motion of the planets by the principle of regularity. For they thought it altogether absurd that a heavenly body, which is a perfect sphere, should not always move uniformly. . . . Callippus and Eudoxus, who endeavored to solve the problem by the use of concentric spheres, were unable to account for all the

[19]Rosen, *op. cit.*, p. 64.

[20]Kopernikus, *Gesamtausgabe,* II, p. 148, lines 9-10.

[21]*Ibid.*, p. 20, lines 34-36.

> planetary movements. . . . Therefore it seemed better to employ eccentrics and epicycles, a system which most scholars finally accepted.
>
> Yet the planetary theories of Ptolemy and most other astronomers, although consistent with the numerical data, seemed likewise to present no small difficulty. For these theories were not adequate unless certain equants were also conceived; it then appeared that a planet moved with uniform velocity neither on its deferent nor about the center of its epicycle. Hence a system of this sort seemed neither sufficiently absolute nor sufficiently pleasing to the mind.
>
> Having become aware of these defects, I often considered whether there could perhaps be found a more reasonable arrangement of circles, from which every apparent inequality would be derived and in which everything would move uniformly about its proper center, as the rule of absolute motion requires. After I had addressed myself to this very difficult and almost insoluble problem, the suggestion at length came to me how it could be solved with fewer and much simpler constructions than were formerly used, if some assumptions (which are called axioms) were granted me.[22]

Thus according to Copernicus in the *Commentariolus,* it was the "scandal" of the Ptolemaic equant that triggered his inquiry. He repeats essentially the same explanation in the *Revolutions* (V, 2):

> Thus they [the followers of Ptolemy] concede here that the equality of circular motion can take place with respect to an alien and not the proper center. . . . These and similar things gave us the occasion to consider the mobility of the earth and other means whereby equality and the principles of this art might abide, and the theory of the apparent inequality be rendered more firm.[23]

The puzzle that now arises for us is whether there is in fact, as Copernicus seems to imply, a plausible link between Copernicus' stubborn

[22]Rosen, *op. cit.,* pp. 57-58.

[23]Kopernikus, *Gesamtausgabe,* II, p. 291, lines 25-31.

insistence on uniform circular motion, which we tend to view as pedantic and fussy, and his bold hurling of the earth into motion amidst the stars, which we are likely to think of as signalling a new and liberating vision of the world.

Figure 2 shows the Ptolemaic theory for a superior planet (except for the dotted line, it is the same as that for Venus). Circle FHG is the deferent circle, with center at E, the earth at D, and C the equant point. The epicycle's center H moves with angular uniformity about the point C while remaining on the circular track with center E; consequently its motion on this circular track is non-uniform. In the *Commentariolus* Copernicus says that the planet's motion on the epicycle is also non-uniform; actually it is uniform with respect to the stars, since Ptolemy measures the uniform rotation of the radius-arm HP from the line CHI, which itself rotates uniformly with respect to the line of apsides FG, while FG according to Ptolemy is fixed with respect to the stars. By the time he wrote the *Revolutions* Copernicus had apparently seen his error, for there he merely states that the revolution of the planet on the epicycle ought to be measured, not in relation to the line CHI, but in relation to the line EHL.[24]

In order to eliminate the equant, Copernicus is forced to introduce new epicycles. In the *Commentariolus* he introduces two new epicycles per superior planet. In the *Revolutions* he replaces the biepicyclic scheme of the *Commentariolus* by an eccentric together with a single epicycle. We explain the latter scheme first, because its relation to the Ptolemaic theory is more easily seen.

In Figure 3, the center Q of the deferent circle is at a distance from point D equal to three-fourths of the distance CD in Figure 2 between earth and equant point. The epicycle's radius is equal to one-fourth of this same distance CD. The point with whose rotation on the epicycle we are concerned starts at F when the epicycle is at the top of the deferent, and rotates counterclockwise, its angular rate being the same as that of the epicycle's center in its counterclockwise rotation on the deferent. It can be shown that if the equant point C of Figure 2 is inserted into Figure 3, then the line from C to the point of interest on the epicycle rotates uniformly; thus as far as angular motion is concerned the

[24]*Revolutions*, V, 2.

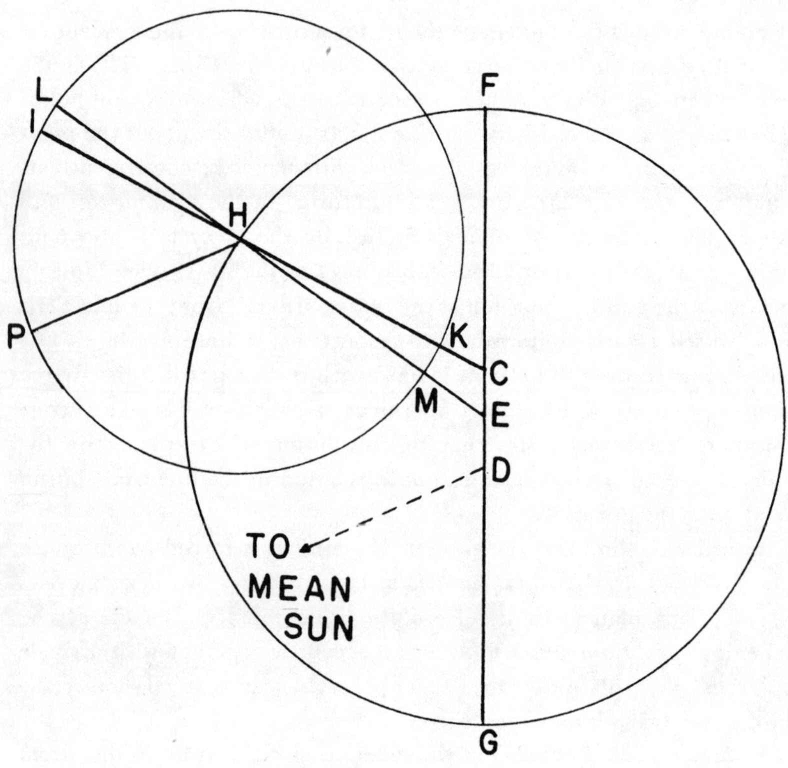

Figure 2. The Ptolemaic Theory for a Superior Planet.

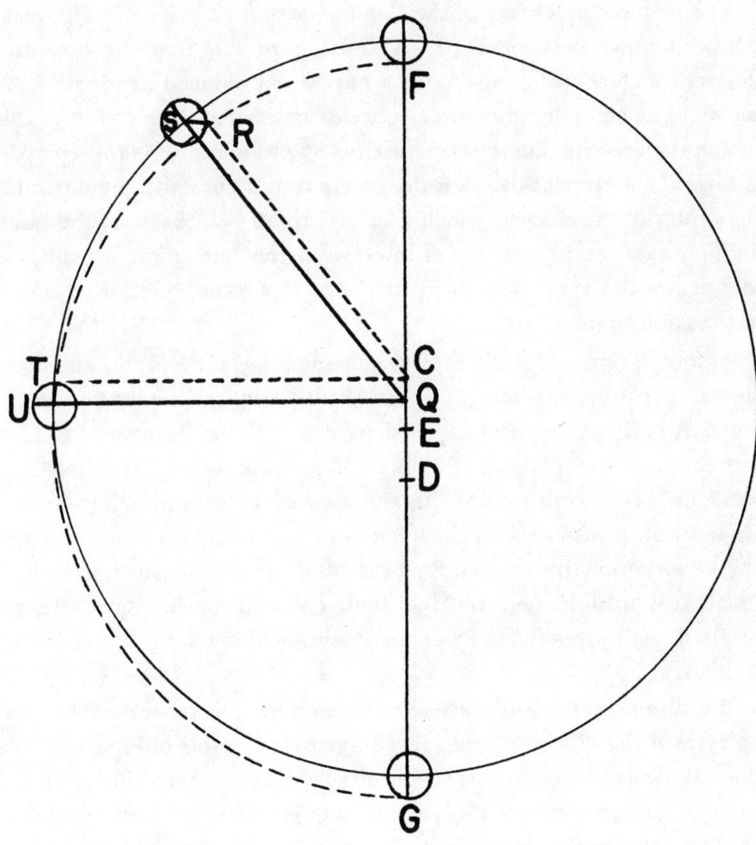

Figure 3. The Replacement for the Equant in Copernicus' *Revolutions.*

scheme of Figure 3 accomplishes the same thing as the equant-point arrangement of Figure 2. The actual path swept out in Figure 3, however, is not quite circular, but bulges out at points to the side such as T.

The biepicyclic scheme of the *Commentariolus* is shown in Figure 4. The difference between Figure 3 and Figure 4 is that the eccentric deferent with center Q of Figure 3 has been produced in Figure 4 by means of an epicycle riding on a concentric deferent. The radius of this new epicycle is equal to the eccentricity DQ of Figure 3. As this epicycle is carried counterclockwise on the deferent at a constant angular rate, the center of the epicycle which it carries rotates clockwise at the same angular rate, and the point of interest on this last epicycle (which is identical with the epicycle in Figure 3) rotates counterclockwise, again at the same angular rate.

As a replacement for the Ptolemaic equant-style theory, we can guess that Copernicus was well pleased with this biepicyclic piece of clockwork. A half century after Copernicus' death Tycho Brahe and Longomontanus will be praising it highly and making use of it in their own work of theory construction; in the view of these anti-heliocentrists, Copernicus' return to the principle of uniform circular motion through the adoption of this mechanism was his most important accomplishment. It is possible, however, that these epicyclic mechanisms were not original with Copernicus, but were obtained by him from Arabic sources.

But suppose now that we cause to be mounted piggyback on the final epicycle of this clockwork the large Ptolemaic epicycle of Figure 2. The motion of the planet on this large epicycle does not keep time with the three angular motions of Figure 4; its rate is rather the average rate of the sun, so that (for instance) in Figure 2 the radius-arm HP remains always parallel to the line from the earth's center D to the mean sun, provided that our theory is that of a superior planet. A certain incoherence or lack of harmony is introduced by mounting this annual epicycle atop the epicyclic mechanism of Figure 4. It is imaginable that, by an easy change of emphasis in the application of the principle that everything celestial should move uniformly about its *proper* center, Copernicus at just this juncture might have begun to consider whether the annual epicycle might not be given a place and center of its own; and he might then have seen in the case of the outer planets the possibility of

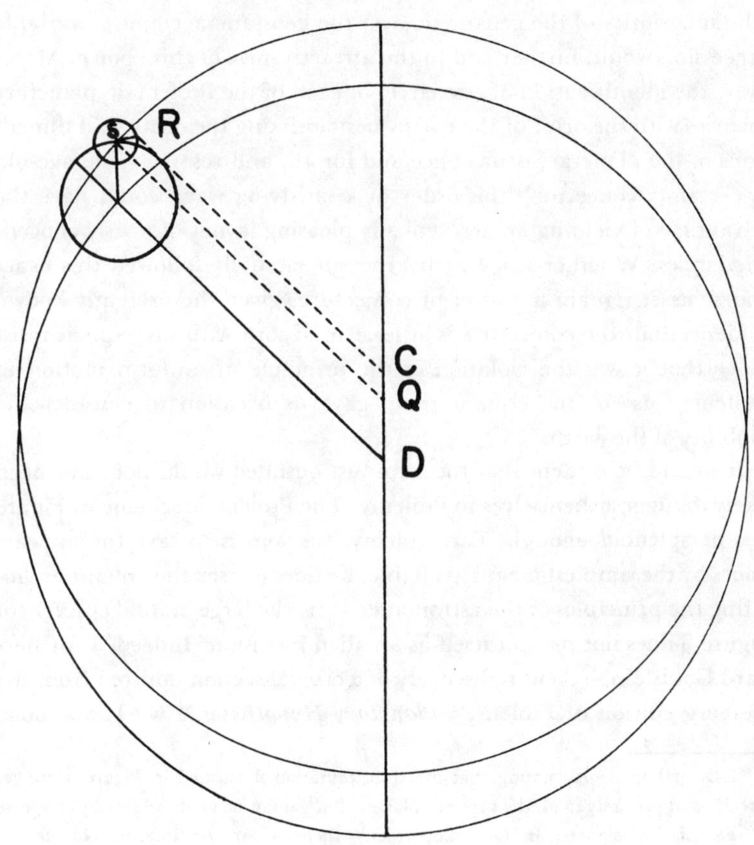

Figure 4. The Replacement for the Equant in Copernicus' *Commentariolus*.

transforming the annual epicycle into the orbit of the sun or earth. Of the two alternatives, the transformation of the annual epicycle into the orbit of the earth is perhaps the simpler; it is illustrated in Figure 5. The superiority of the geokinetic over the geostatic account of variable precession would further add to the attractiveness of this choice. Moreover, the identification of one circle in each of the Ptolemaic planetary theories with the orbit of the earth, besides fixing the order and dimensions of the planetary orbits once and for all, and resolving the age-old uncertainty concerning this order in a satisfying way, would have the advantage of yielding an aesthetically pleasing layout of nearly concentric circles. Whether Copernicus' thought actually followed this exact course must remain a matter of conjecture, given the presently known evidence, but the conjecture is at least in accord with his explicit insistence that it was the violation of the principle of uniform motion in Ptolemy's use of the equant that "gave us occasion to consider the mobility of the earth."[25]

It should be evident that the steps just outlined would not have been likely to suggest themselves to Ptolemy. The Ptolemaic scheme of Figure 2 is perspicuous enough. For Ptolemy, the aim is to save the appearances by the simplest means available; he does not see the equant as violating the principles of the astronomer's art; the large annual epicycle of Figure 2 does not present itself as an alien intrusion. Indeed, with Bernard Goldstein's recent rediscovery of a crucial section omitted from the Heiberg edition of Ptolemy's *Planetary Hypotheses,*[26] we know today

[25]It is perhaps worth noting that the transformation illustrated in Figure 5 applies directly and precisely to all the superior planets, but cannot be easily adapted to the case of the inferior planets. In fact, Copernicus' theories for the inferior planets are encumbered with certain annual motions. Now Copernicus' account of planetary theory in Book 5 of the *Revolutions* starts with the outer planet Saturn and moves inward; while Ptolemy's account in the *Syntaxis* follows exactly the opposite order, moving from the inferior planets outward. This difference could reflect, for one thing, the fact that Copernicus originally discovered his transformation for the outer planets, as suggested in the foregoing account; and it could also reflect a fundamental difference in ways of thinking about the stars, Ptolemy looking up to them as superior and unfathomable beings, as gods, while Copernicus views them from above, as the artifacts of a divine artificer.

[26]Bernard R. Goldstein, "The Arabic Version of Ptolemy's *Planetary Hypotheses,*" *Transactions of the American Philosophical Society,* 1967, *57,* [4].

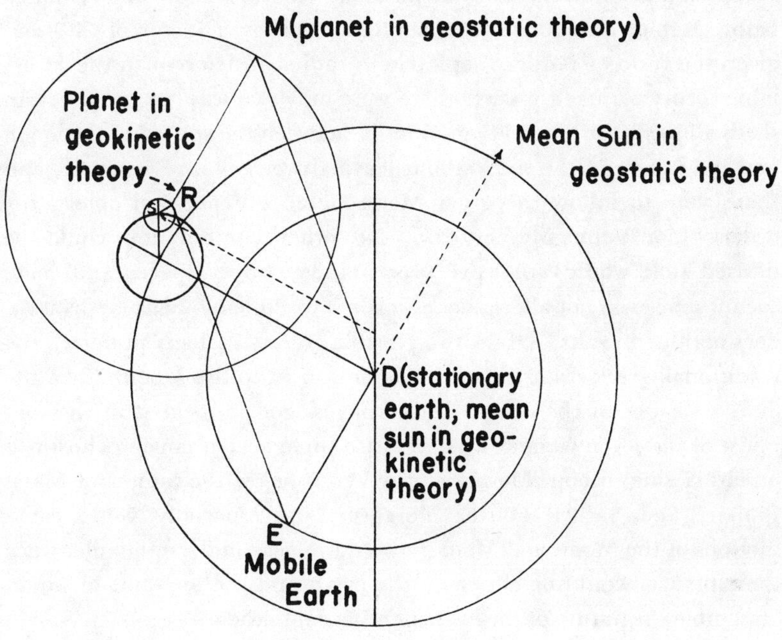

Figure 5. Transformation of the Annual Epicycle of a Superior Planet into the Orbit of the Earth.

that both the Ptolemaic order of the planets, and the fact that into the Ptolemaic theory of each planet other than the sun there entered a heliacal anomaly or motion related to the sun, carried for Ptolemy a certain mythological plausibility and appropriateness. Given the accepted ancient value of solar parallax—it was almost unquestioned before Kepler questioned it, though in error by a factor of 20—and given the ratio of radius of epicycle to radius of deferent in the Ptolemaic theory for each planet, there were only two ways to fit spherican shells allowing for planetary motions in between the sphere of the moon and the sphere of the sun without leaving over any unused space, and these were to follow the order Moon-Mercury-Venus-Sun or else the order Moon-Venus-Mercury-Sun. No other combination could be devised that would not leave over unused, empty space; and mere vacant space was not allowable, according to Ptolemy, because "Nature does nothing in vain." Of the two possible orders, Ptolemy preferred the traditional Greek order that put Mercury next to the Moon. Then the Sun, as leader of the chorus and strongest god, appeared in the very midst of the seven wandering stars, with three accompanying choristers on either side: Moon, Mercury, and Venus below the Sun, and Mars, Jupiter, and Saturn above. Moreover, the especially complicated motions of the Moon and Mercury would appear understandable, since these planets would be closest to the earth and the air, and so would "resemble the nature of the element adjacent to them."

Yet for Copernicus the plausibility of the Ptolemaic scheme has vanished (he dismisses the scheme explicitly in *Revolutions* I, 10). It has vanished precisely because he insists, with a stubbornness that is perhaps not altogether surprising on the part of a provincial convert to the Renaissance ideal of classicist *renovatio,* on the absolutely rigorous application of the ancient principle that the apparent inequalities of the celestial motions are to be "saved" or "demonstrated" by postulating uniform motions:

> . . . it can be seen what sort of argumentation it would be, if we should wish to derive the inequality of the appearances from actual inequalities. For what would we be doing, except offering assistance to those who would detract from this art?[27]

[27]Kopernikus, *Gesamtausgabe,* II, p. 212, lines 29-32 (*Revolutions,* IV, 2).

At the notion of the inconstancy of the celestial motor virtue, Copernicus says, the mind shudders; it is unfitting that such a state of affairs should exist among things that are established in the best system.[28] The reference here to "the things that are established in the best system" may remind us that Copernicus, in contrast to Ptolemy, is a Christian and accepts creationism unquestioningly. So, as the dedicatory epistle of the *Revolutions* shows, he thinks of the world as a *machina mundi,* a machine designed by the most expert artisan of all:

> . . . when I had long considered this lack of certitude in the mathematical tradition concerning the composition of the motions of the spheres of the world, I began to be annoyed that no more certain theory of the motions of the machine of the world, which was created for us by the best and most orderly of all artisans, had been established by the philosophers. . . .[29]

Ptolemy, by contrast, is a pagan whose world has not been created and for whom the highest principle is the first cause of the first motion "up high somewhere near the loftiest things of the cosmos."[30] The planetary movers too, according to Ptolemy, are divine and altogether dissimilar to human things, so that "it is not proper to judge the simplicity of heavenly things by those which seem so to us. . . ."[31] In Book II of the *Planetary Hypotheses* Ptolemy in effect denies that mechanical models of the planetary motions, like those of Eudoxus and Callippus, can have other status than as conceptual aids; he compares the planets to a flock of birds, each moved in accordance with its essence by a life-force within it, in a manner analogous to the action of the will and understanding in humans.[32] The theological vitalism of the *Planetary Hypotheses* may be an after-thought, but it is at least consonant with the positivistic attitude toward astronomical explanation that

---

[28]*Ibid.*, II, p. 13, lines 22-26 (*Revolutions,* I, 4).

[29]*Ibid.*, II, p. 5, lines 18-21.

[30]*Almagest,* I, 1, in *Great Books of the Western World, 16* (Chicago: Encyclopaedia Britannica, cop. 1952), p. 5.

[31]*Ibid.*, p. 429.

[32]Ptolemy, *Opera quae exstant omnia,* vol. II (ed. J.L. Heiberg; Leipzig: B.G. Teubner, 1907), pp. 119-120.

one encounters in the *Syntaxis,* for instance in the following lines of XIII.2:

> it is proper to try and fit as far as possible the simpler hypotheses to the movements in the heavens; and if this does not succeed, then any hypotheses possible.[33]

In a detailed comparison of the Ptolemaic and Copernican mechanisms, these doctrinal differences may tend to fade in significance. The ineptitude of some of Copernicus' constructions, in particular his failure to eliminate annual motions from his theories of Mercury and Venus—a major provocation later on to Kepler; the inaccuracy of some of his calculations; the fact that his detailed system saves the phenomena no more accurately than could the Ptolemaic set of theories with at most some minor revisions in parameters—all this had been recently insisted upon by latter-day contra-Copernicans.[34] By comparison with Ptolemy's sophisticated *salvare apparentia* doctrine, Copernicus' claim to comprehend the machine on the basis of his principle of uniform circular motion has appeared to some as incredibly naive. Neugebauer, in referring to the preface that Osiander added to the *Revolutions,* in which the kinematic models of the work are asserted to be *mere* hypotheses, with no other aim than to serve as guides to computation, remarks: "it is hard for me to imagine how a careful reader could reach a different conclusion."[35]

The fact remains that it was the man with the pedantic principle who instituted the revolutionary cosmology. To Copernicus the principle of uniform circular motion was an ancient and necessary axiom for the astronomical art. Reconstructing the astronomical theories on its basis he was led—not perhaps by what we can regard as logical necessity but certainly a notion of the harmony and order appropriate to a *machina mundi* created by the best and most orderly of artisans—to a new

[33]*Almagest,* ed. cit., p. 429.

[34]Derek de Solla Price, "Contra-Copernicus: A Critical Re-estimation of the Mathematical Planetary Theory of Ptolemy, Copernicus, and Kepler," in Marshall Clagett (ed.), *Critical Problems in the History of Science* (Madison: Univ. Wisconsin, 1959); O. Neugebauer, "On the Planetary Theory of Copernicus," *Vistas in Astronomy,* 1968, *10*: 89-103.

[35]Neugebauer, *op. cit.,* p. 100.

assignment of motions. The actual order in which Copernicus' thought proceeded can only be conjectured, but we can see how the introduction of the epicyclic replacements for the equant might cause him to seek to reassign, either to the sun or the earth, the annual motions included in the Ptolemaic theories of each of the planets. The geokinetic alternative appears, from the standpoint of both planetary and precessional theory, to be the simpler. The appropriateness of immobilizing the sphere of the stars, in order that it might form the container and stationary background against which the motions of the mobile bodies are measured, would also argue for the earth's mobility rather than the sun's. The immobility of the sun, meanwhile, would bring it into an appropriate analogy with a supreme ruler rather than a travelling administrator,[36] and provide a new and stronger justification for the ancient poetic encomia of the sun. However askance we may view the complications in which Copernicus' planetary theory becomes involved because of his initial principle of uniform motion, we cannot deny that it is with Copernicus that the new vision of the world that will inspire Kepler and Galileo emerges. And according to Copernicus' own assertions, and the conjectural extension thereof we have proposed, it emerged because of rather than despite that principle. It is perhaps unlikely, at a time when *innovatio* regularly meant *renovatio*, that it would have happened otherwise.

[36]Kopernikus, *Gesamtausgabe,* II, p. 26, lines 5-10 (*Revolutions,* I, 10).

COMMENTARY

Salomon Bochner
Rice University

Curtis Wilson's paper is technically so accomplished, as was also his previous paper on Kepler, that it would not be possible or even feasible for me to comment on his technical arguments, especially since I am not versed in the technical specifics of the history of astronomy. But I will make some comment, from an approach that is congenial to my way of thinking, on the actual subject matter of the papers of Wilson and Ravetz, which is not only a question of dating the *Commentariolus,* but of forming a picture of how Copernicus intellectually developed, by which challenges and stimuli of his times and by what literary means at his disposal.

Ravetz started a controversy in 1962 by asserting that Copernicus composed his *Commentariolus* even before leaving Kraków for Italy in 1496. This was opposed, by rather brief statements, by Edward Rosen in the annotated Copernicus Bibliography in the 1971 edition of his important book "Three Copernican Treatises." Now Wilson has done the same as Rosen, but very systematically with an impressive array of arguments. The decisive points at issue seem to be whether certain specific astronomical data which seem to underlie the reasoning or rationale in the *Commentariolus* were already in existence before Copernicus left for Italy, or at any rate would have been readily available to him even then. On such points of context Wilson has the better of Ravetz. However important and meaningful as the resolution of such issues may be, the problem of the true intellectual origin of the *Commentariolus* in particular, and of Copernicus' true intellectual motivation in general is not answered thereby. To put it somewhat inelegantly, this question cannot be decided exclusively by examining which "library facilities" were available to him at what times, and with what degree of attentiveness he utilized such facilities.

For my part I would not discard the possibility that Copernicus may well have started nurturing his heliocentric proclivities long before he had the specific knowledge by which to support them. Thus his dissatisfaction with the Ptolemaic assertion that the huge sphere of stars performs daily rotations need not have come about at all by the knowledge of specific astronomical detail from specific astronomical works. It may have been simply the expression of a metaphysical malaise, which was brought about by certain nascent general developments in the scientific relation between mathematical space and cosmic universe, which were destined to profoundly alter astronomy and cosmology in due course.

Let me get to the heart of the matter by commenting, from my point of view, on Copernicus' Motion in Declination, the so-called Third Motion of the Earth. Nowadays, the two motions of the Earth, which Copernicus introduced, namely the rotation around the sun, and the spinning around its axis, are vectorially independent in the underlying mathematical three-dimensional space, say in the absolute space of Newton. The superposition, or addition of these two motions in this space, is commutative, that is either of the two motions can come first. Now, Copernicus did not, or rather did not yet, have such a space, not for decisive operational purposes at any rate. For such purposes, he had only a *machina mundi,* and, in a *machina,* whenever there was a conjunction of the motions of the wheels or gears of the *machina,* there were priorities. In the present conjunction the motion along the ecliptic was deemed by Copernicus to come first, and the daily rotation was deemed to be secondary to this. But, if it was secondary, then, as in a "good" *machina,* the axis of daily rotation "ought" to be firmly affixed to the plane of the ecliptic, and thus form a fixed angle with the polar axis of the ecliptic. However, the axis of the earth does not form such a fixed axis, but remains parallel to itself, and this, in Copernicus' vision, can only come about by an extra motion of the earth, a third one, which brings this about. In this motion "the axis of the Earth describes the surface of a cone in a year, moving in an opposite direction to that of brings this about. In this motion "the axis of the Earth describes the surface of a cone in a year, moving in an opposite direction to that of the Earth's center, that is from east to west."[1]

[1]J.L.E. Dreyer, *A History of Astronomy from Thales to Kepler* (formerly titled *History of the Planetary Systems from Thales to Kepler*), (New York: Dover, 1953), p. 329.

Thus, in a sense, this third motion of Copernicus was nothing but a piece of clumsiness, a testimonial to his incapacity to add up the two motions of the Earth, the daily and the annual, simply and vectorially.[2] However, it is also a testimonial to Copernicus' greatness as astronomer that he was able to extract some very crucial astronomical yield from this piece of clumsiness, once he had installed it. He made the period of the third motion slightly less than a year, and, to quote Dreyer,

> . . . and this slight difference produces a slow backward motion of the points of intersection of the ecliptic and the equator—the precession of the equinoxes. This was now at least correctly explained as a slow motion of the earth's axis and not as hitherto as a motion of the whole celestial sphere, and this almost reconciles us to the needless third motion of the earth, which certainly had its share in the unpopularity against which the Copernican system had to do battle for a long time, as it seemed bad enough to give the earth one motion—but three![3]

But having said this in praise of Copernicus, I must backtrack and point out that even this third motion was not original with Copernicus, but was an imitation of something that Ptolemy had done in connection with the epicycles of the outer planets, in order to account for the fact, astonishing to him, that the planes of these epicycles were parallel to the ecliptic. To quote again Dreyer, from another part of this work,

> As the epicycle of an outer planet was nothing but the earth's annual orbit around the sun transferred to the planet in question, it was of course quite right that the epicycle should be parallel to the ecliptic. In thus remaining parallel to a certain

---

[2]Equally clumsy, and indicative of motion in a *machina mundi* rather than in a "space," is Copernicus' retention of Ptolemy's presumption that in the (epi-)cyclical motion of a celestial body it is the orbis of the body that rotates directly and not the body itself. The body is firmly affixed to the orbis.

Kepler was apparently the first to overcome all such clumsiness. He dispensed with the Third Motion of the Earth (Dreyer, p. 395), and all celestial bodies became "freely floating in space, moved by physical forces acting on them" (Arthur Koestler, *The Sleepwalkers,* (New York: Macmillan, 1959), p. 314.)

[3]Dreyer, p. 329.

> plane the epicycles did what the ancients considered an unusual thing, as they would have thought it natural that the plane of the epicycle should keep at the same angle to the radius joining the centre of the deferent to the centre of the epicycle. The hypothesis therefore demanded the introduction of a small auxiliary circle, the plane of which was perpendicular to the plane of the deferent, the center of which was in the latter plane and which revolved in the zodiacal period of the planet. If we imagine a stud on the circumference of this circle and let it slide in a slot in the epicycle, we see how the latter was kept parallel to the ecliptic.[4]

Also, may I quote the following sentence about Copernicus from another part of Dreyer:

> In his sixth book, the shortest of all, Copernicus deals with the latitudes of the planets, and this is the part of his work in which he keeps closest to Ptolemy,[5]

to which I would like to add the following observation. Unlike Kepler after him, Copernicus neither states nor shows a proper awareness of the fact that the orbit of each planet is in a plane, and that these planes are various planes in a common (three dimensional) overall space. The awareness of such a space, with such various planes in it, would perforce have carried a measure of awareness of a mathematical background space with it. Ptolemy himself did not have such a space either, but it is much easier to "forgive" this failure to Ptolemy than to Copernicus, because once a geocentric view is adopted, the planetary orbits are indeed not planar at all. But Copernicus, from his heliocentric view, was within an ace of "finding" such a space, but he did not find it. In this sense, there never was a "Copernican invention of a planetary *system*," as Derek Price, although militantly opposed to Copernicus, was somehow disposed to concede to him.[6] The glory of having created a

---

[4]Dreyer, pp. 198-199.

[5]Dreyer, p. 339.

[6]Derek J. de S. Price "Contra-Copernicus: A Critical Re-Estimation of the Mathematical Planetary Theory of Ptolemy, Copernicus and Kepler," in Marshall Clagett (ed.), *Critical Problems in the History of Science,* (Madison: University of Wisconsin Press, 1959), pp. 197-218, esp. p. 201.

planetary system belongs to Kepler, and perhaps Tycho Brahe before him.

May I also add that Dreyer's statement above that the ancients would have considered it unusual for the epicycle of an outer planet to remain parallel to a fixed plane in space is already to be found 20 years earlier, in 1887, in the work of Norbert Herz on the history of the determination of orbits of planets and comets, in which he says as follows.

> This uniform motion in an unchanging absolute position in space was a thought that was alien to the ancients, as can be already seen from the fact that they always counted anomalies from instantaneous apogees.[7]

The total absence of a mathematical background space from the astronomy of Ptolemy must not be held against him as an astronomer, because the much vaunted Greek mathematics did not have it either. As I have written elsewhere:

> Most conspicuous, and almost fate-sealing, was the absence from Greek thought of a general conception of space for geometry and geometrically oriented analysis. Greek mathematics did not conceive an overall space to serve as a "background space" for geometrical figures and loci. There is no such background space for the configurations and constructs in the mathematical works of Euclid, Archimedes, or Apollonius, or even in the astronomical work *Almagest* of Ptolemy. When Ptolemy designs a path of a celestial body, it lies in the astronomical universe of Ptolemy; but as a geometrical object of mathematical design and purpose, it does not lie anywhere. In Archimedes, who, in some respects was second only to Isaac Newton, the mathematical constructs were placed in some kind of metaphysical "Nowhere" from which there was "No Exit" into a mathematical "Future."
>
> Such a background space was rather slow in coming. Thus, Nicholas Copernicus did not have it yet. He was an innovator in astronomical interpretation, and not in mathematical

[7]Norbert Herz, *Geschichte der Bahnbestimmung von Planeten und Kometen,* (Leipzig 1887); part IV, Die Breitenbewegungen der Almagest (first page).

> operation. His mathematics was still largely Ptolemaic, and only a bare outline of a *mathematical* background space is discernible in his *De revolutionibus orbium coelestium.* Nevertheless, already a century before Copernicus, Nicholas of Cusa, churchman, theologian, mystic, and gifted mathematician, in Book II of his leading work *Of Learned Ignorance* (*De docta ignorantia*), adumbrated an overall space of mathematics by way of an overall mathematical framework for the space of the universe.[8]

We have observed above that, as far as mathematical background space is concerned, the Copernican Motions in Declination and in Latitude are still conceived entirely in the spirit of the *Almagest.* But in some contexts of *De revolutionibus,* anticipations of a background space are discernible.

First, a whiff of an "absolute" space, but only a whiff, can be detected in Copernicus' insistence on immobilizing the sphere of the stars. In Book I, Chapter 6, when the size of the cosmological Universe is at issue, Copernicus declares that it is "immense" and leaves it at that. Now, whatever this "immense" means—and we will talk about this soon afterwards—, it is certainly not outright "infinite." This being so, if we could assume that Copernicus' "immense" Universe were somehow immersed in an infinite space that surrounded it, then the eighth sphere of the Universe would somehow already belong to the surrounding space, and making this sphere immobile would somehow be in accord with making it "absolute." And it is not at all far-fetched to impute to Copernicus the assumption of the presence of such a space, because some ancient Pythagoreans had already conceived something like this, according to the following famous report of Aristotle in *Physica,* 213b 22-24: "The Pythagoreans, too, held that the void exists and that breath and void enter from the infinite into the Heaven itself which, as it were, inhales."

Secondly, Copernicus' insistence, in Book I, Chapter 6, on his avowal that the Universe is "immense," without any further explanation of what this is meant to be, is of course a kind of refusal, an intentional one, to commit himself as to whether the Universe is finite or infinite. It

[8]Salomon Bochner, "Space" in *Dictionary of the History of Ideas,* (New York: Scribners, 1973), Volume IV, p. 205.

is tempting to take this feature in *De revolutionibus* lightly, or even adjudge it negatively. For my part, I judge it positively, and find it very future-oriented. I adjudge it positively, because, as W. de Sitter stated it so aptly in his book *Kosmos* in 1932: "Infinity is not a physical, but a mathematical conception."[9] In fact, no matter how much you may try to define infinity for some physical object, it turns out that you attach this infinitude to some mathematical structural feature of it. And I find the Copernican "immensity" future-oriented because, as I see things, it was taken over by Newton for his *Principia* which is Newton's counterpart to Copernicus' *De revolutionibus.*

I am fully aware of the prevalent outlook that, say between Copernicus and Newton, a so-called "infinitization of Space," as Koyré has called it, has taken place, and that it somehow culminated in Newton, who was a prime architect of it. Well, I have my opinion on this outlook, which, after quoting a statement of Bishop Barnes of 1931 that "infinite space is a scandal to human thought," I previously expressed thus: "I venture to add that, beginning with Newton, scientists who seemingly acquiesced in infinitude did so not only silently, but also restlessly and even sullenly."[10]

But quite apart of the question of whether Newton was somehow involved in an infinitization of space, I find, no matter how much I go over the *Principia* of Newton, that throughout the length of the *Principia,* and in all three editions of it, Newton *never* states that the Universe, *qua* Universe, is infinite. On the contrary, in the main context, though not the only one, where he touches upon the problem of the size of the Universe, namely in the System of the World, section [57], *On the distance of the stars,* he only uses the word "immense," and nothing stronger. And this seems to be the same "immense" as occurs in Copernicus' *De revolutionibus,* Book I, Chapter 6, as if *with regard to this problem,* nothing had been changed between Copernicus and Newton. When not speaking of the Universe *qua* Universe, but of Space and Time *qua* Absolute Space and Absolute Time, the *Principia* does have once, and only just once, the expression *from infinity to infinity,* and Koyré comments upon this

[9]W. de Sitter, *Kosmos,* (Cambridge, Mass: Harvard University Press, 1932), p. 113.

[10]Salomon Bochner, *Eclosion and Synthesis,* (New York: Benjamin, 1969), p. 256.

> '*From infinity to infinity* retain the same position...' What does *infinity* mean in this place? Obviously not only the spatial, but also the temporal: absolute places retain *from eternity to eternity* their positions in the absolute, that is, *infinite* and *eternal* space, and it is in respect to this space that the motion of a body is defined as being absolute.[11]

But this *infinity* has nothing to do with a cosmological universe. It is a philosophico-theological term and notion that had had currency throughout all the centuries since the earliest Presocratics, especially since the era of Anaximander and Xenophanes. The scholasticism of the High Middle Ages was most intimately familiar with it, however much its "World" may have been "bounded" in the sense of Aristotle and Ptolemy. It is, more or less the same infinity as in Lucretius' argument of the "flying dart," and Lucretius may have had it from a much earlier Pythagorean vision of an "infinite" void surrounding the Heavens. It is this infinity more-or-less that Copernicus had in mind when in Book I, Chapter 8 he suggests that it ought to be left to philosophers of nature to quarrel over whether the World is finite or infinite; whereas in Book I, Chapter 6, when the size of the cosmological Universe is at issue Copernicus declares that it is "immense" and leaves it at that.

But if all this is so, then why is it that Newton somehow creates the impression of advocating infinitude, or at least acquiescing in it, while letting the cosmological Universe be only "immense"?[12] And in a

[11]Alexandre Koyre, *From the Closed World to the Infinite Universe,* (New York: Harper, 1958), p. 166.

[12]In the discussion at the Symposium I was reproved for distinguishing between *mundus* and *caelum,* since in the Middle Ages the two terms were (allegedly) used interchangeably. To which I can only retort that a residue of difference did remain, as the very manner in which Kepler uses the terms in the two contexts demonstrates. Scientists frequently use pseudo-synonyms interchangeably, when no misunderstanding can arise, but become meticulous when the residual difference in meaning is of consequence. This can even be observed in Aristotle. In *De Caelo* he frequently uses *ouranos* (Heaven) and *to pan* (the All) interchangeably. But he decidely prefers *ouranos* when he speculates on the rotation of the Heavens around the Earth, and when he demonstrates that there is only one world ("Space", p. 300). On the other hand, he uses *to pan* in *Physica,* Book 4, Chapter 5, at the end of the essay on *topos,* when raising the question whether the universe as a whole, when viewed as one comprehensive physical system, has a physical *topos* too (ibid.).

parallel feature in Copernicus, why does he say in Chapter 8 of Book I that it should be left to the physiologoi whether the world (*mundus*) is finite or infinite, but in Chapter 6, he only maintains that the heavens (*caelum*) is immense? The reason in both cases is the same. In Newton, what is infinite is the underlying Euclidean Space, and it is infinite as a mathematical object. The fact that Newton fancied to call it "absolute space" does not make the slightest difference. In Copernicus, there is an anticipation of this underlying space, but not as an operational datum, but as a semi-theological one (which Newton's absolute space still in part is), and which was very well known to the scholastics, and was already endowed in Cusanus with purely mathematical features. In fact, Cusanus had a mathematical turn of mind, and the mathematical features which I detect in his Universe are far ahead of his times. To end with a quotation, from J.D. North's important book, *The Measure of the Universe:* "It is easy to speak of the infinite, as every theologian knows, but it is difficult to speak of it meaningfully."[13]

[13](Oxford: Clarendon Press, 1965), p. 23.

III

ON COPERNICUS' THEORY OF PRECESSION

Noel Swerdlow

University of Chicago

Not the least of the problems confronting Copernicus was what to do with observations spread over a period of nearly two thousand years that did not lead to any consistent theory. Using the methods handed down by Ptolemy in the *Almagest* and presented with great clarity by Regiomontanus in the *Epitome of the Almagest,* it generally took a set of three observations to determine the parameters of a geometrical model (since a circle is determined by three points) and a set of two widely spaced observations to determine a mean motion. More than this number would over-determine a problem, and, in the absence of any kind of error theory, lead to inconsistencies that would appear to be long term irregularities in parameters apparently stable over a short period. In the deleted preface to *De revolutionibus,* Copernicus says that one reason that Ptolemy's work is no longer adequate is that "certain motions not yet known to him have been found," and this refers specifically to the long period irregularities resulting from over-determined solutions.[1]

Copernicus considered himself fortunate to know of these additional motions, but it was really the worst thing that happened to him, for the motions unknown to Ptolemy forced upon Copernicus a number of

[1]The preface appears in Copernicus autograph manuscript (henceforth abbreviated Ms.), and this remark is on f. 1v:20. A facsimile of the complete manuscript is reproduced in *Nicolai Copernici Opera Omnia* I, Varsoviae-Cracoviae, 1972. For no apparent reason, the preface was not printed in the first edition, *Nicolai Copernici Torinensis De Revolutionibus Orbium Coelestium, Libri VI,* Nuremberg, 1543 (henceforth abbreviated N). There are a number of facsimiles of N; I have used one from New York, 1965.

exceedingly difficult and altogether fictitious problems. Copernicus was the heir to many sound observations with sound theory to account for them, but he also received observations of varying degree of error leading to theories of varying degree of complexity and the necessity of attacking problems for which there were no rigorously demonstrated methods of solution. He had two choices. The first was to reject all earlier observations until he could first establish a fairly accurate representation of the motions of the heavens using only extremely accurate observations from his own time. Ancient observations could then be measured against this standard to see whether they fit well enough to confirm or correct this preliminary theory, or were so far out of line that they could safely be eliminated as erroneous. This was done by Tycho Brahe and Kepler, and was possibly the intention of Regiomontanus when he began his program of observations at Nürnberg. For a multitude of reasons, most of all his isolation, Copernicus could not, and therefore would not adopt this method. And so the alternative was to trust in the accuracy of his predecessors (for were they not "equal in dedication and carefulness"?),[2] and develop a theory that would, in so far as possible, account for all their observations.

The principal motions not yet known to Ptolemy that brought these troubles to Copernicus were an irregularity in the rate of precession with a (presumably) corresponding irregularity in the length of the tropical year, and a gradual diminution of the obliquity of the ecliptic. Copernicus had a record of earlier observations and parameters that showed these irregularities unmistakably. He was not about to throw out some 1800 years of observations extending back to Timocharis nor some 1650 years of theory extending back to Hipparchus, so he tried to account for all of it, assuming optimistically that each observation and theory was accurate for its time, and that now, after the passage of hundreds of years, it was possible to determine long-period irregularities unknown to earlier astronomers. This was really nothing new, Thābit ibn Qurra, az-Zarqāl, the Alphonsine astronomers, and Johann Werner—to name only those examples known to Copernicus—had tried to do the same thing. Considering the substantial number of attempts, each must have

[2] *De rev.* III, 20 (Ms. f. 100r:7; N f. 90v).

been judged a failure within a couple of centuries, and Copernicus was only one of the last in a succession of failures.

Still more unfortunately, the irregular precession was also one of the first problems that Copernicus was forced to attack. Now, the motions of the moon and planets are all related to the motion of the sun, and thus the subjects treated in Books IV, V, and VI of *De revolutionibus* could not be taken in hand until a correct solar theory had been established. And because Copernicus believed that the tropical year was non-uniform and the sidereal year uniform, he could not establish the length of the sidereal year—to which the motions of the moon and planets are related—until he had first worked out a complete theory of the irregular precession, which is, of course, the difference between the non-uniform tropical year and the uniform sidereal year. There was no way he could just measure the sidereal year directly; he had to correct two widely spaced observations of the sun *with tropical coordinates* by the value of the true precession at each observation in order to obtain the sidereal motion of the sun. Thus, he was forced to derive his precession theory first, and nothing that he attempted presented him with more insurmountable difficulties.

The following study of Copernicus' precession theory will probably raise more questions than it answers. Copernicus seems not to have hesitated to (ever so slightly) falsify observations to fit his completed theory. Is this done throughout *De revolutionibus*? He appears to conceal his derivation of certain critical parameters, indicating that it was done in one way, but in reality using a completely different method about which he says nothing. Tacit assumptions are made in order to simplify a problem, and then later a new section is added, "proving" on the basis of a completed theory what was earlier assumed in order to derive the theory in the first place. Sometimes it is possible to reconstruct what Copernicus in fact did; sometimes it is only possible to show that he could not have done what he claims. I hope that someone will be able to carry this investigation farther, and find solutions to problems that have eluded me.

There has been a previous study in some depth of Copernicus's precession theory by K.P. Moesgaard of the University of Aarhus.[3] While I

[3]"The 1717 Egyptian Years and the Copernican theory of precession," *Centaurus* 13 (1968), pp. 120-138.

reach a number of conclusions differing from Moesgaard, his paper, which I have found extremely helpful, deserves careful examination. The history of precession theory up to and including Copernicus is given in a paper by J. Dobrzycki which I have been able to use only in its English summary.[4] A brief history of the theory of precession in so far as it was known to Copernicus when he wrote the *Commentariolus* is given in my translation of that work along with a description of Copernicus' early (and rather confused) precession theory.[5] Finally, some mention should be made of J.R. Ravetz's proposal that it was the theory of precession that led Copernicus to assert the movement of the earth in the first place. Ravetz's argument, if I understand it correctly, seems to be based on the belief that there is something in a complex, irregular precession that necessitates assigning it to the earth rather than the heavens.[6] That this is untrue can be seen by examining the precession theory of Thābit ibn Qurra or Johann Werner in which all the irregular motions that Copernicus assigns to the earth are given to the heavens with no ill effects other than the general wrong-headedness of the whole theory, a wrong-headedness shared by Copernicus. There is simply no reason whatever to grant the precession to the earth rather than the heavens until *after* the earth is set in motion around the sun. The same thing, incidentally, is true of the diurnal rotation. Further, the precession theory in the *Commentariolus* is extremely crude and in fact probably wrong (the earth's axis is carried about in the wrong direction), so the granting of the precessional motion to the earth was probably the last and most uncertain part of Copernicus' early astronomical investi-

[4]"Theoria Precesji w Astronomii Średniowiecznej," *Studia i Materialy z Dziejów Nauki Polskiej,* C 11 (1965), pp. 3-47 with English summary.

[5]"The Derivation and First Draft of Copernicus's Planetary Theory: A Translation of the Commentariolus with Commentary," *Proceedings of the American Philosophical Society*, 1973, *117*: 445-450.

[6]*Astronomy and Cosmology in the Achievement of Nicolaus Copernicus,* Wrocław, Warszawa, Kraków, 1965; "The Origins of the Copernican Revolution," *Scientific American,* 1966, *215,* 4:88-98. Ravetz's error seems to occur when he says that if the precession is brought about my moving the ecliptic, then the latitudes of stars will be altered. However, in all such theories, e.g., Thābit's or Werner's, the entire sphere of the fixed stars moves along with the movable ecliptic so the latitudes are not affected. Ravetz's argument, which is presented more clearly and effectively in the *Scientific American* article than in the earlier monograph, is nevertheless ingenious.

gations, and in no way his first reason for considering the motion of the earth.

1. *The Observational Evidence*

In III, 2 Copernicus gives a number of observations of three fixed stars—Spica, Regulus, and β Scorpionis—and then shows that various rates of precession follow from the changes in longitude between different pairs of observations.[7] Thirteen observations at eight different dates are given, from Timocharis in - 293 to Copernicus in 1525, but only a few of these are ever used in deriving the parameters of the precession theory. All the observations and corresponding rates of precession have been tabulated by Moesgaard (although the observation for 1498 is hypothetical, and the one for 1515 is, as we shall see, falsified by Copernicus) so we shall tabulate only the time intervals Δt and differences in longitude $\Delta\lambda$ needed for the derivations. Copernicus rounds Δt to integral Egyptian years (abbreviated *ey*), and here we merely copy his values without correction. The epoch used is Era Philipp (12 November, - 323), which Copernicus calls the Death of Alexander.

		Δt^{ey}	$\Delta\lambda$	π
Timocharis	E.P. 30			
		432	4;20°	$1°/100^{ey}$
Ptolemy	E.P. 462			
		741		
		742	11;30°	$1°/65^{ey}$
al-Battānī	[E.P. 1204] E.P. 1202			
		645(!)	9;11°	$1°/71^{ey}$
Copernicus	E.P. 1849			

[7]Copernicus' sources for these observations were: (1) *Almagest* VII, 2-3 (Gerard of Cremona translation, Venice, 1515); (2) *Epytoma Joannis De monte regio In almagestum Ptolemei* (Venice, 1496) VII, 2-6; (3) G. Valla, *De expetendis et fugiendis rebus,* (Venice, 1501), Book XVII. This last is a printing of Ptolemy's star catalogue with an extraordinary number of typographical errors such as the incorrect longitude of Spica, Virgo

The chronological errors in the Δt column are the result of Copernicus' originally dating al-Battānī E.P. 1204 and then correcting the date to E.P. 1202 without likewise adjusting the intervals (Ms. f. 72r). He is also inconsistent about whether the interval between Ptolemy and Battānī is 741^{ey} or 742^{ey}, but for the later derivations he uses 742^{ey}. It is evident that the rate of precession π was slowest in the first interval, fastest in the second, and has slowed somewhat in the third.

Next Copernicus gives a series of values of the obliquity of the ecliptic. There are some problems and inconsistencies in these, but only the first and last are ever really used for anything. The list is as follows:

Ptolemy	23;51,20°	
al-Battānī	23;36°	(Correctly 23;35°)
az-Zarqāl	23;34°	(Correctly 23;33,30°)
Prophatius	23;32°	
In our times	23;28,30°	

Where Copernicus found the values for Battānī, az-Zarqāl and Prophatius is by no means clear,[8] the value for Ptolemy he also attributes to Aristarchus (owing to the corruption of the name Eratosthenes into *Archusianus* in the Gerard of Cremona translation of the *Almagest*),[9] and the value for Copernicus' own time has, as we shall see, a very interesting history. Nevertheless, it is evident that the obliquity has been decreasing from the time of Ptolemy on.

26;30° instead of 26;40°, that Copernicus mentions and then corrects in III, 2. In connection with this, one should consult J. Dobrzycki, "Katalog gwiazd w De revolutionibus," *Studia i Materialy z Dziejów Nauki Polskiej,* 1963, *C7*: 109-153.

[8]It is not quite clear what Copernicus thought Battānī's value was. We read in III, 2, Ms. f. 73r, 23;36°; N f. 65b, 23;26°, evidently a misprint (*xxvi* for *xxxvi*); in III, 6, Ms. f. 79r 23;27° changed to 23;35° (the correct value); N f. 69b 23;25°, evidently a misprint for 23;35°. Rheticus gives 23;35° in the *Narratio Prima* (in Kepler, *Gesammelte Werke* I, Munich, 1938, p. 91). It is probable that Rheticus knew the correct value from the 1537 Nürnberg edition of Battānī, but it is not certain whether Copernicus ever saw this. In III, 12 Copernicus computes from his theory of the variation of the obliquity an obliquity of 23;38° for the time of Battānī, and seems to consider this satisfactory.

[9]In II, 2, which was written after III, 2, he correctly names Eratosthenes. The name is given correctly in both the Greek *Almagest* (Basel, 1538) and in the George of Trebizond translation (Venice, 1528).

2. *Copernicus' Observations*

Copernicus himself is responsible for three observations—two of Spica and one of the obliquity—and two of the three are falsified to agree with computation from the finished theory. It is necessary to get rid of these observations to show that they had no part in the derivation of the parameters. We shall investigate all three.

(1) Longitude of Spica for 1525.

The observation was done by measuring the altitude of Spica at meridian transit. Since the celestial equator intersects the meridian at right angles, the difference between the elevation of the culminating point of the equator and the altitude of the star immediately gives the star's declination. Copernicus observed an altitude of Spica of 27°. He takes the latitude of Frauenburg to be 54;19,30° so the elevation of the intersection of the equator and meridian is 35;40,30° and the declination of Spica to the south is

$$35;40,30° - 27° = 8;40,30° \approx 8;40° .$$

He assumes Ptolemy's value for the latitude of Spica, 2° south, and assumes further that the obliquity of the ecliptic is 23;28,30°. Given the declination and latitude of Spica, and the obliquity of the ecliptic, the longitude is derived as follows (see Figure 1 drawn from the manuscript).[10]

With the observer at E taken as center, draw the meridian circle $ABCD$, let AEC be the celestial equator, BED the ecliptic, and the axis of the ecliptic FEG where F is the north pole. On the meridian take an arc AM equal to the southern declination of Spica, and draw MN paral-

[10]Ms. f. 72rv. Copernicus' method of finding a longitude from a given declination and latitude is an analemma construction similar to one used by Johann Werner for the same purpose in Proposition 2 of *De motu octavae sphaerae* (see below note 11) although Werner's numerical solution, which takes the versed sine of the longitude, is more direct than Copernicus' somewhat clumsy procedure. I would guess that Copernicus' derivation is adapted from Werner's.

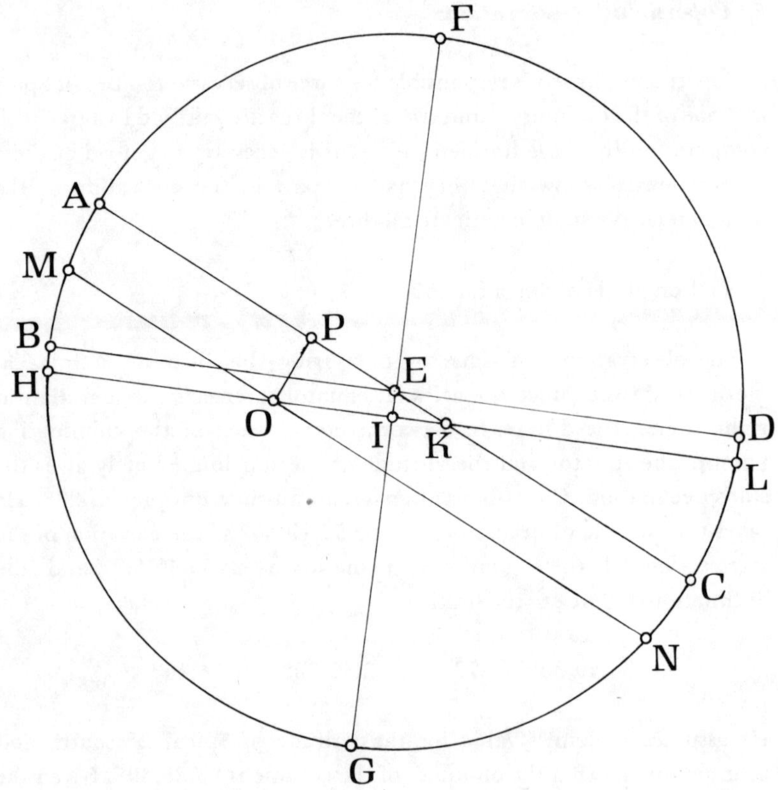

Figure 1

lel to AC. Likewise take an arc BH equal to its southern latitude, and draw HL parallel to BD. HL will intersect the axis of the ecliptic at I, the equator at K, and MN at O which is the location of the star. Draw OP perpendicular to AC and MN. It is required to find OI, the distance of Spica from the autumnal equinox.

Now

arc AM = 8;40°, sin AM = 15069

arc AB = 23;28,30°, sin AB = 39832 (Accurately 39835)

arc BH = 2;0°,

and

arc ABH = 25;28,30°. sin ABH = 43010 (Accurately 43012)

Thus, where

$$BE = 100000\ ,$$

$$HIK = \frac{\sin ABH}{\sin AB} = \frac{43010}{39832} = 107978\ ,$$

and

$$OK = \frac{\sin AM}{\sin AB} = \frac{15069}{39832} = 37831\ ,$$

so that

$$HO = HIK - OK = 70147.$$

But

$$HOI = \cos BH = 99939,$$

and thus

$$OI = HOI - HO = 29792$$ (N and Ms. corr. to 29892 which is an error)

where

$$BE = 100000.$$

However, where

$$HOI = 100000,$$

$$OI = \frac{29792}{99939} = 29810 = \sin 17;20,36°.$$

Copernicus rounds this to 17;21°, and thus the longitude of Spica in 1525 is Libra 17;21°. Aside from the small errors in the sines of AB and ABH, the computation is accurate.

(2) Longitude of Spica for 1515.

Copernicus then says that ten years earlier, in 1515, he found the declination of Spica to be 8;36° to the south and its longitude Libra 17;14°. The computation of this value is not shown, but we may reconstruct it as follows:

The obliquity is again 23;28,30°, and the latitude 2;0° to the south as in the previous derivation. Thus arcs AB, BH, and ABH are unchanged, but the declination

$$\text{arc } AM = 8;36°. \qquad \sin AM = 14954$$

Now, where

$$BE = 100000$$

again

$$HIK = 107978,$$

but

$$OK = \frac{\sin AM}{\sin AB} = \frac{14954}{39832} = 37543\ ,$$

so that

$$HO = HIK - OK = 70435.$$

And since

$$HOI = \cos BH = 99939,$$

thus

$$OI = HOI - HO = 29504$$

where

$$BE = 100000.$$

So where

$$HOI = 100000,$$

$$OI = \frac{29504}{99939} = 29520 = \sin 17;10,11°.$$

Copernicus, however, gives the longitude Libra 17;14°, but we see in the manuscript that he originally wrote 17;10° and then changed the 10 to 14. If now we compute the position of Spica from Copernicus' tables, where λ^* is the sidereal and λ the tropical longitude, we find for January 0, 1515

$$\lambda = \lambda^* + \bar{\pi} + \delta = 170;0° + 26;39,22° + 0;34°$$

$$= 198;13,22° = \text{Libra } 17;13,22°.$$

A date only about two months into the year will of course round to 17;14°, exactly Copernicus' result. He has, therefore, altered the observation to agree with the completed theory of precession.

(3) The Obliquity of 23;28,24° for 1525.

We have seen that for the reduction of the observations of Spica for both 1515 and 1525 Copernicus used the obliquity 23;28,30°. In III, 12 he computes for 1525 an obliquity of 23;28,24° (although even here, on f. 87r of the manuscript he accidentally wrote *secunda xxx* and then changed it to *xxiiii*), and in three other places he claims to have found such a value for the obliquity. However, an examination of the manuscript shows that all references to the obliquity 23;28,24° are later additions. Indeed, in one place Copernicus claims to have found 23;28,30°. The passages in question are as follows:

(a) II, 2: Ms f. 27v; N f. 29a.

After explaining that Ptolemy believed the obliquity to be fixed at 23;51,20°, Copernicus remarks that since Ptolemy's time it has continuously decreased. He then continues (Ms version with N in brackets):

> Reperta est enim iam a nobis et aliis quibusdam coaetaneis nostris distantia tropicorum partium esse non amplius xlvi et scrupulorum primorum lviii fere et angulus sectionis partium xxiii scrupulorum xxix [N partium 23 scrupulorum 28 et duarum quintarum unius], ut satis iam pateat mobilem esse etiam signiferi obliquationem, de qua plura inferius, ubi etiam ostendemus coniectura satis probabili, numquam maiorem fuisse partibus xxiii scrupulis lii nec umquam minorem futuram partibus xxiii scrupulis xxviii.
>
> For it has been found by us and by some of our other contemporaries that the distance between the tropics is not more than about 46;58° and the angle of intersection 23;29° [N 23;28⅖°] so that it is now sufficiently clear that the obliquity of the ecliptic is also movable; concerning this more below where we shall also show by a sufficiently probable conjecture that it was never greater than 23;52° nor will ever be less than 23;28°.

Here we see that Copernicus originally had the value 23;29°. Still more important is his remark that he will show by a *sufficiently probable conjecture* that the obliquity varies between 23;52° and

23;28°. This, as we shall see, is the truth, for Copernicus did in fact originally guess this variation of 0;24°, while the derivation of this value in III, 10 is a later addition and something of a fake.

(b) III, 2: Ms f. 73r; N f. 65b.

After giving the values of the obliquity found by Ptolemy, al-Battānī, az-Zarqāl, and Prophatius, Copernicus says (Ms):

> Nostris autem temporibus non invenitur maior partibus xxiii scrupulis xxviii s [*deleted*: vel xxix secundum aliquos]. . . .
>
> In our times it is found to be not greater than 23;28½° [*deleted*: or 29 [minutes] according to some]. . . .

The deleted part is, of course, not in N. Here Copernicus has the value 23;28,30° which was the obliquity used for the reductions of the observations of Spica.

(c) III, 6: Ms f. 79r; N ff. 69b-70a.

This is a more difficult passage for the manuscript shows two sets of corrections. It is quite important for it allows us to determine Copernicus's source for contemporary values of the obliquity. The first version read:

> Quod denique nostra concernit tempora, Georgius Purbachius anno Christi mcccclx partium ut illi xxiii scrupulorum vero xxviii adnotavit, Dominicus Maria Novariensis anno Christi mccccxci ultra partes integras scrupula xxix et amplius quiddam, nos ab annis xxx frequenti observatione non multum excedentem (?) scrupula xxviii xxiii partes.
>
> And finally, concerning our times, Georg Peurbach in A.D. 1460 observed 23 degrees, as did the former [observers], but 28 minutes, Dominico Maria of Novara in A.D. 1491 [observed] above the whole degrees 29 minutes and some more, we, from 30 years of repeated observation, not much exceeding 23;28°.

The second version, which differs slightly from the first, depends upon two marginal additions and one small deletion from the text:

> Quod denique nostra concernit tempora, Georgius Purbachius anno Christi mcccclx partium ut illi xxiii scrupulorum vero xxviii adnotavit, Dominicus Maria Novariensis anno Christi mccccxci ultra partes integras scrupula xxix et amplius quiddam, Joannes Regiomontanus partium 23 scrupulorum 28 et dimidii, nos ab annis xxx frequenti observatione scrupula xxviii xxiii partes parum differens.
>
> And finally, concerning our own times, Georg Peurbach in A.D. 1460 observed 23 degrees, as did the former [observers], but 28 minutes, Dominico Maria of Novara in A.D. 1491 [observed] above the whole degrees 29 minutes and some more, Johannes Regiomontanus 23;28½°, we, from 30 years of repeated observation, 23;28°, differing slightly.

The third version differs radically from the first two, and is almost identical to the printed edition:

> Quod denique nostra concernit tempora, nos ab annis xxx frequenti observatione, [N invenimus] xxiii partes scrupula xxviii et duas fere quintas unius scrupuli, a quibus Georgius Purbachius et Joannes a monte regio, qui proxime nos praecesserunt, parum differunt.
>
> And finally, concerning our own times, we, from 30 years of repeated observation, [N have found] 23;28° and about ⅖ of one minute [i.e., 23;28,24°], from which our closest predecessors Georg Peurbach and Johannes Regiomontanus differ slightly.

Here Copernicus drops all mention of specific values for his nearest contemporaries and gives his own value as 23;28,24°. In the first version, however, he claimed slightly over 23;28° and in the second 23;28° presumably exactly.

(d) III, 10: Ms f. 76v; N f. 76ab.

In this chapter Copernicus demonstrates that if the obliquity was 23;51,20° in the second year of Antoninus (A.D. 138/39) and 23;28,24° in 1525, then the maximum value is 23;52° and the minimum 23;28°. However, the chapter is a later addition written on a sheet inserted out of order in the manuscript and disturbing the numbering of the chapters from chapter 11 on to the end of Book III. Further, it is written on a sheet of paper of watermark E while all the rest of the section on precession and change of obliquity is written on paper of watermarks C and D. The whole chapter is a fake in which Copernicus pretends to prove what he in fact assumed, and we shall take it up at the end of this paper.

Considering all these passages we see that in every case the obliquity 23;28,24° is a later addition or alteration. It is not clear what value Copernicus himself found "from 30 years of repeated observation." He gives 23;29°, 23;28°, slightly over 23;28°, and 23;28,30°, the value used for reducing the observations of Spica. Now there is an earlier source for this value not mentioned by Copernicus, and it was in fact his source for the other contemporary values of the obliquity cited in the cancelled passage in III, 6.

3. Werner's Record of Recent Measurements of the Obliquity

Johann Werner's *De motu octavae sphaerae* published in Nürnberg in 1522 would probably be all but forgotten today had not Copernicus written a critical review of it in a letter to his friend Bernhard Wapowski dated 3 June, 1524.[11] Copernicus takes Werner to task for the serious error of 11 years in fixing the date of Ptolemy's observation of Regulus on 9 Pharmuthi in the second year of Antoninus. Werner had equated this with 22 February, 150; Copernicus corrects this to 22 February, 139; it is in fact 23 February, 139. Then Copernicus criticizes Werner for using the slowest motion of the precession as his "mean" precession. Werner did this so that the correction due to trepidation would always be positive, but Copernicus believes that a "mean" motion should be a *mean* motion so corrections must then be both positive and negative. Finally he objects to Werner's suspicion of the accuracy of ancient observations that do not agree with computation from his theory. Here Copernicus points out that it is more likely that the fault lies with Werner's theory than with the ancient observations. While Copernicus'

[11] *Libellus Joannis Werneri. . . de motu octavae Sphaerae. Tracatus duo,* Nürnberg, 1522. Copernicus' letter is printed twice in L. Prowe, *Nicolaus Coppernicus,* II (Berlin, 1883, repr. Osnabruck, 1967), pp. 145-153, 169-183. There is a translation in E. Rosen, *Three Copernican Treatises,* 3rd ed. (New York, 1971), pp. 93-106.

second objection is but a quibble, the first and third certainly hit the mark and show that there are grave problems in Werner's theory.

Now the letter to Wapowski was not the end of Copernicus' concern with Werner's book. Although he judged Werner's theory of precession to be unsatisfactory, nevertheless he found some information given by Werner of great use. In Proposition 25, Werner gives the following report on his own and various other recent measurements of the obliquity:

> Quarto supponendum est circa annos a nativitate domini 1514 completos, maximam solis ab aequatore declinationem esse graduum xxiii primorum minutorum xxviii secundorum xxx. Tantam praedictis annis a domini nativitate perfectis et circiter Nurembergae diligenti observatione per regulas Ptolemaei depraehendi, eandem denique Ioannes de Regio monte et plures alii post eum Nurembergae regulis quibusdam aeneis invenerunt. Et in Italia praecipue Bononiae quidam Dominicus Maria Novariensis circa annos domini 1491 accurata inspectione sua invenit maximam solis declinationem graduum xxiii primorum minutorum xxix. Georgius denique Peurbachius, Ioannes de Regio monte praeceptor, in Vienna Pannoniae superioris circa annos domini 1460 reperit eandem maximam solis declinationem graduum xxiii primorum minutorum xxviii. Et quoniam hae tam excellentium mathematicorum inventiones considerationi meae, plurimum conveniunt, ideo libenter credo maximam solis declinationem pro annis domini 1514 completis esse graduum xxiii primorum minutorum xxviii secundorum xxx.
>
> Fourth it is to be assumed that at about 1514 completed years from the birth of Our Lord the maximum declination of the sun from the equator is 23;28,30°. I found this value in the aforesaid completed years from the birth of Our Lord near Nürnberg by careful observation with Ptolemy's rulers,[12] and

[12]This is the so-called parallactic instrument, used to measure zenith distances, and described by Ptolemy in *Almagest* V, 12. Copernicus himself made such an instrument (*De rev.* V, 15) which was later acquired by Tycho Brahe (*Description of his Instruments and Scientific Work as given in Astronomiae instauratae mechanicae,* trans. H. Raeder, E. Strömgren, B. Strömgren (Copenhagen, 1946), pp. 45-46.

> in fact Johannes Regiomontanus and many others after him in Nürnberg found the same value with certain bronze rulers. And in Italy, principally in Bologna one Dominico Maria of Novara at about 1491 years of Our Lord found by his accurate observation a maximum declination of the sun of 23;29°. Finally, Georg Peurbach, the teacher of Johannes Regiomontanus, in Vienna in Upper Pannonia at about 1460 years of Our Lord found the same maximum declination of the sun of 23;28°. And since these findings of such excellent astronomers agree very well with my observation, therefore I willingly believe that the maximum declination of the sun for 1514 completed years of Our Lord is 23;28,30°.[13]

Comparing the first two versions of Copernicus' account in III, 6 with this passage from Werner, we see that the agreement is perfect except that to Dominico Maria of Novara's 23;29° Copernicus adds "and some more." Presumably this is because Copernicus, having worked with Dominico Maria in Bologna, knew something about Dominico's value for the obliquity that Werner did not know, that is, knew that he found it somewhat greater than 23;29°. Werner's attribution to Regiomontanus of 23;28,30° is, to my knowledge, attested nowhere else. In the *Tabulae directionum* and *Tabula primi mobilis* Regiomontanus uses 23;30°. Either Werner is incorrect or he has learned this value from Regiomontanus' papers in Nürnberg. In any case, Copernicus' account agrees with Werner's.

We have seen that Copernicus uses the obliquity 23;28,30° for computing the longitude of Spica. It seems likely from the passages cited above that he himself found values of anywhere from 23;28° to 23;29° during his 30 years of repeated observation, and then decided to use 23;28,30° for his own work on the basis of Werner's testimony that Regiomontanus and many other Nürnberg astronomers found this value.

[13]Werner gave an earlier review of recent values of the obliquity in his *In primum librum Geographiae Claudii Ptolemaei argumenta, paraphrases, et annotationes* (Nürnberg, 1514), cap. III, annot. vi, but the value 23;28,30° is not mentioned and this was not Copernicus' source.

4. Description of the Model

In Chapter 3, Copernicus describes a model for the irregular precession and the change of obliquity. The model is based on some assumptions made by Copernicus on the basis of the foregoing observations. Note that the precession was slowest at some time before Ptolemy, rather rapid between Ptolemy and Battānī, and somewhat slower between Battānī and Copernicus. The obliquity, however, has continuously decreased between Ptolemy and the time of Copernicus. Copernicus now makes three assumptions:

(1) The obliquity will not continue to decrease, but varies within fixed limits.

(2) The period of one oscillation of the obliquity from maximum to minimum and back to maximum is exactly twice the period of the variation of the velocity of precession from slowest to fastest to slowest.

(3) The cycles of the obliquity and precession begin at exactly the same time.

We shall also see that Copernicus made some further assumptions for deriving the parameters of the model.

Copernicus must now develop a model that will produce a uniform precession, an irregular variation of this precession, and a variation of the obliquity. All of these motions are envisioned as applied to the earth's axis of rotation and bringing about corresponding motions of the earth's equator, and hence of the celestial equator, with respect to the ecliptic. We shall consider each component separately.

(a) The Uniform Precession, $\bar{\pi}$.

If one wanted to allow a motion of the earth to represent the precession under the assumption that the earth was fixed at the center of the celestial sphere, it would only be necessary to let the earth's axis slowly describe a circle about the pole of the ecliptic. However, the earth is not fixed, but is revolving in a circle about the mean sun. One could now say that the earth's axis remains parallel to itself as the earth moves about the sun, and further, that it slowly describes a circle about a line parallel to the axis of the ecliptic. But Copernicus does not do this. That the axis of the moving earth remains parallel to itself requires in itself an additional motion in the direction opposite to the motion about the sun, and therefore one must add this motion to the slow precessional motion, both of which take place in the direction opposite to the earth's motion about the sun. The model is shown in Figure 2.

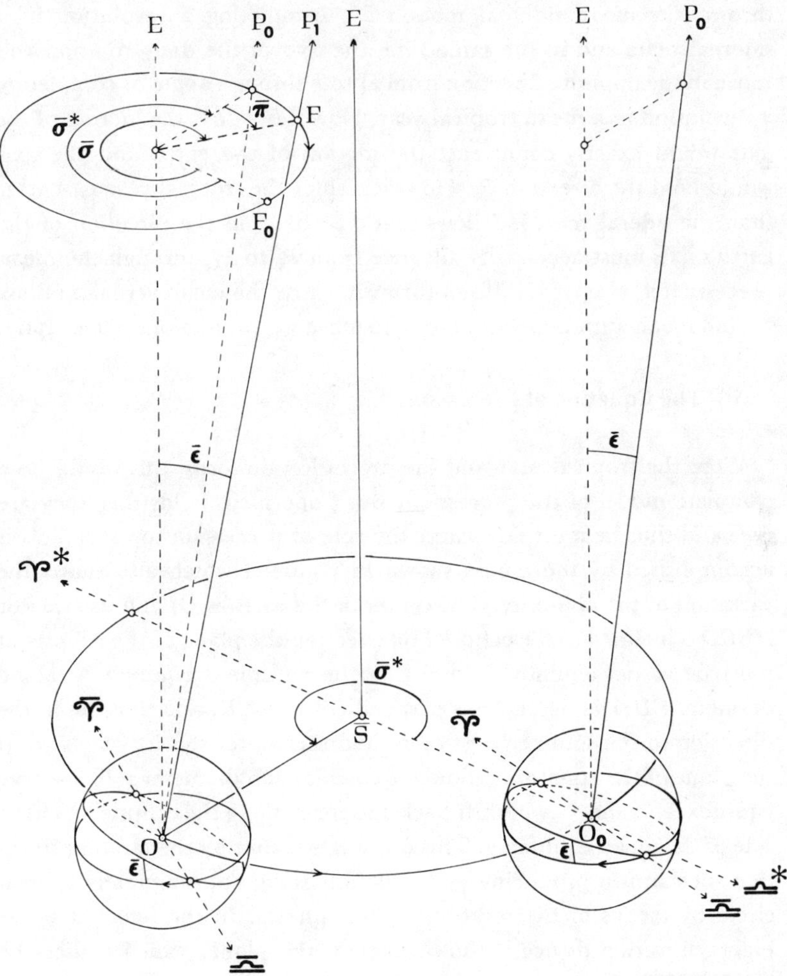

Figure 2

The center of the earth's orbit, the mean sun, is $\bar{S}$, the first star of Aries, from which Copernicus measures sidereal longitudes, is in the direction $\bar{S}$♈*, and the earth is initially in the direction $\bar{S}$♎* at O_0. The pole of the ecliptic holds the fixed sidereal direction E while the earth's axis of rotation is inclined to E by the angle $\bar{\varepsilon}$ of the mean obliquity and lies in the initial direction P_0. The earth now moves from O_0 to O through its mean sidereal motion $\bar{\sigma}^*$, completing a revolution in a sidereal year, and in the same time the axis of the daily rotation will move in the opposite direction from F_0 to F through angle $\bar{\sigma}$, completing a revolution in a mean tropical year. Now if $\bar{\sigma} = \bar{\sigma}^*$, the motion of the axis would exactly counteract the motion of the earth and the axis would hold the direction P_0. However, since the tropical year is shorter than the sideral year, it follows that $\bar{\sigma} > \bar{\sigma}^*$, and the direction of the earth's axis must necessarily advance from P_0 to P_1, through the mean precession $\bar{\pi} = \bar{\sigma} - \bar{\sigma}^*$. This in turn will cause the mean vernal equinox ♈̄ and mean autumnal equinox ♎̄ to move westward along the ecliptic.

(b) The Equation of Precession, δ.

Were the tropical year and the precession uniform, this would be a complete model of the precession. But Copernicus holds that they are not, and thus he must now cause the rate of precession to vary. This is accomplished by the model shown in Figure 3 which also shows the variation of the obliquity. The center of the earth is O, and its equator $ABCD$ is inclined to the ecliptic, intersecting the plane of the ecliptic at the true vernal equinox ♈ and the true autumnal equinox ♎ along diameter CD. Diameter AB is perpendicular to CD, and thus lies in the direction to the summer solstice ♋ and the winter solstice ♑. Now, if the plane of the equator is allowed to oscillate on diameter AB, the true equinoxes ♈ and ♎ will shift back and forth along the ecliptic on either side of the mean equinoxes. This is brought about by the familiar libration mechanism producing an oscillation along approximately a great circle by means of the rotation of two spheres. In the figure it is the higher libration device,[14] shown inset in the plane, which causes the

[14] Copernicus does not specify the order of the libration mechanisms, and it does not really matter. In the *Narratio Prima* (in Kepler, *Werke* I, pp. 108-111, esp. p. 110:27; trans. E. Rosen, *Three Copernican Treatises,* pp. 153-159, esp. p. 157) Rheticus

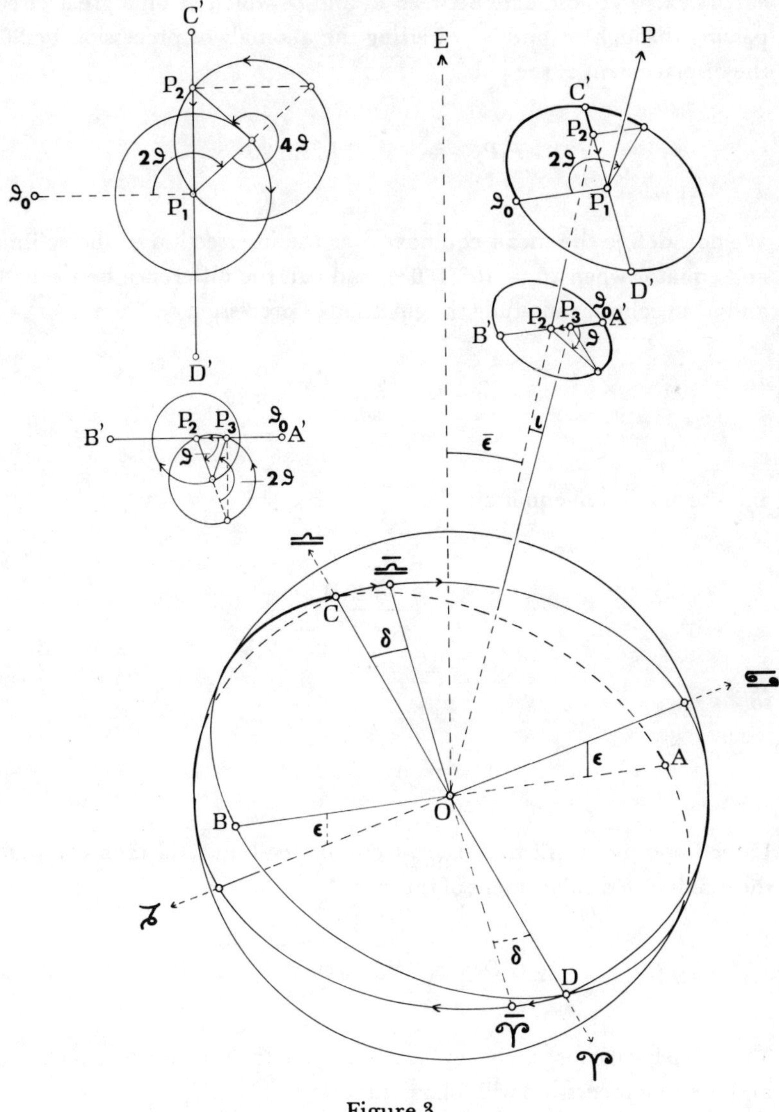

Figure 3

definitely places the device for changing the obliquity above, but I am doing it the other way around for didactic reasons. Descriptions of the libration device can be found in Moesgaard and elsewhere.

earth's axis P_2 to oscillate between C' and D' which lie on a great circle passing through $\bar{♈}$ and $\bar{♎}$. Letting the anomaly of precession be 2ϑ, the displacement of the pole

$$P_1P_2 = \tfrac{1}{2}\, C'D' \sin 2\vartheta\,.$$

We now define the mean equinox $\bar{♈}$ as the intersection of the ecliptic and equator when $2\vartheta = (0°, 180°)$, and call the difference between $\bar{♈}$ and ♈ in ecliptic longitude the equation of precession δ. Then

$$\delta = \frac{\tfrac{1}{2}\, C'D'}{\sin \varepsilon} \cdot \sin 2\vartheta\,,$$

and the maximum equation

$$\delta_{max} = \frac{\tfrac{1}{2}\, C'D'}{\sin \varepsilon}\,,$$

so

$$\delta = \delta_{max} \sin 2\vartheta\,.$$

Later Copernicus will find δ_{max} from observation, and then compute the maximum displacement of the axis

$$\tfrac{1}{2}\, C'D' = \delta_{max} \sin \varepsilon\,.$$

The model will cause an irregular variation in the velocity of precession, and the true precession will follow from

$$\pi = \bar{\pi} \pm \delta\,. \qquad \begin{array}{l} - \text{ for } 0° \leqslant 2\vartheta \leqslant 180° \\ + \text{ for } 180° \leqslant 2\vartheta \leqslant 360° \end{array}$$

(c) The Variation of the Obliquity, ι.

Now the true obliquity ε is not constant, but varies on either side of a mean obliquity $\bar{\varepsilon}$ over a small range ι. This is brought about by a second libration device, also shown inset in the figure, that will cause the plane of the equator to oscillate along diameter CD, thereby changing the angle between diameter AB and the ecliptic by causing the axis P_3 to oscillate between A' and B' which lie on a great circle perpendicular, not to $C'D'$, but to CD. Thus the displacement of the pole P_2P_3 is directly the variation of the obliquity ι, and since it is assumed that the period of the variation of the obliquity is exactly twice the period of the anomaly of precession and the epochs of the anomalies are the same, thus

$$\iota \quad = \quad P_2P_3 \quad = \quad \tfrac{1}{2}\, A'B' \cos \vartheta \quad = \quad \iota_{max} \cos \vartheta \, .$$

We define the mean obliquity $\bar{\varepsilon}$ as the angle of intersection of the ecliptic and equator when $\vartheta = \pm\, 90°$, and then

$$\varepsilon \quad = \quad \bar{\varepsilon} \pm \iota \qquad \begin{array}{l} + \text{ for } 270° \leqslant \vartheta \leqslant 90° \\ - \text{ for } 90° \leqslant \vartheta \leqslant 270° \end{array}$$

Note that the displacement of the pole along $C'D'$ and the maximum equation of precession were related by

$$\delta_{max} \quad = \quad \frac{\tfrac{1}{2}\, C'D'}{\sin \varepsilon} \, .$$

Since ε is not constant, and $C'D'$ must remain constant, the variation of ε will cause a slight change in δ_{max}. Using Copernicus' parameters, the variation of δ_{max} can amount to about 0;1°, but taking this into account would create such a nuisance that Copernicus, who must have been aware of it, neglects it. As long as δ_{max} is taken to be constant, we may dispense with consideration of the motion of P_2 along $C'D'$ and instead represent the equation of precession as a libration of ♈ along the diameter of a circle about $\bar{♈}$ with radius δ_{max}. This is the way that Copernicus treats the equation in deriving the parameters of the model.

5. *Derivation of the Parameters*

In chapters 6-11 Copernicus derives and modifies the parameters for the precession and change of obliquity. The crucial parameter is the anomaly of precession 2ϑ or the anomaly of the obliquity ϑ, which is of course, the same parameter. Only by first isolating ϑ can he then determine the mean precession $\bar{\pi}$ and the equation of precession δ. The problem is by no means simple because he does not yet know even an approximate period for ϑ and π. He must therefore make some further assumptions, and indeed, we shall see that there are two more made tacitly in his derivations.

(a) The Anomaly of Precession, ϑ.

This is the first parameter Copernicus derives. However, he does not show how he did it; instead he is not a little deceptive. His account is about as follows (see Figure 4).

Let *ABCD* be the circle of the anomaly of precession, let the precession be slowest when the anomaly is at *A*, fastest at *C*, let the increasing mean be *B* and the decreasing mean *D*. Between the Spica observation of Timocharis in -293 and Copernicus' Spica observation of 1525 there are about 1819 Egyptian years (all periods are rounded to integral numbers of years). If it is assumed that the period of the anomaly is 1819 years, then, dividing the circle into 360°, the intervals between the observations of Timocharis, Ptolemy, Battānī, and Copernicus will be:

	Δt	$\Delta 2\vartheta$
Timocharis		
	432^{ey}	85;30°
Ptolemy		
	742^{ey}	146;51°

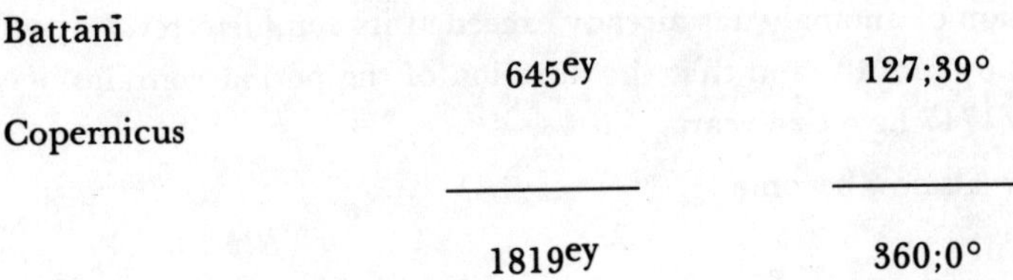

Battānī	645^{ey}	127;39°
Copernicus		
	1819^{ey}	360;0°

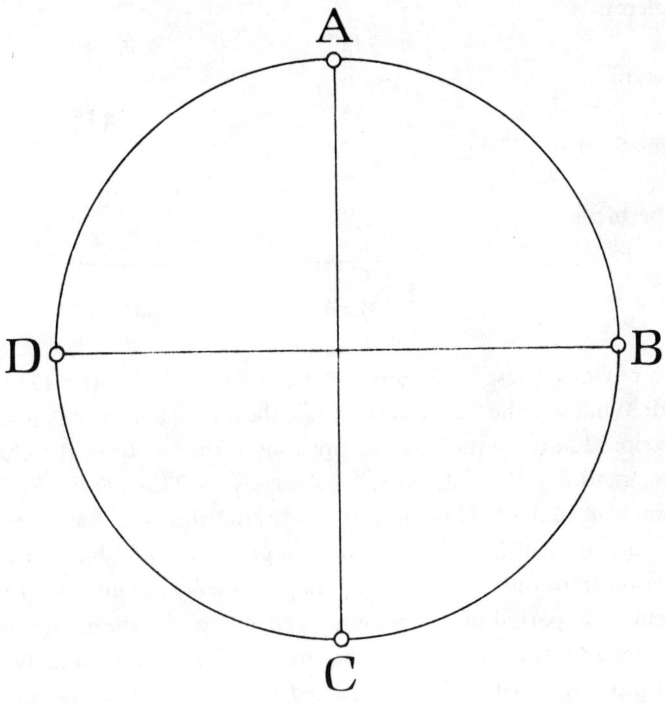

Figure 4

He then continues:

> We have obtained these easily and by a simple guess (*simplici coniectura*), but going over [them] again with a more careful calculation in order that they may agree more precisely with the observations, we find that in 1819 Egyptian years the

> motion of anomaly has already exceeded its complete revolution by 21;24°, and that the duration of the period contains only 1717 Egyptian years. . . .

The intervals now become:

	Δt	$\Delta 2\vartheta$
Timocharis		
	432^{ey}	90;35°
Ptolemy		
	742^{ey}	155;34°
Battānī		
	543^{ey}	113;51°
Timocharis + 1717^{ey}		
	102^{ey}	21;24°
Copernicus		
	1819^{ey}	381;24°

The obvious question is how did Copernicus derive the 1717 year period? What was the "more careful calculation?" He certainly gives the impression that it comes from the precession, that is, from the changes in the longitudes of the fixed stars observed by Timocharis, Ptolemy, Battānī, and himself. However, after working this over for more time than I care to admit, I cannot see how Copernicus could have extracted this period from the observations. Five parameters are mixed up in the problem—the period of the anomaly and its epoch, the period of the precession and its epoch, the maximum equation of the anomaly—and there is no way he could have extracted the first without making some arbitrary assumptions about the other four. Now, we shall see that in order to derive the maximum equation, he does make a nearly correct assumption about the epoch of the anomaly, and in order to then correct the epoch of the anomaly he uses pure trial and error, but no methods of this sort that I have tried will produce the 1717 year period of the anomaly.

There is, however, a different motion from which the mean anomaly can be isolated far more easily, and that is the variation of the obliquity of the ecliptic. Here there is but a single motion, the anomaly itself, and two other parameters—the epoch of the anomaly and the maximum equation. If any one of these parameters is assumed, the other two can be derived from two observations of the obliquity. This, I believe, was Copernicus' method. Back in II, 2 he had remarked about the variation of the obliquity that "...we shall also show by a sufficiently probable conjecture (*coniectura satis probabili*) that it was never greater than 23;52° nor will ever be less than 23;28°." This statement cannot apply to the demonstration of the maximum and minimum values in III, 10 because that chapter is a rigorous proof, not a "sufficiently probable conjecture," and, as I mentioned earlier, III, 10 is a later insertion and something of a fake in that Copernicus proves what he had earlier assumed. Thus, Copernicus' fourth assumption is:

(4) The obliquity varies within a range of exactly 0;24°, from 23;52° to 23;28°.

The derivation of the mean anomaly is as follows:

We assume that the obliquity at the time of Ptolemy was 23;51,20°, the obliquity at the time of Copernicus or Werner 23;28,30°, and that the total variation of obliquity is 0;24°. Now, in Figure 5, we let A be

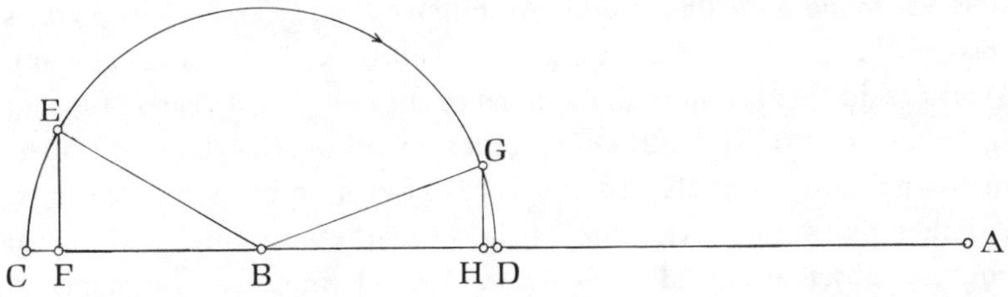

Figure 5

the pole of the ecliptic and taking a point B such that $AB = 23;40°$, we let B be the center of the circle of the anomaly of the obliquity of diameter $DC = 0;24°$ so that the least obliquity $DA = 23;28°$, and the greatest obliquity $CA = 23;52°$. Let the obliquity in the time of Ptolemy be $AF = 23;51,20°$ and thus $FB = 0;11,20°$; and at the time of Copernicus the obliquity will be $AH = 23;28,30°$ so that $BH = 0;11,30°$. Draw perpendiculars at F and H intersecting the circle at E and G. Since the radius of the circle $BC = BD = 0;12°$, it follows that

$$\frac{FB}{BC} = \cos CBE = \frac{0;11,20}{0;12} = \cos 19;11°$$

and

$$\frac{HB}{BD} = \cos DBG = \frac{0;11,30}{0;12} = \cos 16;36° .$$

Therefore

$$\text{arc } EG = 180° - (\text{arc } CE + \text{arc } DG) = 144;13°$$

is the motion of the anomaly between Ptolemy and Copernicus.

Now we must find the time that has ellapsed between these two observations of the obliquity, and since Copernicus gives no particular information about this, we must look for probable dates to assign to each value of the obliquity. For Ptolemy's obliquity, let us take the date of his observation of Regulus, reviewed by Copernicus in II, 14. The observation was made on 9 Pharmuthi, Antoninus 2, and Copernicus equates this (with an error of one day) with 24 February, 139 (j.d. 1771 882). Werner said that his obliquity applied to the completed year 1514, that is, January 0, 1515 (j.d. 2274 411), and we shall use this date for Copernicus's obliquity. The elapsed time is 1376 Egyptian years and 289 days. Another possibility is the interval between the autumnal and vernal equinox observations of Ptolemy in 139/40 and of Copernicus in 1515/16. Copernicus gives these intervals as 1376 Egyptian years, 332 days, and ½ hour for the interval between the autumnal, but 16⅓ hours between the vernal equinoxes. Since Copernicus' observations are for 1515/16, they are at about the time of Werner's observation of the obliquity.

Now, in II, 2, Copernicus gave the interval of time between Ptolemy and al-Battānī as 741^{ey}, so from Ptolemy to 1515 the interval would be

$741^{ey} + 635^{ey} = 1376^{ey}$ instead of about 1377^{ey}. If he used this value for computing the period of the anomaly of precession, he would in fact get 1717 Egyptian years. Letting $\Delta\vartheta = 144;13°$, we have

$$\frac{144;13°}{1376^{ey}} = \frac{180°}{1717;24^{ey}},$$

or if $\Delta\vartheta$ is rounded to 144;15°, then

$$\frac{144;15°}{1376^{ey}} = \frac{180°}{1717;1^{ey}}$$

Either way the result rounds to 1717 Egyptian years. The whole method is thus very simple, and uses only the reasonable assumption that the total variation of the obliquity is 0;24°, that is, what Copernicus calls his "sufficiently probable conjecture."

I believe that Copernicus did indeed derive the period of the anomaly in this way, but there are three problems that should not be overlooked:

1. The intervals taken for the computation are all closer to 1377 than to 1376 Egyptian years. All the intervals between the observations in III, 2 were rounded to integral years, but would Copernicus leave off the 289 or 332 days of the correct intervals when actually computing a parameter? I do not know, but considering that the period is so long, he may not have been concerned about the difference of about a year that the accurate interval would make in the complete period. Of course, we do not really know what terminal dates he used, and he could have chosen dates giving an interval closer to 1376 Egyptian years.

2. On f. 70 of the manuscript there is a different table of the anomaly of precession that must have been accompanied by a table of the mean precession on the preceding folio which has been removed from the manuscript. The table is crossed out, and is doubtless prior to the table incorporated into the printed text which appears on f. 81. The anomaly in the cancelled table is $0;6,17,29,36°/^{ey}$, leading to a period of 1716^{ey} 214^{d}. If this were derived from $\Delta\vartheta = 144;13°$, then $\Delta t = 1375^{ey}\ 122^{d}$, or if $\Delta\vartheta = 144;15°$ then $\Delta t = 1375^{ey}\ 238^{d}$. Neither of these periods fits any plausible dates for pairs of observations of the obliquity, so I have

no idea what the origin of this table could be. It could be based on a miscomputation—hence its deletion—but there seems no way of knowing this.

3. Finally there is the problem of why Copernicus writes as though he derived the period from the precession, not from the variation of the obliquity, and why he does not show how he carried out the derivation. Certainly, his original statement in II, 2 that the maximum variation of the obliquity is a conjecture and the fact that the "proof" of the maximum variation in III, 10 is a later addition argue that he did begin by assuming the variation of 0;24°, and so could have carried out this derivation of the mean anomaly. But why should he want to conceal this? Since he openly assumes that the period of the obliquity is exactly twice the period of the anomaly of precession, there would seem to be nothing wrong with then openly using the obliquity to derive the period. Possibly he thought that a derivation based on these two assumptions would be found objectionable, but then the announcement of a result without showing how it was reached is at least as objectionable. I really have no solution to this problem.[15]

(b) The Mean Precession, $\bar{\pi}$.

Copernicus then says that since, in the 1819^{ey} between Timocharis and his own observation of Spica in 1525 the apparent precession was 25;1°, so in the 102^{ey} after Timocharis the apparent precession must have been about 1;4°, leaving an apparent and mean precession of 23;57° in 1717^{ey}. How Copernicus derives this has been shown by Moesgaard as follows:

[15]Nor am I altogether satisfied with this reconstruction of Copernicus' derivation. The problem is chronological, depending upon the interval of time between the dates assigned to Ptolemy's obliquity and Werner's. Copernicus' account contains two intervals between Ptolemy and Battānī, 741^{ey} and 742^{ey}, while the correct interval is really 740^{ey}. From Battānī to 1525 are really 647^{ey}, but Copernicus' erroneous interval of 645^{ey} leads naturally to 635^{ey} to 1515 instead of the correct 637^{ey}. Combine all this with the cancelled table that leads to something over 1375^{ey} from Ptolemy's obliquity to the date for the obliquity 23;28,30°, and we are left with intervals anywhere from 1375^{ey} to 1377^{ey} while only 1376^{ey} will lead to the period of 1717^{ey}. I can see no way out of these difficulties, yet, since Copernicus says that he originally assumed the 0;24° variation of obliquity, it seems likely that this was his method of deriving the period of the anomaly.

We shall see that Copernicus later assumes that Timocharis and Ptolemy are symmetrical to the zero point of the anomaly, and thus an estimate of the apparent precession in the 102^{ey} preceding Ptolemy may be assumed equal to the precession in the 102^{ey} following Timocharis. Ptolemy notes that the precession between the observations of Menelaus and himself is 0;25° in 40^{ey}, and thus

$$\frac{0;25^\circ}{40^{ey}} = \frac{1;4^\circ}{102^{ey}}.$$

This is probably what Copernicus did, however even here there are problems. When we examine the manuscript (ff. 78v-79r), we see that Copernicus originally said that the apparent precession in 1819^{ey} is 25;3° and in the 102^{ey} after Timocharis 1;6°, leaving as before 23;57° in 1717^{ey}. The 25;3° was then altered to 25;1° and the 1;6° altered to 1;4° (in the latter case he even altered *unum et decimam* to *unum et decimam quintam*), but the 23;57° was not changed, indicating that he had it to begin with and would alter his other numbers to fit it. I can offer no explanation of the 25;3° or the 1;6°, and believe they may be mistakes since Moesgaard's demonstration of the derivation of 23;57° is so convincing.

Copernicus then claims to derive $\bar{\pi}$ and the period of the mean precession from a mean precession of 23;57° in 1717^{ey}. He first says that the period is 25816^{ey}, but

$$\frac{23;57^\circ}{1717^{ey}} = \frac{360^\circ}{25808;46^{ey}}.$$ [16]

Later he says that a precession of 23;57° in 1717^{ey} leads to an annual motion of 0;0,50,12,5°, but

$$\frac{23;57^\circ}{1717^{ey}} = 0;0,50,12,55,46\ldots {}^\circ/ey$$

[16]It is of interest to note that in the 1854 Warsaw edition of *De rev.*, 25816 was changed to 25809, evidently after checking this computation.

Note, however, that

$$\frac{360^\circ}{25816^{ey}} = 0;0,50,12,5,7,54\ldots{}^{\circ/ey} \approx 0;0,50,12,5^{\circ/ey}$$

and that

$$1717 \cdot 0;0,50,12,5^\circ = 23;56,35,47^\circ \approx 23;57^\circ .$$

It appears that first he had either the period of 25816^{ey} or the motion of $0;0,50,12,5^{\circ/ey}$; from this he computed the motion of 23;57° in 1717^{ey}, and then took the difference of 1;6° or 1;4° between this and the apparent precession of 25;3° or 25;1°. The only alternative is that 25816^{ey} is a miscomputation for $25808;46^{ey}$ and $0;0,50,12,5^{\circ/ey}$ is an error for $0;0,50,12,55,46\ldots{}^{\circ/ey}$. Copernicus is not this careless a computer, so I think that he must have derived $\bar{\pi}$ and the period of the mean precession in some other way, but I have been unable to reconstruct such a derivation.

(c) The Maximum Equation of the Anomaly, δ_{max}.

The remaining parameters are the maximum equation of the anomaly, that is, the radius of the circle of anomaly, the epoch of the anomaly, and the epoch of the mean precession. We shall see later that the first two can, indeed must, be found together if they are to be found by rigorous geometry rather than by tinkering. Copernicus, however, did not know how to find them rigorously, so he tinkered, but he tinkered quite well and his results are in close agreement with those found by a rigorous derivation.

In Figure 6, let the circle of the anomaly be described about center O, let the zero point of the anomaly be A, and let the diameter through the points of the maximum equation be DOB. It has been established from the 1717^{ey} period that in the 432^{ey} between Timocharis and Ptolemy $\Delta 2\vartheta = 90;35^\circ$. However, we do not yet know the value of the anomaly at either observation. Very well, says Copernicus, let us assume that A exactly bisects $\Delta 2\vartheta$. So we now have Copernicus' fifth assumption:

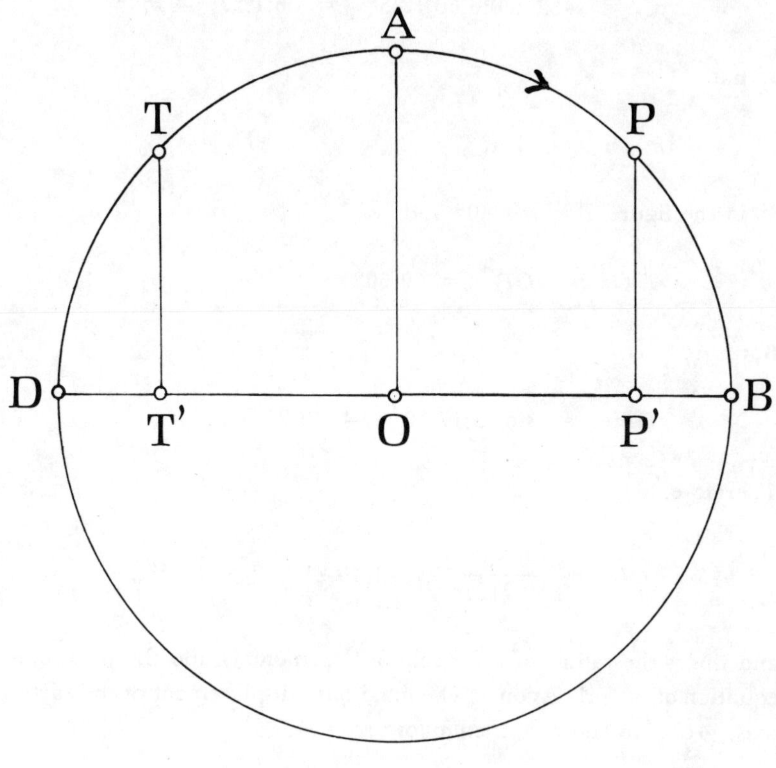

Figure 6

(5) The zero-point of the anomaly of precession fell exactly midway between Timocharis and Ptolemy.

Therefore, we let the anomaly at the time of Timocharis be T, the anomaly at the time of Ptolemy P, and

$$\text{arc } TA \quad = \quad \text{arc } AP \quad = \quad 45;17,30° \ .$$

The apparent precession π between the observations of Timocharis and Ptolemy was 4;20°, and the mean precession

$$\bar{\pi} = 432 \cdot 0;0,50,12,5° = 6;1,27° \approx 6;0°$$

so that

$$\bar{\pi} - \pi = 1;40° .$$

So in the figure $T'P'$ is 1;40° and

$$T'O = OP' = 0;50° .$$

But

$$OP' = \sin 45;17,30° = 0.7107 .$$

Therefore,

$$OB = \frac{0;50°}{0.7107} = 1;10,21\ldots° \approx 1;10° ,$$

and this is the radius of the circle of the anomaly and the maximum equation of the precession.[17] The maximum displacement of the earth's axis, ½ $C'D'$ in Figure 3, then follows from

$$\tfrac{1}{2} C'D' = \delta_{max} \sin \varepsilon .$$

Letting

$$\varepsilon = \bar{\varepsilon} = 23;40° ,$$

$$\tfrac{1}{2} C'D' = 1;10° \cdot 0.4014 = 0;28,5\ldots° \approx 0;28° .$$

It should be noted that Copernicus originally wrote this chapter in quite a different way. In the original version, first, from one-half the equation between Timocharis and Ptolemy, 0;50°, he found the cor-

[17]Here Copernicus wrote in the margin and then deleted "*vel scrupula xi* [i.e., 1;11°] *uti inferius*," and in the following two chapters (Ms. f. 83r:16, f. 85r:13) the radius was written 71′ and then changed to 70′. I do not know what this could indicate.

responding displacement of the axis 0;20°. Then he derived δ_{max} = 1;10°, and after that ½ $C'D'$ = 0;28°. This was also the order in the printed edition, but in the correction sheet the chapter was altered to the form shown here. The manuscript was marked to show the change in order of presentation, but the changes in numbers on the correction sheet were not entered. Since the alterations were made after N was printed, we may assume that whoever made up the correction sheet (Rheticus ?) was responsible, and that the chapter originally went to the printer substantially as it appears in the manuscript.

(d) The Epoch of the Anomaly, ϑ_0.

Copernicus' assumption that the zero point of anomaly bisects the anomaly between Timocharis and Ptolemy amounts to assuming the epoch of the anomaly. Thus, the epoch occurred 45;17,30° of anomaly before Ptolemy. In III, 9, he shows that this assumption was not strictly correct. In Figure 7, let the anomaly at the time of Timocharis be T, at the time of Ptolemy P and at the time of al-Battānī B. It is known from III, 6 that

$$\text{arc } TP = 90;35° \qquad \text{and} \qquad \text{arc } PB = 155;34° .$$

Now in the 742^{ey} between Ptolemy and Battānī the increase in the longitude of Regulus was 11;35° and of β Scorpionis 11;30°, and Copernicus takes 11;30° as the apparent precession in this interval. The mean precession

$$\bar{\pi} = 742 \cdot 0;0,50,12,5° = 10;20,49\ldots° \approx 10;21° ,$$

so that

$$\pi - \bar{\pi} = 1;9° .$$

Thus, in the figure

$$T'P' = 1;40° ,$$

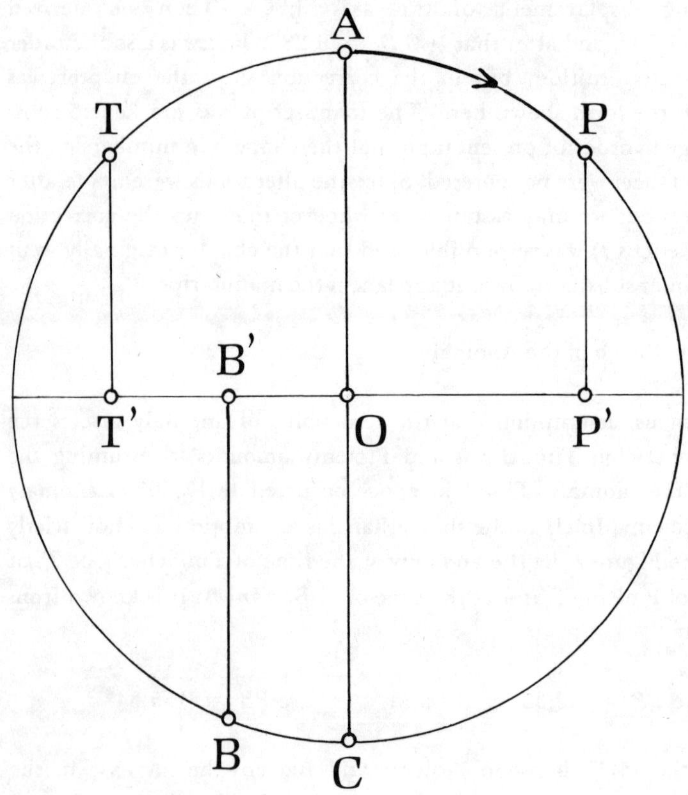

Figure 7

$$P'B' = 1;9° ,$$

and since

$$P'O = 0;50° ,$$

therefore

$$B'O = 0;19° .$$

But since

$$\text{arc } APCB = 45;17,30° + 155;34° = 200;51,30° ,$$

so

$$\text{arc } BC = 20;51,30°$$

and

$$\sin 20;51,30° \approx 0.356 .$$

Therefore

$$B'O = 1;10° \sin BC = 0.356 \cdot 1;10° \approx 0;24° .$$

(Accurately 0;24,55,12°)

But correctly $B'O = 0;19°$ so there is an error of 0;5° which, had Copernicus computed correctly, would have been closer to 0;6°.

Now, Copernicus says, it is necessary to rotate the circle (or move the zero point A which is the same thing) 2;47,30° so that AP is reduced from 45;17,30° to 42;30°, and then everything will be right. I assume that he did this by trial and error, and various alterations in the manuscript (f. 85r) show some of his errors. The results are, for the time of Timocharis

$$APCT = 311;55° ,$$

for the time of Ptolemy

$$AP = 42;30° ,$$

and for the time of al-Battānī

$$APCB = 198;4° .$$ [18]

[18]The correct solution to this problem is shown in Appendix I.

We must next find the anomaly at some epoch and the date that the anomaly was zero. Here we want to know the simple anomaly, that controlling the obliquity, rather than the double anomaly of the precession. Since the double anomaly at the time of Ptolemy was 42;30°, the simple anomaly was 21;15°. Note that in the derivation of the 1717^{ey} period from the change of obliquity, the simple anomaly at the time of Ptolemy is 19;11°, but Copernicus makes no use of this. Copernicus dates Ptolemy's observation of Regulus as $138^{ey}\ 89^{d}$ from the beginning of the Christian Era (1 January, A.D. 1). Thus

$$\Delta\vartheta \quad = \quad 138^{ey} \cdot 0;6,17,24,9^\circ \ + \ 89^{d} \cdot 0;0,1,2,2^\circ \quad = \quad 14;29,23\ldots^\circ$$

so that

$$\vartheta_0 \quad = \quad 21;15^\circ - 14;29^\circ \quad = \quad 6;46^\circ$$

Copernicus gives 6;45°, but it can be seen on f. 86r of the manuscript that he originally wrote 6;46°. If on 1 January, A.D. 1ϑ = 6;45°, then ϑ = 0° on 11 August, A.D. - 65 (j.d. 1697 539).

(e) The Epoch of the Mean Precession, $\bar{\pi}_0$.

Copernicus measures sideral longitudes from γ Arietis which, in the time of Ptolemy, was 6;40° east of the vernal equinox. Thus the apparent precession at this time was 6;40°. Since the anomaly of precession was 42;30°, the equation of the anomaly was

$$\delta \quad = \quad 1;10^\circ \sin 42;30^\circ \quad = \quad 0;48,11\ldots^\circ \approx 0;48^\circ \ ,$$

and since the correction was negative, the mean precession was

$$\bar{\pi} \quad = \quad \pi + \delta \quad = \quad 6;40^\circ \ + \ 0;48^\circ \quad = \quad 7;28^\circ \ .$$

In the interval of $138^{ey}\ 89^{d}$ between Ptolemy's observation of Regulus and the beginning of the Christian Era

$$\Delta\bar{\pi} = 138 \cdot 0;0,50,12,5° + 89 \cdot 0;0,0,8,15°$$

$$= 1;55,36\ldots° \approx 1;56° ,$$

and thus at epoch

$$\bar{\pi} = 7;28° - 1;56° = 5;32° .$$

(f) The "Demonstration" of the Variation of the Obliquity.

It has been mentioned that in II, 2 Copernicus called the variation of the obliquity from 23;52° to 23;28° a *sufficiently probable conjecture*, and then it was shown that the 1717 year period for ϑ could be derived from this assumption. We also mentioned that in III, 12, he computed an obliquity of 23;28,24° for 1525 from the completed theory, and also changed earlier references to the obliquity in the manuscript to this computed value. In III, 10 Copernicus goes through the motions of deriving the 0;24° variation of obliquity. This chapter was a later addition written by Copernicus in what must have been an attempt to cover his tracks by proving what he had in fact assumed. As stated before, it is written on a sheet of paper of water mark E, is inserted out of order in the manuscript, and disturbs the numbering of the remaining chapters of Book III. Further, Copernicus made an error in the value of $\Delta\vartheta$ between Ptolemy's obliquity in 139 and his own obliquity for 1525. This error was rectified in the printed edition. The derivation is as follows (see Figure 8).

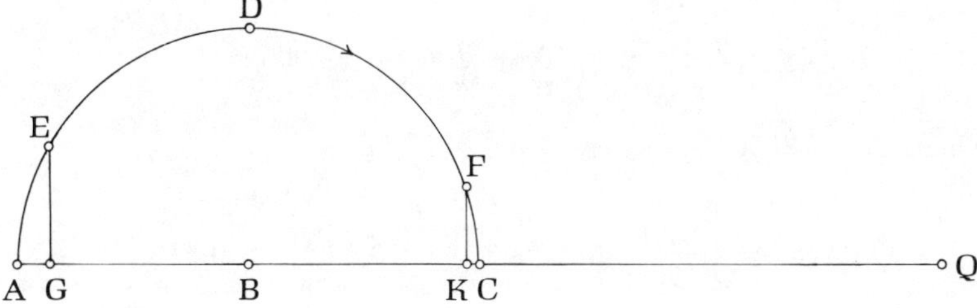

Figure 8

Let Q be the pole of the ecliptic, B the center of the circle of the anomaly of the obliquity, and ADC a semicircle of the anomaly. The maximum obliquity is AQ, the minimum CQ, and the arc of libration of the earth's axis is AC. (Copernicus, by a strange error, calls AC an "arc of the ecliptic," *circumferentia zodiaci,* but AC is of course the path of the earth's axis, is perpendicular to the ecliptic, and is removed from the pole of the ecliptic by the minimum obliquity CQ.) At the time of Ptolemy, in the second year of Antoninus, the obliquity is GQ = 23;51,20°, and at the time of Copernicus, in 1525 it is KQ = 23;28,24°. Thus, the difference

$$GK \quad = \quad GQ - CQ \quad = \quad 0;22,56^\circ \ .$$

At the time of Ptolemy, the anomaly AE = 21;15°, and, Copernicus says, in the intervening 1387^{ey}, the anomaly EF = 144;4°. This is an error, for using the tables in III, 6 for 1387^{ey}

$$\Delta\vartheta \quad = \quad 145;24,16,9^\circ \approx 145;24^\circ \ .$$

In the printed edition EF is corrected to 145;24°. Now, using the correct numbers from N rather than the errors that follow from the incorrect anomaly in the manuscript,

$$\text{arc } ED = 68;45^\circ \qquad GB = \sin ED = 932$$

$$\text{arc } DF = 76;39^\circ \qquad BK = \sin DF = 973$$

and thus

$$GK \quad = \quad GB + BK \quad = \quad 1905$$

where AC = 2000. But where GK = 0;22,56°

$$AC \quad = \quad \frac{2000}{1905} \cdot 0;22,56^\circ \quad = \quad 0;24,1^\circ \approx 0;24^\circ \ ,$$

and thus, Copernicus has proved what he earlier assumed. Naturally, since the whole thing is circular. The incorrect anomaly in the manuscript leads to $AC = 0;24,9°$ which also rounds to 0;24°.

Appendix I

Viète's Criticism and Solution

Appendicula II of François Viète's *Apollonius Gallus* is called "Concerning problems for which the astronomers do not explain a geometrical solution so that they solve them poorly."[19] He begins with an introductory paragraph criticizing the mathematical competence of Ptolemy and, still more, of Copernicus, whose treatment of the precession is singled out as especially inept.

> Indeed Copernicus not only confesses a lack of skill [in geometry], but proves it in Chapter Nine of Book Three of the Revolutions when, from the observations of Timocharis, Ptolemy, and al-Battānī, he attempts to find the maximum equation of the equinoxes and the positions of the anomaly from the limit of slowest velocity. For now, a master not of the science [of geometry] but rather of the dice, he commands the circle to be revolved until the error, which he does not realize to have originated from his ungeometrical method, might at last, if luck allows, be corrected.

The first problem in the appendix is applicable to Copernicus' precession theory. We have seen that in order to find the maximum equation of the precession Copernicus assumed that the values of the anomaly at the time of Timocharis and Ptolemy were exactly symmetrical to the zero point of anomaly. This assumption then led to a 0;5° error in the equation of precession at the time of Battānī, so Copernicus shifted the zero point by trial and error until he found an acceptable result. Viète considered this a very sloppy procedure, and in the first

[19] *Opera Mathematica,* ed. F. van Schooten (Leiden, 1646, repr. Hildesheim, 1970), pp. 343-346.

problem shows how both the diameter of the small circle and the position of the zero point can be found rigorously (see Figure 9).

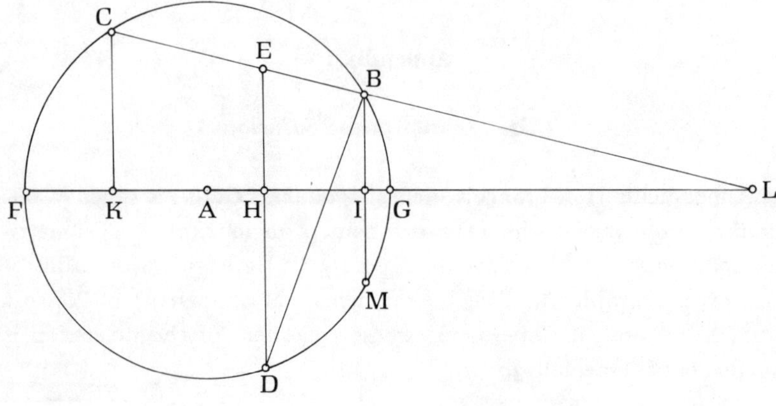

Figure 9

Problem I

Given a circle and three points on its circumference, to find the diameter in which, when perpendiculars descend from the given points, the segments of the diameter intercepted by the perpendiculars will have a given ratio.

In a given circle *BCD* with center *A*, let points *B, C, D* be designated. It is required to find the diameter of the circle in which, when lines fall perpendicularly from points *B, C, D,* the segments of the diameter intercepted by them will have a given ratio. Let the ratio of the segments intercepted by the perpendiculars dropped from *B, C, D* be as *S* to *R*. Let *CB* be divided in *E* so that

$$\frac{CB}{BE} = \frac{S}{R},$$

and join *DE* which will cut *FG* drawn through the center at right angles [to *DE*] in *H*. I say that *FG* is the required diameter in which, when *BI*, *CK*, *DH* fall perpendicularly,

$$\frac{KI}{HI} = \frac{S}{R}.$$

For lines *CB* and *FG* will either be parallel or not. And if they are parallel, then

$$CB = IK$$

and

$$EB = IH ,$$

and therefore by construction

$$\frac{KI}{HI} = \frac{CB}{EB} = \frac{S}{R}.$$

But if *CB* and *FG* will meet each other, let them meet in *L*. Therefore,

$$\frac{LC}{LK} = \frac{LE}{LH} = \frac{LB}{LI},$$

and *dividendo* [by subtraction

$$\frac{LC-LB}{LK-LI} = \frac{LE-LB}{LH-LI} = \frac{CB}{KI} = \frac{EB}{HI}\Bigg],$$

and *permutando* [alternately]

$$\frac{KI}{HI} = \frac{CB}{EB} = \frac{S}{R}.$$

As was required.

In this way, given *KH* and *HI* together with arcs *BC, CD, BD,* one may find arc *BG*, and thus the positions from the limits, and diameter *FG* itself in the parts of *KH* and *HI*. For let triangle *BDC* be constructed. Now it will be of given angles on account of the given arcs, and therefore it will also be of given sides in parts of diameter *FG*. Therefore triangle *EBD* will have sides *EB* and *BD* given in the same parts together with angle *EBD*. Therefore angle *EDB* or *DBM* will be given which, subtracted from arc *BMD,* leaves arc *BM* which is twice *BG*.

Given the motion of the anomaly and the equations between any three observations, we may use Viete's solution to find the diameter of the circle and the location of the zero point of anomaly. The numerical solution is as follows (see Figure 10).

Repeat the previous figure somewhat modified and let *B* be the anomaly at the time of Timocharis, *C* the anomaly at the time of Ptolemy, and *D* the anomaly at the time of al-Battānī. On diameter *FG*, the equation between Timocharis and Ptolemy is *IK*, and between Ptolemy and Battānī *HK*. Draw *AL* perpendicular to *FG*. By construction

$$\frac{BC}{EC} = \frac{IK}{HK} .$$

We are given

$$\text{arc } BC = 90;35^\circ ,$$

$$\text{arc } CD = 155;34^\circ ,$$

$$\text{arc } DB = 113;51^\circ ,$$

and

$$IK = 1;40^\circ , \qquad HK = 1;9^\circ .$$

Now,

$$BC = 2 \sin 45;17,30^\circ = 1.4214 ,$$

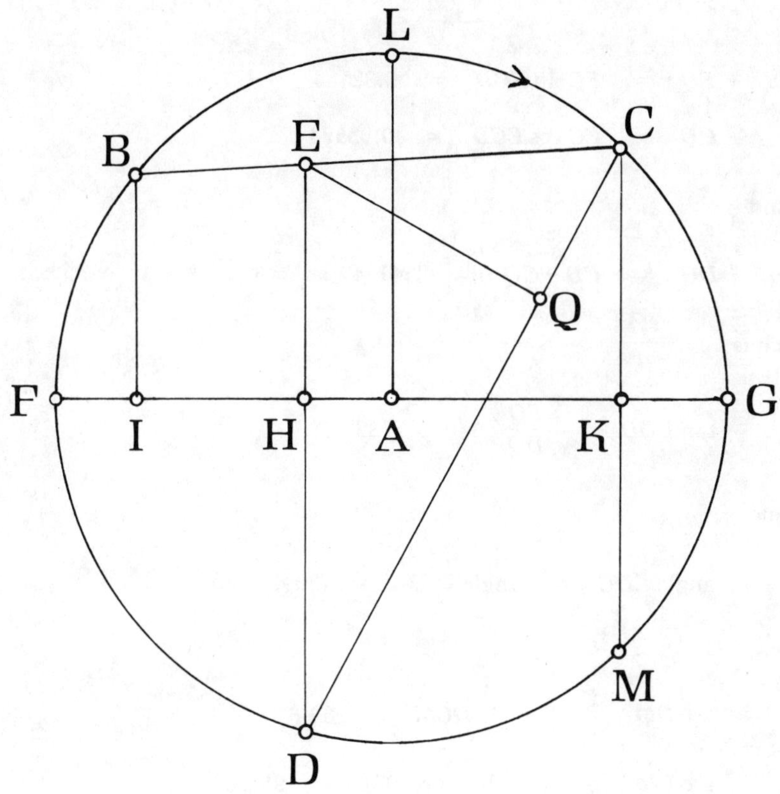

Figure 10

$$CD = 2 \sin 77;47° = 1.9547 ,$$

and

$$EC = \frac{HK}{IK} \cdot BC = \frac{1;9}{1;40} \cdot 1.4214 = 0.980766 .$$

Since

$$\text{angle } ECD = 56;55,30° ,$$

so

$$EQ = EC \sin ECD = 0.82184 ,$$

$$CQ = EC \cos ECD = 0.53523 ,$$

and

$$DQ = CD - CQ = 1.41947 .$$

Thus

$$\tan EDC = \frac{EQ}{DQ} = 0.57898 ,$$

and

$$\text{angle } EDC = \text{angle } DCM = 30;4° .$$

Now,

$$\text{arc } DM = 2 \text{ angle } DCM = 60;8° ,$$

$$\text{arc } CM = \text{arc } CD - \text{arc } DM = 95;26° ,$$

and

$$\text{arc } CG = ½ \text{ arc } CM = 47;43° .$$

Therefore, at the time of Ptolemy

$$\text{arc } LC = 42;17° ,$$

at the time of Battānī

$$\text{arc } LGD = 197;51° ,$$

and at the time of Timocharis

$$\text{arc } LGB = 311;42° .$$

So Copernicus' error in correcting the location of the zero point was only 0;13°.

The maximum equation is the radius of the circle, which Copernicus found to be 1;10°. In Figure 10,

$$IA = \sin 48;18° = 0.74664 ,$$

$$AK = \sin 42;17 = 0.67280 ,$$

so that

$$IK = 1.41944 .$$

But where $IK = 1;40°$

$$AG = \frac{1;40°}{1.41944} = 1;10,27° ,$$

so Copernicus' value for the maximum equation is also satisfactory. Therefore, although Viète's criticism of Copernicus' ungeometrical method is sound, there can be no objection to Copernicus' numerical results.[20]

[20]Editor's Note: It is worthwhile mentioning that Viète's solution was known and accepted by Johannes Praetorius. A group of manuscript folios inserted into Praetorius' heavily-annotated copy of *De revolutionibus* (located in the Schweinfurt Stadtarchiv) shows Praetorius working through Viète's criticism of Copernicus in Appendicula II of the *Apollonius Gallus*. Praetorius comments: "Franc. Vieta reprehendit Copernicum quod non geometricè progressus sit. Proposuit is libr. III. cap. ix. tale problema." And elsewhere, ". . . nos ad methodem Vietae tractabimus." Praetorius' result is, exactly as in the above computation, that the anomaly at the time of Ptolemy is 42;17° and that the radius of the small circle is 1;10°. Ernst Zinner found a similar reference to Viète's solution in a copy of *De revolutionibus* (1543) in the Munich Universitätsbibliothek (since destroyed in World War II) where the annotator—probably Hans Georg Herwart von Hohenberg, the Bavarian chancellor and prolific correspondent of Kepler—adds the following comment:

Appendix II

On the Chronology of the Manuscript of De Revolutionibus

The recent facsimile edition of the autograph of *De revolutionibus* contains an interesting and well argued consideration of the chronology of the manuscript by J. Zathey of the manuscript division of the Jagiellonian Library in Cracow.[21] Zathey's analysis should be read in the Polish or Latin edition since the English translation has been very incompetently edited, altering Zathey's arguments in a most uniformed way and doing his introduction a great disservice. This study of Copernicus' precession theory provides some additional evidence for dating parts of the manuscript and necessitates a considerable revision of its chronology.

It was explained near the beginning of this paper that, because the length of the sidereal year must be determined prior to the demonstration of lunar and planetary theory in Books IV, V, and VI of *De revolutionibus,* and because the sidereal year must be derived from the true tropical motion of the sun by means of the true precession, therefore the whole theory of precession must have been worked out by Copernicus prior to working out the solar, lunar, and planetary theory as presented in *De revolutionibus.* This is not to say that he could not have done preliminary investigations of, say, the eccentricities and the tropical longitudes of the apsidal lines of the planets before the precession theory was complete. It is just that whatever he may have done was necessarily modified before he wrote these books in the manuscript, since all conversions of tropical to sidereal longitudes are computed from the finished precession theory.

Now, we have seen that in III, 6 Copernicus quoted contemporary values of the obliquity of the ecliptic from Johann Werner's *De motu octavae sphaerae.* Werner's treatise was published in 1522, and Coper-

---

"Graviter reprehenditur Copernicus à Vieta, quod in problemata quae proponitur lib. III. Cap. ix non geometricè progressus est" (*Enstehung und Ausbreitung der Coppernicanischen Lehre,* [Erlangen: Sitzungsberichte der physikalisch-medizinischen Sozietät zu Erlangen, No. 74, 1943], p. 452).

[21]*Nicolae Copernici Opera Omnia* I, Varsoviae-Cracoviae, 1972.

nicus's letter concerning it is dated 3 June, 1524. Further, Copernicus uses Werner's value of the obliquity 23;28,30° for the reduction of his observation of Spica made in 1525, and this observation of Spica is then used for determining the mean rate of precession. Thus, the precession theory, Book III, 1-12, could not have been worked out, let alone written in the manuscript before 1525, and all the following parts of *De revolutionibus*—solar theory, lunar theory, planetary theory—must be later still.[22]

The one notable change that Copernicus made in the exposition of the precession was the addition of chapter 10 in which he "demonstrates" the maximum variation of the obliquity, and this addition went hand in hand with the other alterations of the text, the insertion of the obliquity 23;28,24° as a substitute for some other value. This group of alterations can be approximately dated, and it is surprisingly late. Now, chapter 10 is inserted out of order in the manuscript on a sheet (ff. 76-77) of paper bearing the watermark conventionally designated E. Paper E is dated from its use by Copernicus for one letter in 1537 and two letters in 1539. Further, Copernicus' statement in II, 2 (f. 27v) that

[22]See "The Holograph of *De revolutionibus* and the Chronology of its Composition," *Journal for the History of Astronomy,* 1974, *5*: 186-198, where I argue that none of the manuscript was written before 1525 and that, in fact, it all could be several years later. It should be noted that our principal evidence for dating the manuscript is the dating of paper E, and we really do not know how long before using paper E Copernicus was using papers C and D.

In a comment following the presentation of this paper, Professor E. Rosen remarked that in III, 1 Copernicus refers to the invention of an "eleventh sphere" (Ms. f. 71v:4), which appears to be a reference to Werner, while in I, 11 (Ms. f. 11v:9) he refers to modern astronomers adding a tenth sphere, indicating that Copernicus wrote this before he received Werner's book. This is an ingenious analysis, and would serve to establish a part of *De rev.* written prior to 1525. There are, however, two problems that render it untenable. First, Werner uses only an eighth, ninth, and tenth sphere in his theory of irregular precession and change of obliquity. While there is an additional sphere for the diurnal rotation of the entire universe, it is always called the *primum mobile,* and would not be counted as an eleventh sphere in considering the theory of precession. Thus, the passage in I, 11 must itself be a reference to Werner's theory, and it is difficult to know what the eleventh sphere in III, 1 means, unless it is a slightly sarcastic remark about the great number of spheres employed to account for these motions. Second, f. 11 is a sheet of paper of watermark D, but f. 71 is of watermark C. Therefore, the passage in I, 11 must be later than or perhaps contemporary with III, 1, but cannot be earlier.

the variation of the obliquity between 23;52° and 23;28° is a "sufficiently probable conjecture" is also on a sheet of paper E. Thus Copernicus was already using E while he was still content to call the maximum variation a conjecture and before he decided to write III, 10. This would be, say, middle or late 1530's.

In fact, he had not yet made the change when Rheticus came to visit him in 1539. Rheticus' *Narratio Prima* can be exceedingly useful for discovering details of Copernicus book as of 1539 for there are a number of cases of Rheticus' reporting numerical values which were then altered in the manuscript or in the printed edition. The obliquity is just such a case. The *Narratio Prima* says nothing of either the obliquity 23;28,24° nor of the demonstration of the maximum variation that was based on this value. Rheticus says only that

> In our time it appears not greater than 23;28,30°. . . . Moreover, [my] learned teacher assumes that the minimum obliquity will be 23;28°, the difference of which from the maximum is 0;24°.[23]

This agrees perfectly with Copernicus' statements in III, 2 and II, 2, and could not have been written if III, 10 and its accompanying changes, which Rheticus could hardly have missed, had already been written. One may wonder if Copernicus wrote III, 10 and inserted the value 23;28,24° on Rheticus' suggestion. If Copernicus did not tell Rheticus that he had derived the period of the anomaly by assuming a variation of 0;24°, Rheticus might have suggested to Copernicus that he add a proof of the maximum variation which could easily be derived from the period of the anomaly, from Ptolemy's obliquity for 139, and from the obliquity of 23;28,24° computed in III, 12 for 1525. This, of course, is pure speculation, but III, 10 was certainly not present in the manuscript when the *Narratio Prima* was written.

[23]Kepler, *Werke* I, p. 91:28. . .42-44.

COMMENTARY: REMARKS ON COPERNICUS' OBSERVATIONS

Owen Gingerich

Harvard University and

Smithsonian Astronomical Observatory

In our modern scientific age, there is an undeniable mystique attached to "observations." Somehow they are the precious words of the Book of Nature, clearly allied with facts, and on the direct path to truth. Only occasionally are we reminded how tortuous and treacherous this path is.

Werner Heisenberg, in his autobiographical *Physics and Beyond,*[1] recounts a charming episode, a conversation with Einstein:

" 'But you don't seriously believe,' Einstein protested, 'that none but observable magnitudes must go into a physical theory'?"

" 'Isn't that precisely what you have done with relativity'?" Heisenberg asked in some surprise.

" 'Possibly I did use this kind of reasoning,' Einstein admitted, 'but it is nonsense all the same. Perhaps I could put it more diplomatically by saying that it may be heuristically useful to keep in mind what one has actually observed. But on principle, it is quite wrong to try founding a theory on observable magnitudes alone. In reality the very opposite happens. It is the theory which decides what we can observe. You must appreciate that observation is a very complicated process.' "

Mankind's love affair with the efficacy of specific observations has evolved as gradually as science itself. Hipparchus may have discovered the precession of the equinoxes from observations of the star Spica during the lunar eclipses of 21 April -145 and 21 March -134; if so, it is the first known astronomical discovery based on specific observations. Claudius Ptolemy, in his *Almagest,* gives the first recorded example of a

[1]Werner Heisenberg, *Physics and Beyond* (New York: Harper and Row, 1971), p. 63.

theory based on specific data points. His data present us with a perplexing problem, first described in some detail by Delambre in the "Discours Préliminaire" of his *Histoire de l'Astronomie Ancienne.*[2] Delambre suggested that Ptolemy had not observed at all and that his supposed observations were fabricated from theory. More recently, R.R. Newton has reiterated these arguments, concluding with a statistical confidence level of 10^{200} to 1 that Ptolemy's parallax and solar observations were fudged.[3]

The results of Ptolemy's parallax theory (admirably discussed here by Dr. Henderson) cannot be understood more than superficially unless we realize that the data are chosen to agree with the "plenum" theory of the planetary system. That is, Ptolemy's numbers for the distance to the sun and moon agree perfectly with a scheme in which the Venus apogee nests perfectly within the solar perigee, the Mercury apogee within the Venus perigee, and the lunar apogee within the Mercury perigee.

How can we reconcile the idea that Ptolemy fabricated his observations with the fact that he not only gives graphic descriptions of his instruments, but he repeatedly informs his reader that numerous other observations confirm those given? And if the data are computed from theory, on what does the theory rest, and why did it continue to give rather acceptable predictions for over a millenium?

Certainly one aspect to the answer—and something equally relevant to Copernicus—is the absence of any systematically developed error theory. As Professor Swerdlow has indicated, if three observations suffice to specify the required parameters, then a fourth, redundant, observation is likely to be an embarrassment. This, I believe, accounts for the paucity of observations both in the *Almagest* and in *De revolutionibus*. Perhaps Ptolemy actually used scores of observations, but in preparing the book, it is quite possible he generated "ideal" observations for pedagogical purposes. Alternatively, he could have been very selective in choosing his data from a multiplicity of observations avail-

---

[2]J.B.J. Delambre, *Histoire de l'Astronomie Ancienne* (Paris: Mme. V. Courcier, 1817; New York: Johnson Reprint, 1965).

[3]R.R. Newton, "The Authenticity of Ptolemy's Parallax Data—Part I," *Quarterly Journal of the Royal Astronomical Society,* 1973, *14:* 367-388.

able to him. Pannekoek[4] has shown rather convincingly how Ptolemy could have done this to achieve his convenient but erroneous value for precession, one degree per century.

Copernicus had available only a few ancient observations, the handful preserved in the *Almagest.* To these he added only a minimum number of his own. Professor Swerdlow has carefully analyzed the three Copernican observations used in the precession theory. It is instructive to examine some of the other new observations, and in Table 1 I have listed them for the five planets. I think that any astronomer initially coming onto this list can only be startled by its brevity. Copernicus uses three oppositions for each superior planet and three elongations of Mercury in order to derive the positions of their apsidal lines; it is his own discovery that these lines have changed since Ptolemy's day. In addition, he gives another observation for each superior planet in order to establish its distance. Like Ptolemy, Copernicus does not list any specific observations to establish his latitude theory.

Let us look more closely at the oppositions given. Two of them fall in the daytime, when the planet is below the horizon! This abruptly reminds us that a stated planetary opposition is not an observational event devoid of theory. Precisely when and where the opposition occurs depends on the solar theory, for one must know the position of the sun as well as the position of the planet. (In fact, for Copernicus as well as for Ptolemy, it was necessary to reckon the position of the fictitious mean sun and not the true sun; this is related to the circumstance that Copernicus used the center of the earth's orbit, rather than the physical sun, as the center for his system.) Hence, all these "observations" are at least one step removed from the actual sighting of the planet in the sky. Thus it is not surprising that the positions or times are sometimes altered in the *De revolutionibus* manuscript. For example, on the initial recent Mercury observation, Copernicus states first a position of 13° and "almost 25′," and then 13°40′, and finally he settles on 13½°.[5] The

[4]A. Pannekoek, "Ptolemy's Precession," in A. Beer, ed., *Vistas in Astronomy,* Vol. 1 (London: Pergamon, 1965), pp. 60-66.

[5]Copernicus manuscript f. 180v. Walther's observation was printed in *Scripta clarissimi mathematici M. Joannis Regiomontani, de torqueto, astrolabio armillari*. . . (Nürnberg: Johannes Montanus and Ulrich Neuber, 1544), f. 55; presumably Copernicus had prior access to a manuscript copy.

Table 1. Copernicus' Planetary Observations in *De revolutionibus*

Date, Frauenberg time	Precessed position (Copernicus observed)	Actual position (Tuckerman)	Δ O-T	Computed (Copernicus)	Δ O-C	Reference (*De revolutionibus*)
SATURN						
5 May 1514 10:48 p.m.	232°39′	232°18′	+21′	232°36′	+3′	V, 6
13 July 1520 12:00 noon	300 43	300 37	+6	300 37	+6	V, 6
10 Oct. 1527 6:24 a.m.	27 29	27 33	−4	27 26	+3	V, 6
24 Feb. 1514 5:00 a.m.	236 14	235 43	+31	236 8	+6	V, 9
JUPITER						
30 Apr. 1520 11:00 a.m.	227 36	226 55	−41	227 43	−7	V, 11
27 Nov. 1526 3:00 a.m.	75 55	76 14	−19	75 58	−3	V, 11
1 Feb. 1529 6:00 p.m.	141 8	140 43	+25	141 5	+3	V, 11
19 Feb. 1520 6:00 a.m.	232 27	231 57	+30	232 28	−1	V, 14
MARS						
5 June 1512 1:00 a.m.	262 46	262 26	−20	262 55	−9	V, 16
12 Dec. 1518 8:00 p.m.	90 19	89 2	+1°17′	90 28	−9	V, 16
22 Feb. 1523 4:00 a.m.	160 40	158 24	+2 16	160 42	−2	V, 16
1 Jan. 1512 6:00 a.m.	218 41	218 39	+2	218 37	+4	V, 19
VENUS						
12 Mar. 1529 7:30 p.m.	36 35	36 59	−24	36 38	−3	V, 23
MERCURY	(Walther)					
9 Sept. 1491 5:00 a.m.	163 30	163 33	−3	163 10	+20	V, 30
9 Jan. 1504 6:30 a.m.	273 20	273 35	−15	273 23	−3	V, 30
18 Mar. 1504 7:30 p.m.	26 30	26 31	−1	26 58	−28	V, 30

position is, however, borrowed from Bernard Walther, who gave 13°23′. On the other hand, the isolated planetary positions used to establish the distances are presumably straight observations and are not so altered.

In this connection, it is fascinating to note Kepler's handling of oppositions nearly a century later. In his *Astronomia nova,* we see for the first time a scientist coping with the ambiguities of redundant data. In a sense, astronomy comes of age in this great masterpiece. Kepler gives us, for example, twelve Martian oppositions where four suffice to establish the parameters, and he calculates the error in each position. Eventually his route to the ellipse is a matter of picking an acceptable curve out of data whose scatter is almost as large as the subtle discrepancies he is seeking.[6] The standard story is that Kepler spares us nothing in his extended commentary on Mars; but in reality, he is almost as silent as Copernicus about the derivation of his opposition positions. Contrary to the general opinion, these are not simply copied from Tycho's observing books. Each one cost Kepler a great deal of agony, as only an examination of his manuscript notes reveals. His notes show, for example, that the time and position given in Chapter 15 of the *Astronomia nova* for the Martian opposition of 8 June 1591 actually rest on no less than twelve real observations made by Tycho between 13 May and 16 July, a laborious exercise never mentioned in Kepler's book.[7]

When the Copernican planetary observations in Table 1 are compared with the actual planetary positions computed by Tuckerman, yet another astonishing feature comes to light. This comparison reveals appalling inaccuracies—especially in the observations for Mars, where the position given for his most important date, 22 February 1523, errs by more than two degrees.[8] Nevertheless, Copernicus has matched the

[6]See Curtis Wilson, "Kepler's Derivation of the Elliptical Path," *Isis,* 1968, *59:* 5-25, and D.T. Whiteside, "Keplerian Planetary Eggs—Laid and Unlaid, 1600-1605," *Journal for the History of Astronomy,* 1974, *5:* 1-21.

[7]Leningrad Kepler manuscripts XIV, f. 214; see also Owen Gingerich, "Kepler's Treatment of Redundant Observations," in F. Krafft *et al.*, eds., *Internationales Kepler-Symposium, Weil der Stadt 1971* (Hildesheim: Gerstenberg, 1973), pp. 307-314.

[8]Copernicus uses this Mars observation of 22 February 1523 to establish its mean motion (*De revolutionibus,* f. 158v). The daily value of 27′41″ 40‴ 22′^{V} is given in the mean motion table on f. 146 of the manuscript; this is a replacement page with watermark E, glued in later, as the new Warsaw facsimile clearly shows. A few pages before, on f. 143v,

parameters of his planetary model very successfully to the poor data (with the exception of Mercury). All this indicates that Copernicus failed to test his theory against other observations, which he no doubt had available. This, in turn, suggests that the entire exercise was carried out primarily to show that the heliocentric cosmology was compatible with reasonable planetary predictions rather than to reform the accuracy of astronomical predictions. The evidence illuminates the mentality of a gifted theoretician for whom the observational foundations of his science held only a secondary interest. This conclusion strongly reinforces the view of Copernicus' attitudes implicit in Professor Swerdlow's preceding analysis of the precession.[9]

Copernicus gives an earlier and more correct value, 27′41″ 40‴ 8′ᵛ. I have not yet established if this latter value was based on one of the other Mars observations that he reports.

[9]An important by-product of Professor Swerdlow's examination of the Copernican precession is to show how late in the astronomer's career the actual writing of *De revolutionibus* occurred.

Another by-product of Swerdlow's analysis is to remind us of the difficulties Copernicus had with chronology, e.g., the dating of the observation of 2 Antonine 9 Pharmuthi. L.A. Birkenmajer, in his *Mikolaj Kopernik* (Cracow, 1900), has already documented the immense effort Copernicus spent in merely organizing the Egyptian months. As Swerdlow tells us, Copernicus converts the above date to 22 February 139 in the *Letter Against Werner;* in *De revolutionibus* (Nuremberg: Petreius, 1543), f. 46, however, he gives it as 24 February 139. It is interesting to notice that Erasmus Reinhold inscribed the correct date, 23 February 139, in his own copy, which is now preserved at the Royal Observatory in Edinburgh.

It is also interesting to note that Reinhold was fully aware of the origin of the *De revolutionibus* value of the obliquity, 23°28′30″, which Swerdlow has deduced as coming from Werner. On f. 70, Reinhold has written across the top of the page:

Dominicus Maria Bononiae anno Domini 1491 reperit maximam Solis declinationem part 23.29′.

Wernerus dicit se ex alios [sic] circa annum Domini 1514 reperisse eandem partium 23.28′.30″.

Idem Wernerus narrat Almonem seu Anglum quendam ad annum Dni 1323 eam reperisse 23.33.30 fere.

Item fuisse maximum Solis declinatione Alphonsi aetate ad annum Dni 1252 23.35.45-fere:

In Epitome Regiom. annotata est partium 23.28′.

Despite Reinhold's acuity in such matters, he accepts Copernicus' poor observations unhesitatingly and without challenge, and proceeds to fit his tables to them with even more precision than Copernicus did!

In the past year, new evidence has come to light that Copernicus was not unaware of major discrepancies between prediction and observation. Bound in the back of Copernicus' personal copy of the 1492 edition of the *Alfonsine Tables* are sixteen extra leaves; at the bottom of the last page, following the record of a pair of conjunctions observed in 1500 in Bologna, there is written in highly abbreviated Latin:[10]

Mars superat numerationem plus quam gr.ij

Saturnus superatur a numeratione gr.1½

The translation is: "Mars surpasses the numbers by more than two degrees; Saturn is surpassed by the numbers by 1½ degrees."

Although this passage has been known for nearly a century, only recently, with the aid of modern computers, has it become possible to fathom its meaning. If we compare the planetary longitudes predicted by the *Alfonsine Tables* with the actual positions, distinctive error curves arise, as shown in Figure 1. Errors of the stated size for Saturn occurred between 1496 and 1506, and again beginning in 1525. Two-degree errors of the opposite sign occurred for Mars in 1504, 1506, and 1508 and again in 1521, 1523, and 1525. If we further assume that the absence of information about Jupiter implies that Jupiter agreed well with the tables on the date in question, then the sole possibility for all three planets is the winter of 1503-1504.[11] Because this corresponds to a series of important conjunctions between Mars, Jupiter, and Saturn, and because they could be observed with considerable precision without special instruments, it seems highly likely that Copernicus actually witnessed these conjunctions;[12] in any event, he must have become aware of

[10]This page has been reproduced various times, most recently on page 88 in Owen Gingerich, "Copernicus and Tycho," *Scientific American,* 1973, *229:* 86-101.

[11]I am much indebted to Jerzy Dobrzycki, who helped alert me to these possibilities; my conclusion differs slightly from his, which is published in "Uwagi o syewdzkich zapiskach M. Kopernika," ("Notes on Nicholas Copernicus's Swedish Scripts"), *Kwartalnik Historii Nauki i Techniki,* 1973, *18:* 485-494.

[12]These conjunctions were accurately observed and recorded by Bernard Walther, but their publication came after Copernicus' death (see note 5). There is, of course, the possibility that Copernicus had access to them in manuscript, just as he had to the Mercury positions.

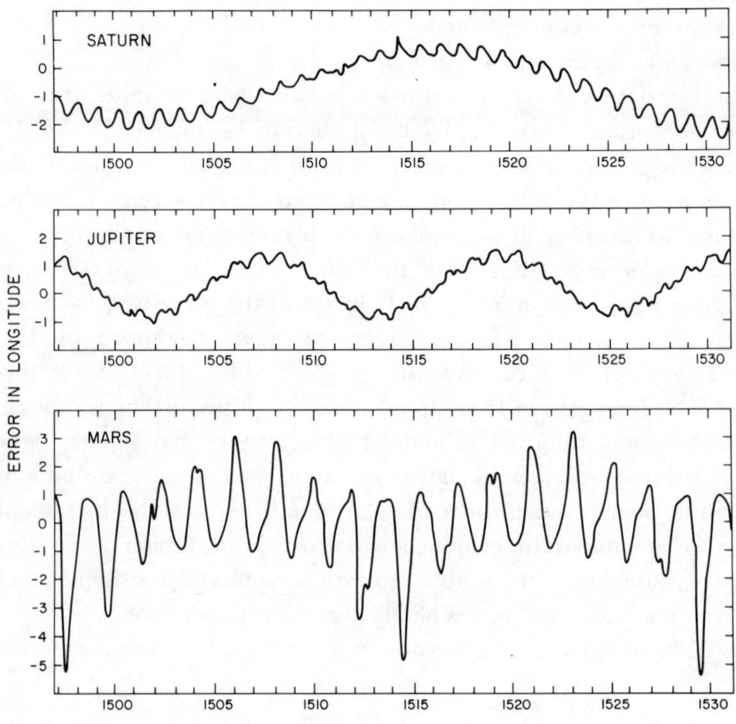

Figure 1. The graphs show for Copernicus' principal observing years the errors in planetary longitudes when the positions are computed from the *Alfonsine Tables*. The figure has been constructed by Barbara Welther.

the discrepancies with the tables. That he has not mentioned such a motivation for reform in his *Commentariolus,* written shortly thereafter, again reinforces the conclusion that Copernicus was interested more in cosmology than in predictive accuracy.

IV

ERASMUS REINHOLD'S DETERMINATION OF THE DISTANCE OF THE SUN FROM THE EARTH

Janice Henderson

Queen's College, City University of New York

Soon after the publication of *De revolutionibus* certain astronomers who read Copernicus' work closely noted a series of errors—inconsistencies, obvious computational errors, and discrepancies between observations and positions computed from the tables in *De revolutionibus.*[1] One of the careful readers of Copernicus' work was Erasmus Reinhold (1511-1553),[2] the Wittenberg astronomer who compiled the Prutenic

[1]See, for example, the letter of Matthias Lauterwalt to Georg Rheticus, dated 17 February, 1545, in K.H. Burmeister, *Georg Joachim Rhetikus, 3* (Wiesbaden, 1968), pp. 59-67.

[2]Erasmus Reinhold was born on October 21, 1511, in Saalfeld in Thuringen, where he died of the plague on February 19, 1553. He spent most of his life in Wittenberg, where he was a mathematics professor at the University after 1536, a dean of the University in 1549, and rector in 1550. Reinhold's published works, in addition to the Prutenic Tables (Tübingen 1551), include his edition of and commentary on Peurbach's *Theoricae novae planetarum* (Wittenberg, 1542), which was reprinted numerous times; an edition of the Greek text of Book I of the *Almagest*, with a translation into Latin and a commentary by Reinhold (Wittenberg, 1549); a short work on spherical astronomy, *Themata quae continent methodicam tractationem de Horizonte rationali ac sensibili*. . .(Wittenberg, 1541); ephemerides for use in Wittenberg for the years 1550-1551, *Ephemerides duorum annorum 50, et 51, supputatae ex novis tabulis astronomicis* (Tübingen, 1550); and a trigonometric work, essentially a new edition of the *Tabulae directionum* of Regiomontanus, *Primus liber tabularum directionum*. . .(Tübingen, 1554).

Tables (1551). Reinhold noted that the very observations in *De revolutionibus* from which Copernicus derived his parameters and tables could not be reproduced from the tables themselves. He undertook the task of rederiving all the parameters and recomputing the tables, an endeavor to which he devoted seven years.[3]

Even a cursory examination of the Prutenic Tables reveals the extent of Reinhold's revisions. Not only are the tables computed for smaller intervals (every degree rather than every three degrees in the correction tables, for example) and to more sexagesimal places, but a comparison of individual entries indicates that the numerical results obtained by Reinhold differ from the values of Copernicus. This gives rise to two questions: how does one account for the differences in the numerical values of Reinhold and Copernicus, and, given the differences, in what sense are the Prutenic Tables Copernican? One should keep in mind that the issue of heliocentricity plays no role in these questions. It is certainly irrelevant to the construction of tables, which, if they are to be at all practical, must be designed for an observer on the earth. One cannot determine from the Prutenic Tables, for example, whether or not Reinhold subscribed to a heliocentric system.[4]

The two questions just raised are answered by Reinhold himself, who left a detailed account of his derivation of solar, lunar and planetary parameters, preserved in an unpublished holograph manuscript entitled "Commentarius in opus revolutionum Copernici." In the preface to the Prutenic Tables Reinhold states that he brought together the observations of Copernicus with those of Ptolemy and others, and, aside from the bare observations and traces (*vestigia*) of the explanations, he took nothing else from Copernicus. He goes on to say "I have explained the causes and the computation of the individual tables in our commentary which I wrote on the work of revolutions of Copernicus." There are in fact three references to this commentary in the Prutenic

[3]Erasmus Reinhold, *Prutenicae tabulae coelestium motuum* (Tübingen, 1551), α4r.

[4]Reinhold gives no indication of his opinion of the heliocentric theory of Copernicus, either in his printed works or in his manuscripts. There are two passages in his commentary on *De revolutionibus* which seem to imply Reinhold's adherence to a Tychonic system, but in my opinion, the evidence is slim and hardly sufficient for a definitive assessment of his philosophical outlook.

Tables,[5] but it was thought lost already by the end of the sixteenth century. Maestlin in 1596 expressed regret that Reinhold was not able to produce the commentary which he had planned.[6] It was rediscovered early in the present century by Valentin Rose and Ludwig A. Birkenmajer.[7] The manuscript also contains an incomplete treatise on spherical trigonometry, a commentary on a portion of the *Almagest* (dated 1 January 1549), and a commentary on Book 2 of Pliny.[8] Reinhold's name does not appear anywhere in the manuscript, but he is known to have lectured on the *Almagest* in 1549,[9] and agreement between the contents of the commentary and the Prutenic Tables leaves no possible doubt.[10]

The contents of the commentary on *De revolutionibus* correspond to Reinhold's description in the Prutenic Tables. Taking the observations cited in *De revolutionibus* and made not only by Copernicus but by Ptolemy and Hipparchus as well, and using models structured exactly like those of Copernicus, Reinhold derives anew all the essential solar, lunar and planetary parameters. These parameters are precisely those used in the computation of the Prutenic Tables. There is some doubt as to the completeness of the surviving manuscript. It begins with a discus-

---

[5]Reinhold, *Prutenic Tables,* α2r, α4r, β2v.

[6]Johannes Kepler, *Gesammelte Werke, 1* (Munich, 1938), p. 85.

[7]A. Birkenmajer, "Le commentaire inédit d'Erasme Reinhold sur le *De Revolutionibus* de Nicolas Copernic," *Actes du Colloque international sur la science au seizième siècle* (Paris, 1960), pp. 171-177.

[8]The complete contents of the manuscripts, bound together in Berlin Deutsche Staatsbibliothek, Latin 2° 391, are as follows; *Commentarius in opus revolutionum Copernici* (Books III and IV) ff. 1-63; spherical trigonometry, June 4, 1543, ff. 64-124; theorems on the theory of circles, ff. 125-171; On Ptolemy's *Almagest* Book V (chapters 12-16), January 1, 1549, ff. 172-186; the remainder of the commentary on *De revolutionibus* (Book V), ff. 192-294; *Tabulae in capita mathematica Plinii,* ff. 295-324; an astrological work in a different hand, dated 1575, ff. 325-356. The manuscript is described in V. Rose, *Verzeichniss der Lateinischen Handschriften der Königlichen Bibliothek zu Berlin* v. 2, part 3 (Berlin, 1905), pp. 1366-1367. At least one other manuscript of Reinhold's survives, also in Berlin, Deutsche Staatsbibliothek, Latin 4° 32. This contains lecture notes on Euclid, dated between November 1543 and August 1550.

[9]E. Zinner, *Entstehung und Ausbreitung der Coppernicanischen Lehre* (Erlangen, 1943), p. 513.

[10]In addition, Prof. Gingerich attests to the handwriting as that of Reinhold. See Owen Gingerich, "The Role of Erasmus Reinhold and the Prutenic Tables in the Dissemination of Copernican Theory," *Studia Copernicana 6* (Wroclaw, 1973), pp. 43-62; 123-126.

sion of chapter 13 of Book III of *De revolutionibus*, and covers the rest of Book III, as well as Books IV and V. In general, it deals with the contents of Copernicus' work chapter by chapter, omitting those chapters which do not entail mathematical derivations. I believe this may well account for the absence of a commentary on the first two books of *De revolutionibus*. The reason Reinhold begins only with chapter 13 of Book III may well be that he could not understand Copernicus' mathematical derivation of the precession. On the other hand, it seems quite likely that Reinhold would have commented on Book VI, as it contains material with which Reinhold is quite capable of dealing, and which is essential for the Prutenic Tables. There is also a section missing from the beginning of the derivation of Jupiter's parameters, where Reinhold has indicated in the manuscript "Initium vide in alio libro."[11]

Reinhold's commentary on *De revolutionibus* contains nothing original of a purely theoretical sort, nor does it provide insight into the obvious technical problems of Copernicus' work. Using the models, observations and mathematical methods of *De revolutionibus*, Reinhold displays in detail the computation of each parameter. His results differ from those of Copernicus because he is a careful, meticulous computer. He uses seven and sometimes eight place trigonometric functions, and he carries all the digits throughout each computation. He very rarely makes an arithmetical mistake. Copernicus, on the other hand, made numerous computational errors which affect the parameters in *De revolutionibus*.

One section of Reinhold's commentary on *De revolutionibus* of particular interest is his analysis of the sizes and distances of the sun and moon. Reinhold's normal procedure is to begin with the observations stated by Copernicus and derive the parameters as Copernicus should have, had he carried out the derivations accurately. His treatment of the solar distance, however, is the only portion of his commentary which is not strictly modelled on *De revolutionibus*. It appears that Reinhold was not satisfied with the solar distance resulting from Copernicus' data, and he altered the data in order to arrive at a solar distance close to Ptolemy's value.

[11]Erasmus Reinhold, *Commentarius in opus revolutionum Copernici*, Deutsche Staatsbibliothek, Berlin, Latin 2° 391, f. 208r.

At the beginning of the section of the commentary on the lunar and solar distances, Reinhold refers to "calculations of Ptolemy which I have written elsewhere for M. Johanne Aurifaber."[12] In subject matter, this corresponds exactly to the notes on the *Almagest* found in the Berlin manuscript, which cover only Book V, chapters 12 to 16, dealing precisely with the sizes and distances of the sun and moon. It appears to be a self-contained unit, with an epilogue, and Book V may very well be the only portion of the *Almagest* on which Reinhold commented, other than his separate publication of Book I.

My intent in what follows is to illustrate 1) how Reinhold arrives at values for the lunar distances which differ slightly from those of Ptolemy and Copernicus, and 2) how Reinhold's derivation of the solar distance, which is a function of the lunar distance, defers to Ptolemy's value rather than to that of Copernicus. Reinhold's copy of the *Almagest* was undoubtedly the Basel (1538) edition of the Greek text.[13] He himself published in 1549 an edition of Book I of the *Almagest*, identical to the Basel edition, with a translation into Latin and a commentary. Reinhold was using the Nürnberg (1543) edition of *De revolutionibus*.[14] The surviving manuscript in Copernicus' own hand,[15] though it is not the manuscript from which the printed edition was prepared, serves to throw light on some of the internal problems of this work. In the manuscript, for example, one finds Copernicus frequently changing or correcting some, but not all, of the numerical values in a given compu-

[12]Reinhold, *Commentarius,* f. 50v. There are two contemporaries of Reinhold named Joannes Aurifaber, both of whom studied at Wittenberg. The first Joannes (1519-1575), from Mansfeldt, was Luther's private secretary, known for his publication of Luther's *Letters.* The second Joannes (1517-1568), from Breslau, is probably the one to whom Reinhold refers. He graduated from the University of Wittenberg in 1538, was a docent there until 1550, and was a lifelong friend of Melanchthon, Reinhold's colleague and supporter. Cf. Anon. "Aurifaber," *The Encyclopaedia Britannica*, eleventh edition, v. 2 (New York, 1910).

[13]Claudii Ptolemaei, *magnae constructionis, id est, perfectae coelestium motuum pertractationis lib. xiii* (Basel, 1538).

[14]Nicolaus Copernicus, *De revolutionibus orbium caelestium.* Facsimile edition of the Nürnberg 1543 edition, *Copernicus de revolutionibus orbium caelestium libri vi* (Amsterdam, 1943).

[15]Nicolaus Copernicus, *De revolutionibus orbium caelestium.* Facsimile edition of Copernicus' manuscript, *Nikolaus Kopernikus Gesamtausgabe* v. 1 (Munich, 1944).

tation, thus creating inconsistencies. Copernicus' derivation of the solar distance, as revealed by an analysis of his manuscript, proves to be very interesting in itself, but, while I occasionally refer to the manuscript in what follows, I have not attempted a detailed reconstruction here.[16]

The earliest surviving attempt to determine the solar distance is that of Aristarchus of Samos.[17] His method was to measure the angle subtended at the earth by the moon and sun when the moon appeared exactly halved. In the triangle formed by sun, moon and earth, with a right angle at the moon, the cosine of the angle at the earth is the ratio of the lunar distance to the solar distance. Both the measurement of this angle, which is very close to ninety degrees, and the determination of the moment when the moon is exactly halved are very difficult to perform. The result found by Aristarchus, that the solar distance is between 18 and 20 times the lunar distance, is cited by Archimedes in the *Sand-Reckoner.*[18] Archimedes himself estimated the solar distance as 30 times the lunar distance, although his estimate appears not to be based on any geometric method. It is difficult to tell how seriously Archimedes should be taken, since he is interested in over-estimating the size of the universe.

Hipparchus invented a new geometric method for relating the distance of the sun to that of the moon such that when one is given the other can be calculated. Given the ratio of the solar and lunar distances, there remains the problem of determining one of them in absolute terms, i.e., in earth radii. This can be done by a direct observation of parallax, since the reciprocal of the sine of the parallax of an object seen on the horizon is equal to the distance of the object in terrestrial radii. Parallax is measured by comparing the apparent position of an object with the true position predicted by theory, and thus requires a theory which can adequately predict true positions. The lunar distance can be obtained with some degree of accuracy from the parallax of the moon,

[16]For an analysis of Copernicus' derivation of the solar distance, and a more detailed discussion of Reinhold's computation of the solar and lunar distances, see Janice Henderson, *On the Distances between Sun, Moon, and Earth according to Ptolemy, Copernicus and Reinhold,* Yale Doctoral Dissertation [unpublished], 1973.

[17]T.L. Heath, *Aristarchus of Samos, the Ancient Copernicus* (Oxford, 1966).

[18]Archimedes, *The Works of Archimedes,* ed. T.L. Heath (New York, 1953), p. 223.

but solar parallax is so small that the distance of the sun cannot be determined in this way. Hipparchus, however, appears not to have attempted a direct measurement of either lunar or solar parallax. Ptolemy and Pappus report that Hipparchus *assumed* values for the solar distance and from these derived two different sets of lunar distances, presumably arising from the two different assumptions about the solar parallax. This may indicate that Hipparchus did not have sufficient confidence in the accuracy of his lunar theory to use it for a direct measurement of parallax. Indeed, his lunar model may have been adequate for syzygy, and thus for the prediction of eclipses, but not for positions of the moon at arbitrary elongation from the sun. Hipparchus' value for the solar distance was 490^{tr} (terrestrial radii), which corresponds to a parallax of exactly 0;7°.[19]

Ptolemy used the geometric method of Hipparchus, but derived the solar distance from the lunar distance. For Ptolemy, and for those who followed his example, which includes al-Battānī, Copernicus and Reinhold, the lunar distance and the lunar model were essential features of the determination of solar distance. Ptolemy made several changes in the lunar model of Hipparchus to improve its accuracy at positions other than syzygy. In particular, he had noted that the simple model of Hipparchus produced longitudes near quadrature which did not agree with observations. Ptolemy's complex lunar model accounts for these deviations by bringing the center of the lunar epicycle closer to the earth at quadrature, increasing the apparent size of the epicycle and thus the effect of lunar anomaly. The ratio of Ptolemy's maximum lunar distance at syzygy to the minimum distance at quadrature implies that the diameter and parallax of the moon would appear almost twice as large at the moon's closest approach to the earth as at its furthest distance. While nothing of the kind should occur, and Ptolemy makes no reference to this problem,[20] the parallax measurement reported in the

[19]Noel Swerdlow, "Hipparchus on the Distance of the Sun," *Centaurus* 1969, *14*: 287-305.

[20]In the *Planetary Hypotheses* Ptolemy states that when the moon is at its mean distance from the earth (48^{tr}), the apparent lunar diameter is 4/3 times that of the sun. This follows from the assumption that solar and lunar diameters appear equal when the moon is at its maximum distance (64^{tr}), since (64/48) = (4/3). Thus Ptolemy does in fact use his lunar model and lunar distances to determine the apparent lunar diameter. Ptolemy's

Almagest curiously confirms the close approach of the moon to the earth near quadrature.

Copernicus' lunar model was specifically designed to correct the exaggerated variation in distance and parallax produced by Ptolemy's model without impairing the quality of its predictions of lunar longitudes at quadrature, and was in fact successful in this respect. His double epicyclic model increases the equation of anomaly at quadrature while keeping the center of the epicycle at a constant distance, and thus the relative distances remain within reasonable limits. Copernicus describes two lunar parallax observations in *De revolutionibus* IV. 16, both made near quadrature. Though his computatons are filled with errors, the results effectively disprove the close approach of the moon to the earth implied by the Ptolemaic model.

The procedure for finding the lunar distances is as follows:

1) one observes the apparent zenith distance of the culminating moon under certain specified conditions,

2) one computes the true zenith distance of the moon at the time of the observations,

3) the parallax is then the difference between the true and apparent zenith distances,

4) the true lunar distance in terrestrial radii at the time of the observation is determined from the parallax,

5) the relative lunar distance at the time of the observation is found from the model,

6) the ratio of true and relative distances is then used to convert the extreme distances of the model, i.e., greatest and least distances in syzygy and quadrature, into true distances in terrestrial radii.

The observation itself merely involves a statement of the time of the observation and the apparent zenith distance. Ptolemy cites an observation he himself made in A.D. 135, when the moon was within 12° of quadrature. Copernicus describes two observations of lunar zenith distance, one just before and one just after the moon was at first quarter.

---

assumptions imply that at the minimum lunar distance (32^{tr}), the apparent lunar diameter is twice (64/32) that of the sun. See Bernard R. Goldstein, "The Arabic Version of Ptolemy's *Planetary Hypotheses*," *Transactions of the American Philosophical Society*, 1967, new series, *57*: 8 and 11.

In his report of the first observation, the time has twice been altered in the manuscript. For the second observation, three different values of the observed zenith distance are given in the manuscript, plus a fourth value in the printed Nürnberg edition. It does not appear that these changes were made in an effort to produce preconceived results. Some changes are too small to appreciably affect the results, while others have an adverse effect on the lunar distance, assuming Copernicus' aim was to maintain Ptolemy's lunar distance in syzygy while increasing the distance in quadrature.

The computation of the true zenith distance is rather lengthy and involves many intermediate steps. It entails computing the lunar longitude and latitude at the time of the observation, using the mean motion and correction tables in the *Almagest* or *De revolutionibus.* When Reinhold computes the true zenith distance for Ptolemy's observation, he is using the parameters and tables of the *Almagest*, and his result, that the moon is 49;48° from the zenith in Alexandria, is identical to Ptolemy's. In calculating the true zenith distance at the time of Copernicus' observation, Reinhold is not using the tables and parameters of *De revolutionibus,* but his own recomputed parameters of the lunar model. Even so, Reinhold's results are virtually identical to those of Copernicus, differing by 0;2° for the first observation and by only 0;0,36° for the second. In individual steps, Copernicus and Reinhold may differ by as much as 0;15°, as in the case of lunar anomaly, but the effect on the final result is small.

The computation of lunar parallax from the true and observed zenith distances involves only a simple subtraction. Thus the differences between Reinhold's results and those of Ptolemy and Copernicus remain constant. I would venture to say that it is at this point that Copernicus makes the changes in his apparent zenith distances, i.e., after obtaining the true zenith distance, he altered the apparent zenith distance to produce different values of the parallax.

The true lunar distance at the time of the observation is a function of the parallax and the apparent zenith distance. Copernicus and Reinhold find the distance by a direct solution, applying the law of sines to the triangle through the moon, the observer, and the center of the earth. The three known elements in the triangle are the parallax, the apparent zenith distance, and the terrestrial radius, assumed to be one.

Ptolemy's solution is somewhat longer, since he divides the triangle into two right triangles and solves each triangle separately. Reinhold's result differs from ptolemy's by $0;5^{tr}$. This discrepancy arises from a step early in the computation, where Ptolemy finds

$$\text{chord } 2 \,.\, 1;7^{\circ} \quad = \quad 2;21$$

instead of 2;20,19. The results of Reinhold and Copernicus differ by $2;24^{tr}$ for Copernicus' first observation, but by only $0;31^{tr}$ for the second.

The determination of the relative lunar distance requires a lengthier geometric procedure. Here the distance is a function of the parameters of the lunar model and the particular configuration of the model at the time of the observation. While Copernicus found the true lunar distance for both his observations, he finds the relative lunar distance for his second observation only. Copernicus and Reinhold find 93 998 and 93 930 parts, respectively, for the relative lunar distance, where the radius of the lunar deferent is 100 000 parts. Ptolemy and Reinhold differ by only 0;1 parts, where the deferent radius is 49;41 parts.

The relative lunar distance at any given moment can be determined from the model. Given the values of the true and relative distances for any one configuration, the ratio of these distances can be used to convert relative distances into true distances. The greatest, least and mean distances in syzygy and quadrature are listed below for Ptolemy and Copernicus, along with Reinhold's recomputed values. The last line indicates the ratio of greatest to least lunar distances.

TRUE LUNAR DISTANCES IN TERRESTRIAL RADII – PTOLEMY, COPERNICUS, REINHOLD

	Ptolemy	*Reinhold's* Recomputation of Ptolemy	*Copernicus*	*Reinhold's* Recomputation of Copernicus
Syzygy				
Greatest	$64;10^{tr}$	$64;19;49^{tr}$	$65;30^{tr}$	$64;57^{tr}$
Mean	$59;\ 0^{tr}$	$59;\ 9,15^{tr}$	$60;18^{tr}$	$59;48^{tr}$
Least	$53;50^{tr}$	$53;58,41^{tr}$	$55;\ 8^{tr}$	$54;39^{tr}$
Mean	$48;51^{tr}$	$48;59,58^{tr}$	$60;18^{tr}$	$59;48^{tr}$
Quadrature				
Greatest	$43;53^{tr}$	$44;\ 1,15^{tr}$	$68;20^{tr}$	$67;47^{tr}$
Mean	38 $38;43^{tr}$	$38;50,41^{tr}$	$60;18^{tr}$	$59;48^{tr}$
Least	$33;33^{tr}$	$33;40,\ 7^{tr}$	$52;17^{tr}$	$51;49^{tr}$
S/s	1;54,45	1;54,38	1;18,25	1;18,29

Reinhold's results differ from those of Ptolemy by less than $0;10^{tr}$, and Copernicus and Reinhold differ by approximately half a terrestrial radius. While these differences may seem small, the solar distance is extremely sensitive to small changes in the lunar distance. Assuming Ptolemy's value for the diameters of the moon and the earth's shadow, Swerdlow[21] has shown that the solar distance becomes infinite when the lunar distance is $60;57^{tr}$. If Ptolemy, like Hipparchus, had assumed that the solar and lunar diameters appeared equal when the moon was at its mean distance in syzygy (59^{tr} for Ptolemy), he would not have been able to obtain the solar distance using Hipparchus' method.

The geometric method of Hipparchus and Ptolemy for determining solar distance utilizes the near equality of the apparent diameters of the sun and moon and assumes an arrangement of the sun, moon and earth as illustrated in Figure 1. The plane of the ecliptic passes through the centers of these three bodies cutting off great circles. ABC is the circle of

[21]Noel Swerdlow, *Ptolemy's Theory of the Distances and Sizes of the Planets: A Study of the Scientific Foundations of Medieval Cosmology,* Yale Doctoral Dissertation [unpublished], 1968, p. 65.

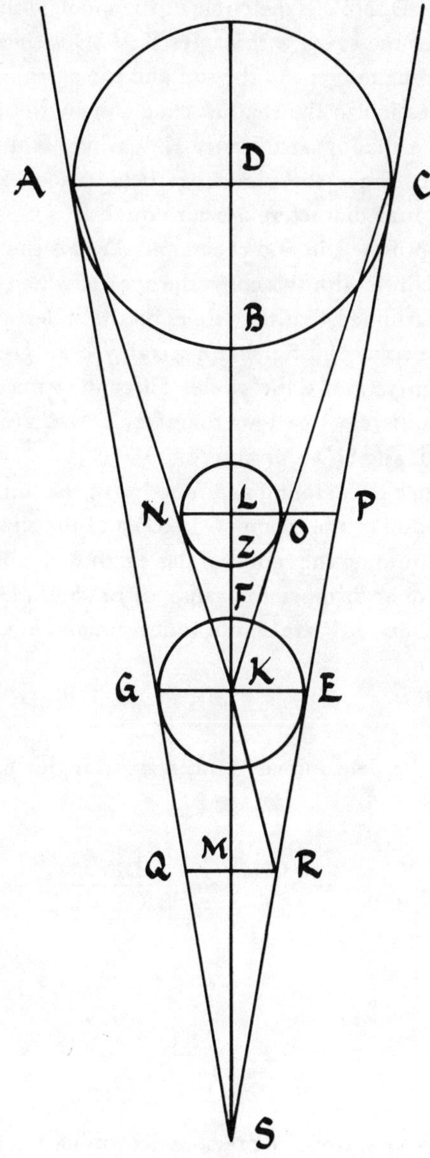

Figure 1. The Solar Distance.

the sun with center D, NZO is the circle of the moon with center L, and GFE is the circle of the earth with center K. ASC is the intersection of the plane with a cone tangent to the sun and the earth. GES is the portion of this intersection in the shadow cone cast by the earth. QMR is the diameter of the shadow at distance KM, where KM is equal to the lunar distance KL. Since the moon has been placed at the distance where solar and lunar diameters appear equal, it is possible to form a cone tangent to both the sun and the moon. AKC is the intersection of the plane of the ecliptic with this cone, the apex of which is at the center of the earth. It is assumed that the circles on the spheres of tangency of the cones with the spheres differ inappreciably from great circles, and thus that the diameters of these circles differ inappreciably from the diameters of the spheres. The four diameters, ADC, NLO, GKE and QMR, are parallel, and NO is produced to P.

The solar distance DK is found as a function of the lunar distance KL and the angular radii of the moon (∡ LKO) and the shadow (∡ MKR) at this distance, assuming the radius of the earth is one. From the angular radii and the lunar distance the values of LO and MR can be determined. Since LOP and MR are equidistant from and parallel to KE,

$$2KE = 2^{tr} = MR + (LO + OP)$$

and thus OP can be determined.[22] Then in triangles KCE and KCD, where KE is 1,

$$\frac{1}{OP} = \frac{CK}{CO} = \frac{DK}{DK - KL},$$

and the solar distance is

$$DK = \frac{KL}{1 - OP}.$$

This method thus requires the previous determination of

[22]This follows from Euclid I.15, I.29, and I.34. Reinhold offers a proof in his commentary on the *Almagest*, f. 181r.

1. the lunar distance at which the solar and lunar diameters appear equal,

2. the angular diameters of the sun and moon at that distance,

3. the ratio of the radii of the moon and the shadow of the earth where it intersects the lunar orbit at that distance.

Ptolemy found that the solar and lunar diameters appeared equal when the moon was at its greatest distance at syzygy. He determined this distance in terrestrial radii, found the angular diameter of the moon at this distance, which was then equal to the angular diameter of the sun, and found the angular diameter of the earth's shadow at this distance.

Between the time of Ptolemy and Copernicus, the only attempt well-known in the West to redetermine the solar distance was that of al-Battānī,[23] who adopted Ptolemy's lunar model, lunar distance, apparent solar diameter, and the ratio of lunar and shadow radii. He redetermined the apparent diameter of the moon, however, which necessitated recomputing the lunar distance at which solar and lunar diameters appear equal. If al-Battānī had used his value for this lunar distance to find the solar distance, he would have found that his parameters produced a solar distance 200 times greater than that of Ptolemy, a result which may well have seemed unreasonable and surprising. He circumvents this problem by simply multiplying his lunar distance by the ratio of Ptolemy's solar and lunar distances, obtaining a result closer to Ptolemy's, namely 1146^{tr}.[24]

Copernicus appears to have original values for the apparent solar and lunar diameters and for the ratio of the lunar and shadow radii. I believe, however, that he has merely adjusted the values of Ptolemy and al-Battānī to fit the relative distances in his solar and lunar models. Copernicus' solar diameter at apogee follows from Ptolemy's value of the solar diameter, and his lunar diameter follows from al-Battānī's value of the lunar diameter. There are two different values for the ratio in Copernicus' manuscript. The first is an adjustment of Ptolemy's value

[23] Al-Battānī, *Al-Battānī sive Albatenii, Opus Astronomicum,* ed. C.A. Nallino, Pubblicazioni del Reale Osservatorio di Brera in Milano, no. 40 (Rome 1903); reprint, Frankfurt, 1969, v. 1, pp. 57-61.

[24] Noel Swerdlow, "Al-Battānī's Determination of the Solar Distance," *Centaurus,* 1972, *17*: 97-105.

of the ratio. Copernicus, it must be understood, does not reveal the source of his parameters, but merely states them as if he had determined them anew. In fact, he has done nothing of the kind. This is particularly apparent in his derivation of the radii of the moon and the shadow. I believe that the two observations which Copernicus says he used (*De revolutionibus* IV.18) were fictitious.

From the manuscript of *De revolutionibus* one can see that Copernicus began his derivation of the solar distance with a value of 79:30 for the ratio. He did not find the lunar distance at which solar and lunar diameters appear equal, but merely used the greatest distance at syzygy ($65;30^{tr}$) to derive a solar distance of 1179^{tr}. Later, he returned to this section of the manuscript and changed the value of the ratio to 403:150. At this time, he also computed the lunar distance (62^{tr}) at which solar and lunar diameters appear equal. While he recomputed various intermediate values in the derivation, he did not bother to adjust the final result. The Nürnberg edition of *De revolutionibus* contains the value 403:150 and the correctly computed lunar distance, but retains the solar distance of 1179^{tr}, which thus simply does not follow from the stated parameters.

The solar distance itself is not constant, but varies between a greatest distance in apogee and a least distance in perigee. In the *Almagest,* Ptolemy considered only one solar distance, 1210^{tr}; since he nowhere in the *Almagest* specified that this was the mean distance, as he does in the *Planetary Hypotheses,* some later astronomers assumed it to be the greatest distance.[25] Al-Battānī and Copernicus, by estimating the solar diameter at apogee and determining the lunar distance at which solar and lunar diameters appear equal, found the solar distance in apogee. The Copernican solar model, with its slowly changing eccentricity, introduces a further variation in distance such that there is actually a greatest and least solar distance in apogee and perigee. In Figure 2, the sun is at B and the center of the earth's orbit moves on circle AGD in a period of 3434 years, while the center of the circle F moves slowly in the opposite direction in a period of about 53,000 years (*De revolutionibus* III.20). The minimum, mean and maximum eccentricities are repre-

[25]See, for example, Rheticus' statement in E. Rosen trans., *Three Copernican Treatises* (New York, 1959), p. 162.

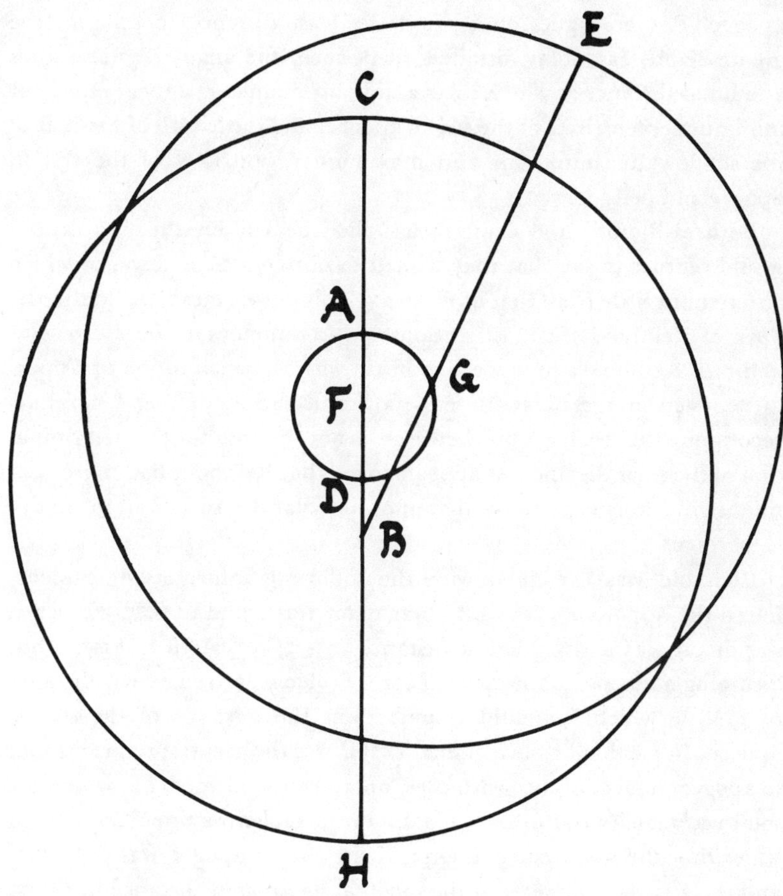

Figure 2. Copernicus' Solar Model for the Changing Eccentricity and Apsidal Line.

sented by BD, BF and BA. If the earth is at E, G is the center of its orbit, and CD equals GE. Copernicus himself does not consider the effect of the changing eccentricity on the solar distance, but Reinhold treats it at length. After calculating the greatest solar distance in apogee, Reinhold goes on to find the least distance in apogee, the greatest and least solar distances in perigee, the angular solar radius and lunar distance at which solar and lunar diameters appear equal for minimum eccentricity of the sun in apogee, and the length of the axis of the shadow for minimum and maximum eccentricity of the sun in apogee and perigee.

Both al-Battānī and Copernicus relied heavily on Ptolemy, and I would venture to say that they wished to obtain a solar distance which differed but little from that of Ptolemy. This is even more explicit in the work of Reinhold. In most sections of his commentary on *De revolutionibus,* Reinhold's practice is to begin with the assumption of Copernicus, such as the observational data and the model, and carefully recompute the results. This, however, is not the case for the determination of the solar distance. It appears to me that Reinhold had more confidence in Ptolemy's determination of the solar distance than in that of Copernicus.

Reinhold was presented with the following information. Ptolemy found the apparent solar and lunar diameters equal to 0;31,20° when the moon was at its greatest distance in syzygy, 64;10^{tr}. From this, assuming a moon-shadow ratio of 5:13, Ptolemy derived a solar distance of 1210^{tr}, which Reinhold assumed was the distance of the sun at apogee. In particular, he assumed 1210^{tr} was the greatest solar distance in apogee, in accordance with the Copernican solar model in which the solar eccentricity was near its maximum in Ptolemy's time. Copernicus states that the apparent diameter of the sun in apogee was 0;31,40°, and the angular diameter of the moon at its greatest distance in syzygy (65;30^{tr}) was 0;30°. Ostensibly assuming a moon-shadow ratio of 150:403, he derived a solar distance of 1179^{tr}. Since the solar eccentricity was near its minimum in Copernicus' time, this should correspond to the least solar distance in apogee.

Reinhold combines Ptolemy's information about the sun with Copernicus' information about the moon to obtain a solar distance in Ptolemy's time. He accepts Copernicus' value of 0;30° for the apparent diameter of the moon at its greatest distance in syzygy and uses his own recomputed value for this distance, based on the Copernican lunar model. He accepts Ptolemy's value of 0;31,20° for the apparent diameter of the sun and derives the greatest solar distance in apogee. The one variable which he does not specify is the apparent diameter of the shadow. Given his assumptions, this is determined by his choice of a moon-shadow ratio. Reinhold's procedure is to try different values of the moon-shadow ratio in an essentially trial-and-error attempt to determine the solar distance.

Reinhold begins by assuming Ptolemy's value for the ratio (5:13) and obtains a solar distance nearly three times as large as that of Ptolemy (3108^{tr}). He remarks that this value is too large and that Copernicus was correct in assuming a different value for the ratio. Using Copernicus' value of 150:403, Reinhold next computes a solar distance of 1301^{tr}, which he considers still too large. He makes one more attempt, with a ratio of 7:19, obtaining 1187^{tr}. While this value is close to that of Copernicus, he is obviously still not satisfied with the result.

Reinhold's next step, in a section which he entitles *analysis,* is to assume a value for the solar distance and from this to derive a value for the ratio. This may seem contrived, since the ratio of the lunar and shadow diameters has a definite value and it can be derived from the comparison of two lunar eclipses. Reinhold may have felt, however, that Copernicus' derivation of the ratio in *De revolutionibus* IV.18, on which Reinhold chose not to comment, was not based on real eclipse observations. But neither did Reinhold make the observations necessary to provide an independent confirmation of this quantity.

Reinhold assumes a greatest solar distance of 1200^{tr}, which he says, does not differ too much from Ptolemy's value. His own recomputed value of Ptolemy's solar distance, in his commentary on *Almagest* V.15, is 1194^{tr}, which may explain why he chose a value less than 1210^{tr}. According to Reinhold, the value of the true lunar radius (LO) is fixed at $0;17^{tr}$ and the lunar distance (KL) at which solar and lunar diameters appear equal is $62;10^{tr}$. In order to compute the ratio, one must find the true shadow radius (MR) that corresponds to a solar distance of 1200^{tr}. This can be easily accomplished by computing OP, since

$$\text{MR} = 2 - (\text{LO} - \text{OP}) ,$$

and LO is known. Since

$$\text{DK} = \frac{\text{KL}}{1 - \text{OP}} ,$$

it follows that

$$\text{OP} = \frac{\text{DK} - \text{KL}}{\text{DK}} .$$

Instead of finding OP directly, i.e.,

$$\text{OP} = \frac{1200^{\text{tr}} - 62;10^{\text{tr}}}{1200^{\text{tr}}} = 0;56,53,30^{\text{tr}} ,$$

Reinhold finds KL under the assumption that DK is 1^{tr}:

$$\text{KL} = \frac{62;10^{\text{tr}}}{1200^{\text{tr}}} \cdot 1^{\text{tr}} = 0;3,6,30^{\text{tr}} .$$

However, he has made a mistake at this point, for the manuscript indicates a value of $0;3,5,40^{\text{tr}}$ for KL. This accounts in part for a solar distance of 1208^{tr}, rather than 1200^{tr}, which Reinhold obtains when he works the analysis backwards (his following *synthesis*). In triangles KCE and KCD, where KE is 1,

$$\frac{1}{\text{OP}} = \frac{\text{DK}}{\text{DK} - \text{KL}} .$$

Thus if DK is 1,

$$\text{OP} = 1 - \text{KL} = 1^{\text{tr}} - 0;3,5,40^{\text{tr}} = 0;56,54,20^{\text{tr}} .$$

The true shadow radius is then

$$MR = 2^{tr} - (0;17^{tr} + 0;56,54,20^{tr}) = 0;46,5,40^{tr} .$$

Reinhold reduces the ratio as follows:

$$\frac{LO}{MR} = \frac{0;17}{0;46,5,40} = \frac{61200}{165940} = \frac{30600}{82970} = \frac{3060}{8297} .$$

At this point he "throws out" one from the 8297 and proceeds to find

$$\frac{3060}{8296} = \frac{1530}{4148} = \frac{765}{2074} = \frac{45}{122} .$$

The small error introduced by changing 8297 to 8296 also helps account for the difference between 1208^{tr} and 1200^{tr}.

At this point in the manuscript, Reinhold draws a line and begins a new section entitled *synthesis*. The small difference between analysis and synthesis is immediately apparent in the calculation of MR. The ratio 45:122 was obtained from a value of $0;46,5\ 40^{tr}$ for MR. Working backwards, however,

$$MR = 0;17^{tr} \cdot \frac{122}{45} = 0;46,5,20^{tr} .$$

From this it follows that

$$LOP = 2^{tr} - 0;46,5,20^{tr} = 1;13,54,40^{tr}$$

and

$$OP = 1;13,54,40^{tr} - 0;17^{tr} = 0;56,54,40^{tr} .$$

The solar distance is thus

$$DK = \frac{62;10^{tr}}{1^{tr} - 0;56,54,40^{tr}} = 1208^{tr} .$$

Reinhold's aim, quite obviously, is to obtain a solar distance in agreement with Ptolemy, rather than Copernicus. The value he finally accepts is 1208^{tr}, compared to Ptolemy's value of 1210^{tr}. It is surprising to me that Reinhold did not notice the error in his analysis which produced 1208^{tr}, rather than the 1200^{tr} which he assumed. Reinhold's value for the least solar distance in apogee, 1197^{tr}, is considerably larger than Copernicus' value of 1179^{tr}. This seems to me a telling example of the strong influence of Ptolemy, even after the time of Copernicus.

Ptolemy's influence on Copernicus is much in evidence throughout *De revolutionibus*, which is modelled very closely on the *Almagest*. Not only does Copernicus adhere to the subject matter of the *Almagest*, presented in roughly the same order, but his mathematical methods, used in the derivation of parameters and the construction of tables, are virtually identical to Ptolemy's. The determination of the solar distance is but one example of Copernicus' use of the actual numerical values in the *Almagest*. The underlying concern throughout *De revolutionibus* is to show that the Copernican models produce results which agree with the Ptolemaic models.[26]

One of the reasons the Ptolemaic system survived without rival until the time of Copernicus is the high quality of Ptolemy's workmanship. The *Almagest* is a model of logical rigor, both in its presentation of the subject matter and in the accuracy of the numerical details. As for the observations, John Britton has suggested, on the basis of his analysis of solar and lunar observations and parameters in the *Almagest,* that Ptolemy's procedure for deriving the parameters from the observations may be considerably more sophisticated than is indicated in the *Almagest*. Ptolemy never overdetermines a model in the *Almagest,* i.e., he only presents as many observations as are needed to derive the parameters. Since Ptolemy's parameters are a great deal more accurate than one would expect on the basis of the average errors in his observations, it appears that he may have been drawing on a large number of observations.[27] After many determinations of the parameters, and perhaps

[26]Otto Neugebauer, "On the Planetary Theory of Copernicus," *Vistas in Astronomy, 10* (Oxford, 1968), p. 103.

[27]John Britton, *On the Quality of the Solar and Lunar Observations and Parameters in Ptolemy's Almagest,* Yale Doctoral Dissertation [unpublished], 1967.

applying some non-rigorous averaging procedure, he may have selected the values which he thought best. There is nothing comparable to this in the work of Copernicus and Reinhold, who limit themselves to the procedures explicitly described in the *Almagest*.

There is a marked contrast between Copernicus and Reinhold in their treatment of numerical detail. Copernicus is quite unreliable and uneven in this respect. In the chapters on the solar and lunar distances, one has many examples of Copernicus altering his observational data, committing computational errors, and making partial corrections which are not carried through, resulting in internal inconsistencies in the printed editions of *De revolutionibus*. I believe Reinhold was well aware of Copernicus' lack of concern for the arithmetical precision in his work, and that this accounts for Reinhold's decision to recompute all parameters before constructing the extensive Prutenic Tables, since nothing short of a complete recomputation could rectify the situation. Reinhold's commentary displays the painstaking care which he lavished on even the smallest detail.

An analysis of Copernicus' manuscript reveals that he was particularly careless in his computation of the solar distance, the basic dimension of the solar system. He did not independently derive the parameters which he used, but merely modified the values of Ptolemy. This has the advantage, of course, of producing a solar distance which does not differ greatly from that of Ptolemy, which was undoubtedly Copernicus' intent. There is no doubt about Reinhold's desire to derive a solar distance approximately equal to Ptolemy's. When he assumes that the solar distance is 1200^{tr}, he explicitly says that this does not differ too much from Ptolemy's value. To my knowledge, this is the only example in the commentary on *De revolutionibus* where Reinhold rejects Copernicus' data. He was then forced to rely on Ptolemy, and perhaps this is why he wrote a commentary on this very subject as it is treated in the *Almagest*. The unwillingness of Copernicus and Reinhold to venture very far from the solar distance of Ptolemy is a good example of the extent to which they both realized the difficulty of improving upon Ptolemy's results.

COMMENTARY: THE ROLE OF ERROR IN ANCIENT METHODS FOR DETERMINING THE SOLAR DISTANCE

Ernest S. Abers

Charles F. Kennel

University of California, Los Angeles

Our U.C.L.A. colleagues in the history of science know the trepidation with which we agreed to present a commentary at this symposium on the Copernican achievement. They also observed our anxiety increase, apparently with some amusement, as the meeting approached. What could physicists say to an audience of professional historians that would not be naive and anachronistic? After considerable discussion, however, we decided to present what we could: the impressions of two practicing physicists about Professor Henderson's paper.

Our first reaction was to be astonished that in the sixteenth century people were arguing about small corrections to a value for the solar distance which, today, we know to be too small by a factor 20. Why were they fudging the fourth decimal place when they were wrong by an order of magnitude? We were therefore naturally predisposed to believe the central thesis of Professor Henderson's paper: that certain astronomers and, in particular, Erasmus Reinhold, had "massaged" their observational data and manipulated their calculations so as to agree with each other and with astronomical tradition. This led us to two further questions. How could astronomers agree to disagree only upon small deviations from the traditional estimate for the solar distance? Secondly, why did they agree on a range of solar distances that was too small?

The situation described in Henderson's paper, it seems, was inherent in the ancient methods for determining the solar distance cited in her paper. Since the sun is so far away, one cannot triangulate from two places on the earth; one has to use the moon. In Antiquity, there were several such techniques for estimating the distance D_M from the earth to the moon or the distance D_S from the earth to the sun, or both together. Aristarchus had proposed that a careful measurement of the angle between the sun and moon when the moon was exactly at quadrature would yield the ratio D_S/D_M. Ptolemy attributed to Hipparchus the method of eclipses described in Henderson's paper, whereby D_S/D_M could be determined as a function of D_M/R_E, where R_E is the radius of the earth. In principle, the results from these methods could be combined to produce estimates of D_S/D_M and D_M/R_E. Moreover, given an independent theory for the lunar position, one could estimate the diurnal lunar parallax, and, from that, D_M/R_E; whereupon the method of eclipses would yield in principle D_S/R_E. Ideally, the computations combining various permutations of these methods should have agreed. Moreover, the problem should have been overdetermined for there were three observations to determine two ratios.

In fact, both the method of lunar quadrature and the method of eclipses are mathematically singular since small errors in measurement lead to large errors in the inferred solar position. In Aristarchus' method, $D_M/D_S = \cos \vartheta$, where ϑ is the angle between the sun and moon at quadrature. There were at least two sources of error in determining ϑ. The precise timing of the moon when it was exactly at quadrature must have been the main difficulty. However, even if the timing of quadrature could have been exact, the modern solar distance makes ϑ rather more like 89°50′ than Aristarchus' 87°, within the traditional estimate of 10′ accuracy for naked-eye observations of 90°, which would be the value of ϑ for infinite solar distance. Finding D_S/D_M therefore depends upon measuring ϑ very accurately. For $\vartheta = 87°$, Aristarchus found $D_S = 20D_M$ whereas the modern value is closer to $D_S = 400R_M$. Aristarchus' method has another peculiar property. As timing accuracy improves, a given uncertainty in measuring ϑ leads to increasingly large errors in D_S. Let $\Delta\vartheta$ be the uncertainty in the measurement of ϑ, and ΔD_S, the resulting uncertainty in the solar distance. Then

$$\frac{\Delta D_S}{D_S} = \tan\vartheta\, \Delta\vartheta$$

Now, when $\vartheta = 87°$, $\tan\vartheta \simeq 20$, while when $\vartheta = 89°50'$, $\tan\vartheta \simeq 400$. Thus, at the true solar distance, a measurement uncertainty $\Delta\vartheta$ produces 20 times the relative error, $\Delta D_S/D_S$, than at 87°. Thus any improvements in timing accuracy would have been defeated by increasingly stringent requirements upon the basic angular measurement. The ten minutes of arc precision attainable with good naked eye measurements, implies that when $\tan\vartheta \simeq 400$, the relative uncertainty $\Delta D_S/D_S$ is about one. Therefore, naked-eye measurements could never determine the true solar distance with Aristarchus' method.

The eclipse method for estimating D_S is discussed extensively in Professor Henderson's paper. Let us translate her notation into the one we have been using; the key formula in her paper is

$$DK = \frac{\overline{KL}}{1 - \overline{OP}}$$

or

$$DS = \frac{D_M}{1 - \dfrac{\overline{OP}}{R_E}}$$

where the various distances are defined in Figure 1 of her paper. Since, in this diagram, $R_E(\overline{KE})$ is the average of MR and $\overline{OP} + R_M$, where R_M is the radius of the moon.

$$D_S = \frac{R_E D_M}{\overline{MR} + R_M - R_E}$$

The "shadow-moon" ratio, r, which is known to be about 8/3 from the duration of central lunar eclipses, is just $\overline{MR}/R_M$, so

$$D_S = \frac{R_E D_M}{(1 + r)R_M - R_E}$$

It is important that R_M cannot be determined independently, but only in terms of D_M and the angle, φ, subtended by the moon at the center of the earth:

$$R_M = \tfrac{1}{2} D_M \sin \varphi$$

Since φ is small (about 30′), we approximate sin φ by φ (in radians), so that $R_M \approx (\varphi/2)D_M$. Then

$$D_S = \frac{R_E D_M}{\tfrac{1}{2}\varphi(1 + r)D_M - R_E} = \frac{D_M}{\frac{D_M}{D_M^*} - 1}$$

where we have defined a "critical distance" D_M^* by

$$D_M^* = \frac{2R_E}{\varphi(1 + r)}$$

Evidently the method is singular. If D_M is near D_M^*, D_S is large and very sensitive to the small difference $D_M - D_M^*$. In fact, D_M differs from D_M^* by one part in four hundred. This sensitivity is no accident, and has nothing to do with the fact the moon is exactly as far away as it is, or that the angles subtended by the sun and the moon are almost the same. This last coincidence makes it possible to draw Figure 1, but the problem is not much harder with a more general geometry. The method for finding D_S is always singular since D_S is in fact much larger than D_M.

Let us estimate the error ΔD_S introduced by errors in the assumed lunar parameters. If we assume that $\vartheta(1 + r)$, and therefore D_M^* are measured perfectly, but that there is an error ΔD_M in D_M (e.g. by diurnal parallax), then

$$\frac{\Delta D_S}{D_S} = \frac{\Delta D_M}{D_M} \frac{D_M^*}{D_M^* - D_M}$$

On the other hand, if D_M is known exactly, but there is an observational error ΔD_M^* (from either ϑ or r), then

$$\frac{\Delta D_S}{D_S} = \frac{\Delta D_M^*}{D_M^*} \frac{D_M}{D_M - D_M^*}$$

The complete fractional error in D_S is the sum of the two contributions

$$\frac{\Delta D_S}{D_S} = \frac{1}{D_M - D_M^*} \left[\frac{D_M}{D_M^*} \Delta D_M^* - \frac{D_M^*}{D_M} \Delta D_M \right]$$

The factor $1/(D_M - D_M^*)$ in front indicates that the *errors* are singular when D_M is near D_M^*. For example, the modern value $1/(1 - D_M/D_M^*) \approx 400$ implies that a relative error in D_M is magnified 400 times in calculating $\Delta D_S/D_S$. Even the ancient value of 20, which led to $D_S \approx 1200R_E$, means that errors in D_M and $\varphi(1 + r)$ would have been magnified by a factor of 20 in D_S.

On the other hand, the eclipse method is inherently stable when used to calculate the distance to the moon, for, inverting the formula for D_S, we get

$$D_M = \frac{D_M^*}{1 - \dfrac{D_M^*}{D_S}}$$

Assuming *any* large value for D_S compared to D_M or D_M^* (say from a quadrature measurement) leads to $D_M \approx D_M^*$. Since this is very close to the right answer, it was confirmed by lunar parallax observations.

Now we are in a position to speculate upon what may have happened. *The calculation was unstable.* Each person who did it found an alarmingly large variation in D_S for apparently minor changes in calculational procedure. Each was faced with the problem of reconciling his own calculations on observations as well as those of others with one another. He chose to rediscover the traditional value, which was not in disagreement with anything he could have known.

In summary, we hope we have made it plausible by our brief analysis of the role of error in the ancient methods of determining the solar distance that the situation described in Henderson's paper was inevitable. Any solar distance exceeding $1200R_E$ was unresolvable. Clearly the social and philosophical influences affecting the strategies for resolving discordant observations and calculations present very interesting historical questions, particularly in this case. We can only speculate upon such influences. For example, the role of scientific authority is considerable even today.[1] It is natural to believe that others may know something you don't. One could try to determine the solar distance using a completely different line of reasoning. O. Gingerich pointed out to us that a D_S of approximately $1200R_E$ was consistent with the principle of plenitude. It would have been a relief to Ptolemaics to know that this estimate was at least not inconsistent with the known lunar distance.

But the simple desire to agree with more ancient authority surely cannot offer a complete explanation of the fact that everyone, Ptolemaic or Copernican, agreed upon a value for D_S 20 times too small. The number $1200R_E$ indeed must have come from somewhere. Someone must have computed it first, and if it was Aristarchus, and if he indeed advocated a heliocentric solar system it is hard to believe that considerations of geocentric plenitude would matter to him. It is a striking fact, however, that people have traditionally been reluctant to increase the

[1]Mercury's diurnal period of rotation was "known" since the 1890's to be 88 days, a result which seemed physically reasonable by analogy with the earth-moon system where the tides have locked together the moon's periods of rotation and revolution. Observations up to 1965 agreed with this estimate. In 1965, radar observations established Mercury's rotation period to be 59 days, exactly 2/3 of its orbital period, far outside the earlier experimenters' estimates of the probable error in their measurements. The radar observations of Mercury provoked a major advance in celestial mechanics, in which it was realized that "spin-orbit" couplings could produce small whole number ratios of orbital and rotation periods.

size of the universe, as the later history of the development of cosmology indicates. The estimated size of the universe has increased with time. $1200R_E$ was just about the smallest solar distance consistent with observations and error.

Acknowledgement

It is a pleasure to acknowledge many illuminating conversations with our U.C.L.A. colleagues, Robert S. Westman, Amos Funkenstein, and Robert G. Frank, Jr.

V

THE ARTS FACULTY AT THE UNIVERSITY OF CRACOW AT THE END OF THE FIFTEENTH CENTURY

Paul W. Knoll

University of Southern California

When "Nicholas the son of Nicholas of Toruń" enrolled in the arts faculty of the *studium Cracoviensis* in 1491, he entered a late medieval university which was older than most in central Europe but yet sufficiently young to be open to the great pedagogical and intellectual changes which were transforming Europe. Particularly in the arts curriculum, he was to be confronted during his three, and possibly four, years in Cracow with a variety of cultural currents. It is difficult to speak with any certainty about what specifically Copernicus gained in these years and it is difficult to determine what influence this stay may have had upon his eventual formulation of the hypothesis put forward first in *De hypothesibus motuum coelestium a se constitutis commentariolus* (1507) and developed more completely in *De revolutionibus orbium coelestium* (1543).[1] Nevertheless, an examination of the university in which Copernicus probably first confronted the problems he was later to

[1]Polish scholars have generally tried to attribute as much of Copernicus' achievement as possible to his Polish background. The best corrective to this has been the precision and specificity of the scholarship of Edward Rosen. See particularly the ninety-six page biography and items 1029 through 1046a in the "Select Bibliography" in his *Three Copernican Treatises,* 3rd edition (New York: Octagon, 1971). Numerous biographical and commemorative studies on Copernicus have of course appeared in connection with the "Rok Kopernika." Some, which are available in English and have received wide distribu-

transform is useful for determining something of the cultural orientation of late medieval Europe. To that modest purpose this short study is devoted.

1. The University in the Fourteenth Century

Heralded by the example of Emperor Charles IV at Prague in 1347, the last Piast ruler of Poland, King Casimir the Great, founded the University of Cracow in 1364.[2] He had taken great pains from 1351 onward to prepare for the creation of the brightest cultural jewel which could adorn a medieval *corona.* After negotiations in which king, counsellors, and Cracovian officials were eventually united, the establishment of the university was proclaimed in May 1364, celebrated at an international congress in August of that year, and confirmed, with modifications, by the pope one month later. The institution which emerged in this process was modeled upon the University of Bologna, with professors chosen by students and the rector being elected by the students from among their number. Although Casimir had originally requested a theological faculty, the Piast university was limited to one chair in liberal arts, two in medicine, three in canon law, and five in civil law. Upon this structure, three brief observations may be made.

At the beginning inclined to grant the whole of the king's petition, Pope Urban V had eventually denied Casimir's request for a "studium generale . . . in quacumque facultate . . . ," reserving for Prague a theological monopoly in central Europe and casting implicit doubts upon the ability of the Poles to preserve and protect the tradition of doctrinal

tion, are useful; but the most reliable short biographical treatment of Copernicus in Polish is still contained in *Polski Słownik Biograficzny* (Cracow [later also Wrocław and Warsaw]: Ossolineum, 1935-), XIV (1968), pp. 3-16. A reliable calendar of the data and documents relating to Copernicus is now contained in Marian Biskup, ed., *Regesta Copernica* (Wrocław, Warsaw: Ossolineum, 1973 [*Studia Copernicana,* VII]).

[2]The documents relating to the founding of the University are most completely discussed, and are printed in their entirety, in Stanisław Krzyżanowski, "Poselstwo Kazimierza Wielkiego do Avinionu i pierwsze uniwersyteckie przywileje," *Rocznik Krakowski,* IV (1900), pp. 1-111. The materials printed in *Codex Diplomaticus Universitatis studii generalis Cracoviensis,* I (Cracow, 1870), are incomplete. Much of what follows immediately in the text above is derived from my "Casimir the Great and the University of Cracow," *Jahrbücher für Geschichte Osteuropas,* 1968, *16:* 232-249.

orthodoxy.[3] Second, the practical needs of the *regnum Poloniae* for men carefully trained in the administration of law is revealed by the heavy concentration upon legal studies in the university. (That chairs in civil law outnumbered those in canon law reflects no royal prejudices; it was a condition dictated simply by the organization of lectures in these two fields.[4]) Third, the chair in liberal arts, which carried an income only half that of three other chairs and only one quarter that of the remaining seven chairs, was clearly inadequate to support enough students for the higher faculties. Casimir apparently expected that the city and the diocese of Cracow would see fit to endow others, and so he specified in his bull of foundation only that which he would support from the royal salt monopoly at Wieliczka. (This hope for privately endowed chairs was to be realized in the following century.)

These were promising beginnings, but the Casimiran foundation did not prosper. The financial basis of the *studium* was compromised in the last years of Casimir's reign, ecclesiastical and political intrigues during the 1370's and 1380's caused attention to be focused away from the university, and after the king's death there arose no Maecenas to continue the tradition of patronage. Although we know the names of some students who may be assigned to the 1360's and early 1370's and are therefore able to determine that some parts of the *studium* were activated, the university soon thereafter fell into a decrepitude which suggests that it had ceased to function. There is record neither of students nor professors, no trace whatsoever of activity. Not until the last decade of the century was the university revived.

Her interest turned to the question of education by the urging of master Bartholomew of Jasło, Queen Jadwiga helped bring about the refounding of the *studium Cracoviensis.*[5] She won her husband,

[3]See the comments of Adam Vetulani, *Początki najstarszych wszechnic środkowoeuropejskich* (Wrocław, Warsaw, Cracow: Ossolineum, 1970), pp. 94-103.

[4]Vetulani, "U progu działalności Krakowskiego wydziału prawa," in *Studia z dziejów wydziału prawa Uniwersytetu Jagiellońskiego* (Cracow, 1964), pp. 27-33, and "L'enseignement universitaire du droit à Cracovie d'après les desseins de Casimir le Grand," in *Études d'histoire du droit canonique dédiées à Gabriel LeBras,* I (Paris: Sirey 1965), pp. 373-383.

[5]The speeches of Bartholomew of Jasło are printed and discussed in Maria Kowalczyk, "Odnowienie Uniwersytetu Krakowskiego w Świetle mów Bartłomieja z Jasła," *Małopolskie Studia Historyczne,* 1964, *6*, 3-4: 23-42. See also Zofia Kozłowska-Budkowa,

Władysław Jagiełło, to her cause, and he obtained permission for a theological faculty from Pope Boniface IX in 1397 and three years later issued a new founding charter. This time, however, the school was organized on the Parisian pattern and was clearly a masters-run university. Another important difference in the *Alma mater Jagiellonica* (taking its name from its second founder) was that in the decades to come it prospered mightily. In sheer numbers alone, the statistics are impressive, for during the fifteenth century 18,338 individual students matriculated at Cracow.[6] The dimensions of this prosperity as they relate to such areas as theology, law, and medicine will not concern us here directly; but before we turn to the arts faculty and its curriculum, it is appropriate to point out briefly that the university functioned in this period in that *Collegium Maius* which is even today historically and architecturally one of the most important of the physical sites of Cracow.

2. *The* Collegium Maius

At first, medieval universities were not places; they were states of being, paper corporations whose reality rested not in some physical locus, but in privileges, in disputations, in persons. Thus the University of Paris functioned wherever the students and teachers gathered; thus the University of Bologna could secede from the town without ceasing to be the university. All who have heard Professor Astrik Gabriel discourse wittily about academic taverns in medieval Paris or have read Professor Charles Homer Haskins' delightful letters from students seeking housing know the transitory physical character of the early universities.

Poland's first *studium* followed this pattern. Despite some previous

"Odnowienie Jagiellońskie Uniwersytetu Krakowskiego (1390-1414)," in Kazimierz Lepszy, ed., *Dzieje Uniwersytetu Jagiellońskiego w latach 1364-1764* (Cracow: Uniwersytet Jagielloński, 1964), pp. 37-71.

[6]The statistic is drawn from the *Album studiosorum Universitatis Cracoviensis,* ed. B. Ulanowski and A. Chmiel, vol. I (Cracow, 1887), by Antoni Karbowiak, "Studya statystycne z dziejów Uniwersytetu Jagiellońskiego 1433/4-1509/10," *Archiwum do Dziejów Literatury i Oświaty w Polsce,* 1910, *12:* 1-82; pp. 77-81 especially. The document relating to the refoundation are most conveniently available in *Codex dipl. Univ. Crac.*, I, pp. 24-25, 118.

royal attempts to build permanent academic quarters in the district of Kazimierz,[7] the Casimiran foundation apparently functioned in private homes, various churches, miscellaneous rooms of the royal castle, and its personnel lived wherever they could individually obtain bed and board.

Physical foundations were not long in coming, however, and by the fifteenth century, educational institutions were increasingly tied to buildings. It was in this tradition that King Władysław Jagiełło decreed in 1400 that the former house of Stephen Pęcherz in St. Anne's Street should become the home of the *studium*:[8]

> We have resolved to designate our house as a place of residence for the masters and for the daily meeting of the students in order that the doctors, masters, licentiates, bachelors, and students might carry on freely their lectures and scientific duties, and we intend that this house should remain forever in the possession of the professors. . . .

In this building space was limited, and as the university grew during the next half century, several adjoining houses were acquired and incorporated into the university structure. A Commons room and an oriel for lectures were built within this complex, as well as quarters for the professors (but not for the students, who still found their housing elsewhere). The large number of Prague professors then in Cracow as well as the generally close ties between the two schools probably accounts for the architectural influence of the Prague *Carolinum* upon the *Collegium Maius* in this period.[9]

Nearly another fifty years passed before the completion of the building of the university. The acquisition of additional houses and their combination into existing structures created a rectangular compound

---

[7]For the polemic connected with this issue, see Henryk Barycz, *Alma Mater Jagellonica* (Cracow: Wydawnictwo Literacki, 1958), pp. 29-38; Jan Dąbrowski, "Czy Uniwersytet krakowski działał na Kazimierzu," *Rocznik Biblioteki Polskiej Akademii Nauk,* V (1959), 53 ff.; and Vetulani, "Les Origines de l'Université de Cracovie," *Acta Poloniae Historica,* 1966, *13:* 19-24.

[8]The royal document is printed by Stanisław Tomkowicz, "Gmach biblioteki Jagiellońskiej," *Rocznik Krakowski,* IV (1900), 115.

[9]*Ibid.,* pp. 117-118; and Karol Estreicher, *Collegium Maius* (Cracow: Uniwersytet Jagielloński, 1968), pp. 58-62.

which was completely self-contained. An architecturally mixed exterior which combines a variety of facades and construction styles becomes on the interior courtyard a purposeful and harmonious whole. After a final building project during the time when Copernicus was in Cracow, only a new library (now the Senate Hall) and minor additions and remodelings were added later.

Equally of note is the transformation of the interior courtyard, onto which the life of the university faced. The chief architect of this transformation, a certain Master John, followed the wishes of the Masters that "the *Collegium Maius* house should be made in the handsomest shape for visitors and for common usage."[10] He created an environment which is Italian in its inspiration and bears close comparison to the Bargello in Florence and to the Collegio de Spagna in Bologna. Its style is clearly humanistic in contrast to the older Gothic influences from Prague and its local adaptations from Cracow.[11]

Styled in the still common Gothic manner between 1515 and 1519, the aforementioned library is of special importance in our present Copernican context. It was built on the site of an older (probably fourteenth century) house, which during the last decade of the fifteenth century held an astronomical observatory. What observational astronomy the young Copernicus might have undertaken in his years in Cracow would probably have been done in this room.

3. *The Arts Curriculum at the Beginning of the Century*

Studies began at the refounded University of Cracow in 1400 with a curriculum in the Arts faculty which was not markedly different from that at institutions elsewhere in Europe. According to the detailed statutes of 1406,[12] in order to obtain the title of bachelor, a student needed in the space of two years to attend eleven courses. His lessons in grammar were limited almost exclusively to the *Doctrinale puerorum* of Alexander of Villa Dei, while rhetoric was largely confined to the rela-

[10] *Conclusiones Universitatis Cracoviensis ab a. 1441 ad a. 1589,* ed. H. Barycz (Cracow: Polska Akademia Umiejętności, pp. 74-75.

[11] Estreicher, *Collegium Maius,* pp. 82-90.

[12] *Statuta nec non liber promotionum Philosophorum ordinis in Universitate studiorum Iagellonica ab anno 1402 ad annum 1849,* ed. J. Muczkowski (Cracow, 1849), pp. i-xxiv.

tively uncommon *Poetria nova* of Geoffrey of Vinsauf, although in some examinations Donatus was substituted here. These works were usually heard in so-called Ordinary lectures. In Extraordinary or Private lectures, which were not required and which were given on an irregular basis, there was some use of the grammatical and rhetorical works of Boethius and Alanus de Lille.

The student's greatest attention was not surprisingly devoted to the logical works of Aristotle. For example, the handbook of Peter Hispanus was to be read for three months, while the *Logica vetus* (that is, the *Predicamenta* or *Categories* and *De interpretacione*) was to be studied for another four months. Boethius' works on the *Logica nova,* the *Libri priorum et posteriorum analeticorum* and *Liber elenchorum,* were utilized as the basis for another ten and one half months of lectures. Finally, the work of an aspiring bachelor was concluded with a brief introduction to Aristotle's *Libri physicorum* and his *Liber de anima.*

Heavy Aristotelian emphasis characterized the studies of a candidate for the master's degree also. Nine works of the Philosopher were studied during a space of some thirty-eight months. The continuation of logical studies was provided by a set of lectures on the *Topics,* while physics (or natural philosophy) was treated in *De generatione,* the *Parva Naturalia, De coelo et mundo,* and *Meteora.* Next came six months of lectures on Aristotle's *Metaphysics,* which were followed by extended sessions devoted to the *Ethics,* the *Politics,* and *Economics.* Despite the universal scope of these works, they did not completely treat the subjects of the *quadrivium.* To conclude this part of the arts curriculum, therefore, there were lectures based upon the *Theorica planetarum* of Gerard Sabionetta of Cremona, on the *De Algorithmo* of John of Hollywood (de Sacrobosco), the musical handbook of Jean de Muris, the optical study *Perspectiva communis* of John Peckham, and to conclude, three books of Euclid's *Geometry.* Finally, to complete the arts course, the student must participate in a specified number of disputations, receive the positive recommendation of the promotor, and formally be promoted or incepted. In sum, then, this was a minimal five-year process which corresponded to older traditions of general studies in the medieval west.[13]

[13]The early arts curriculum at Cracow is discussed by Casimir Morawski, *Histoire de l'Université de Cracovie, moyen age et Renaissance,* 3 vols. (Paris and Cracow: Alphonse

4. Mathematics and Astronomy

In the arts this structure was modified during the course of the next hundred years by four developments, each of which deserves separate attention. The first was the tendency to reduce the requirements of the curriculum. (Although this development is the least important of the four, it had a certain bearing upon the curriculum reforms later in the century by stimulating demands for change.) In particular, exemptions were granted upon request for music, as in fact they regularly were at both Paris and Oxford in this same period.[14] Related to this lessening of academic rigor was the introduction of new texts upon which lectures were based. While some of the new works cited below in connection with astronomy and mathematics do not lie in this category, it is clear that the substitution of such commentaries as Sędzijów of Czechło's brief *Algorismus proportionum* for John of Hollywood's *De algorithmo* or the same author's treatment of optics in place of Peckham's more sophisticated work clearly represents a weakening of the curriculum. As a result, by the end of the century, those degrees in arts which had previously required at least five years now took only four. (Again, this development parallels a reduction in the *pro forma* requirements in the arts at both Paris and Oxford in this period.[15])

Specialization and expertise in mathematics and astronomy at Cracow was a second, and far more positive development. This process was heralded in 1405 by the foundation of a special chair in astronomy and

Picard and G. Gebethner, 1900-1905), I, 82-119 and Kozłowska-Budkowa, "Odnowienie Jagiellońskie Uniwersytetu Krakowskiego," pp. 71-87. For the arts course in the west in this period, see James A. Weisheipl, "The Curriculum of the Faculty of Arts at Oxford in the Early Fourteenth Century," *Mediaeval Studies,* 1964, *26:* 143-185; and Hastings Rashdall, *The Universities of Europe in the Middle Ages,* ed. F.M. Powicke and A.B. Emden (Oxford: Oxford University Press, 1936), I, pp. 446-470; III, pp. 153 ff.

[14]See the comments of Strickland Gibson, ed., *Statuta antiqua Universitatis oxoniensis* (Oxford: Oxford University Press, 1931), p. xciv, no.2.

[15]In addition to the works cited above in note 13, see Weisheipl, "The Place of the Liberal Arts in the University Curriculum During the XIVth and XVth Centuries," in *Arts libéraux et philosophie au moyen âge — Actes du quatrième congrès international de philosophie médiévale* (Montréal and Paris: J. Vrin, 1969), pp. 209-213, and the unpublished thesis by John Fletcher, *The Teaching and Study of Arts at Oxford, c. 1400—c. 1520* (Oxford: 1962).

mathematics by a private citizen in Cracow, Nicholas Stobner.[16] His wish was that there be lectures on mathematics, theoretical astronomy, and that the almanac be kept current.

Yearly lectures were not given until at least 1415, although this chair may have functioned sporadically as early as 1410. One of the earliest names connected with the Stobner chair was Lawrence of Racibórz, who lectured from 1420 to 1427 upon Gerard of Cremona's *Theorica planetarum*. Apparently all of his works have perished.[17] He was succeeded by Sędziwój of Czechło, who was active from 1429 until at least 1431.[18] He wrote commentaries upon the standard mathematical, optical, and astronomical texts of the day; and although many of these were popular enough to be themselves utilized as texts in the *studium,* they are not generally of high quality.

Exceptions should, however, be noted. One was his commentary upon Gerard.[19] In this work, Sędziwój continues the earlier Polish astronomical tradition of Witelo, Franco de Polonia, and others in the thirteenth and fourteenth centuries.[20] He is particularly stimulating in his discussion of the origins of planetary motion, suggesting that he may have had some knowledge of the work of Buridan and Oresme. A sec-

---

[16]The identity of the founder was determined by Aleksander Birkenmajer, "Les astronomes et les astrologues silesiens au moyen age," in *Etudes d'histoire des sciences en Pologne* (Wrocław, Warsaw: Ossolineum, 1972 [*Studia Copernicana,* IV], p. 456, n. 62. The early history of the Stobner chair is discussed briefly in Birkenmajer's article "L'Université de Cracovie, centre international d'enseignement astronomique a la fin du moyen âge," in this same collection of his articles, pp. 485-487 in particular.

[17]He and his writings are discussed by J. Rebeta, "Miejsce Wawrzyńca z Raciborza w najdawniejszym okresie krakowskiej astronomii XV wieku," *Kwartalnik Historii Nauki i Techniki,* 1968, *13:* 553-564. See also the eulogy upon him given by Matthew of Łabiszyń in 1448 in Maria Kowalczyk, *Krakowskie mowy uniwersyteckie z pierwszej połowy XV wieku* (Wrocław, Warsaw: Ossolineum 1970), p. 135, n. 33, p. 186 (MS Cracow B.J. 2231).

[18]Jerry Wiesiołowski, "Sędziwój z Czechła (1410-1476), Studium z dziejów kultury umysłowej Wielkopolski," *Studium Żródłoznawcze,* 1964, *9:* 75-104, provides a short biographical study.

[19]This unedited treatise is contained in MS Cracow B.J. 1929.

[20]The fundamental works on Witelo by Aleksander Birkenmajer are now available in French translation in his *Etudes d'histoire des science en Pologne,* pp. 97-434. Other aspects of the tradition are discussed in Mieczysław Markowski, "Astronomia i astrologia w Polsce od X do XIV wieku," in *Historia Astronomii w Polsce* (in press).

ond work of note is the revision of the Alphonsine tables, which in one version or another had long served throughout Europe as the basis of astronomical calculation. For Cracow, however, these tables were inaccurate and misleading, and Sędziwój prepared a new edition. The observational data which were included in the resulting *Tabulae resolutae de mediis et veris motibus planetarum super meridianum Cracoviensem*[21] was continually up-dated during the fifteenth century by Sędziwój's successors in the Stobner chair and others. These included the names of otherwise unknown figures such as Johannes Orient and Petrus de Zwanów. After about 1431, Sędziwój left the university, but his long and distinguished life as royal diplomat and close associate of the great Polish historian Johannes Długosz kept him in close contact with Cracow. As a result, he was able to see the elaboration of the astronomical and mathematical tradition he had represented.

A chair in applied astronomy and astrology was founded in 1459, and this initiated a period during which, to use Aleksander Birkenmajer's phrase, the University of Cracow "emerged as the international center of astronomical education at the end of the middle ages."[22] The founder of this chair was Marcin Król of Żurawica, who with three others brought Cracow to this pre-eminent position.

Receiving his master's degree at Cracow in 1445, Marcin Król later studied at Prague, Leipzig, and Bologna. Although he was eventually appointed to a chair of medicine at Cracow, his real interests lay in arithmetic, geometry, astronomy, and astrology, which were apparently derived in part from his arts study, in part from his year in Bologna where he attended the astronomy lectures. His most important work was a critique of the Alphonsine Tables entitled *Summa super Tabulas Alphonsi.*[23] In this work he demonstrated that the actual positions of the planets did not correspond to the data given in the tables. In providing corrections, however, he utilized explanations for his more precise data which anticipated the later, and better known, work by the Viennese

[21]See the list of manuscripts cited in Markowski, *Burydaniszm w Polsce w okresie przedkopernikańskim* (Wrocław, Warsaw: Ossolineum 1971 [*Studia Copernicana,* II]), p. 263, n. 96.

[22]Birkenmajer, "L'Université de Cracovie, centre international d'enseignement astronomique à la fin du moyen age," in *Etudes d'histoire des sciences en Pologne,* pp. 483-495.

[23]The work, MS Cracow B.J. 1927, ff. 501-637, remains unedited.

astronomer and mathematician George Peuerbach, the *Theoricae novae planetarum.*[24]

Marcin Bylica, who was born about 1433, began his study at the University of Cracow in 1452. As a student of Marcin Król, he became in 1459 the first incumbent of the newly founded chair. Shortly thereafter, he traveled to Italy, where he studied at Padua and Bologna, and became the close friend of Regiomontanus. Much of the rest of his life was spent outside Poland, particularly in Hungary;[25] but he retained close contacts with the University of Cracow. During his lifetime he sent the *studium* many books, including a heavily annotated copy of Regiomontanus' *Tabulae Directionum* and a copy of Peuerbach's *Theoricae novae planetarum,* which were widely consulted at the University.[26] Sometime before his death in 1494, he bequeathed to Cracow his large library of mathematical, astronomical, and astrological works, as well as his extensive collection of astronomical instruments.[27]

Years before this enrichment, Jan of Głogów had finished the arts course at Cracow in 1468, and had immediately begun to lecture on astronomy and astrology. He remained as a professor there for forty years until his death in 1507.[28] His extensive writings include numerous horoscopes, many astrological handbooks, as well as several pure astronomical treatises. In these latter, he combined increasingly accurate

[24]See the recent work by Zdzisław Kuksewicz, "Marcin Król z Żurawicy," *Materiały i Studia Zakładu Historii Filozofii Starożytnej i Średniowiecznej,* I (1961), pp. 118-140.

[25]See the very useful article of Leslie S. Domonkos, "The Polish Astronomer Martinus Bylica de Ilkusz in Hungary," *The Polish Review,* 1968, *13,* 3: 71-79. The history of Polish-Hungarian cultural relations during the Renaissance is of fundamental importance in understanding the dissemination of Italian humanism in east central Europe. For a general overview, see the cooperative Polish-Hungarian symposium, *Renaissance und Reformation in Polen und Ungarn* (Budapest: Akadémiai Kiadó, 1963). More specifically, see Jerzy Zathey, "Marin Bylica z Olkusza, professor Akademie Istropolitany," *Humanizmus a Renesancia na Slovensku v 15-16 storočí* (Bratislava, 1967), pp. 40-54.

[26]MSS Cracow, B.J., 597 and 599 respectively.

[27]See the discussion of this library and these instruments in Zofia Ameisenowa, *Globus Marcina Bylicy z Olkusza i mapy nieba na Wschodzie i Zachodzie* (Wrocław, Warsaw: Ossolineum, 1959 [also in English translation: *The Globe of Martin Bylica of Olkusz and Celestial Maps in the East and in the West* (1959)]). The most useful short treatment of Bylica is by A. Birkenmajer in *Pol. Słow. Biog.*, III, pp. 166-168 (also in *Études d'histoire des science in Pologne,* pp. 533-536, in French).

[28]For a short biography, see *Pol. Słow. Biog.*, X, 450-452.

observational data with interpretations of celestial phenomena which implicitly challenged accepted views. For example, in three works, his *Introductorium in tractatum sphaerae Joannis de Sacrobusco* (written before 1506), his commentary upon the *Theorica planetarum* of Gerard of Cremona (before 1483), and his *Interpretatio Tabularum resolutarum ad meridianum Cracoviensem* (*ca.* 1488), he raised older Averroist challenges to the Ptolemaic system of cycles and epicycles and suggested that the sun controlled the motion of the other planets.[29] (In this context, the earth was not considered to be a planet.)

Before Jan had written these treatises, he had been the teacher of Wojciech of Brudzewo, who became a master at Cracow in 1474.[30] He taught there until his death in 1495, lecturing on mathematics, optics, and astronomy. In this last field, he did not comment upon the work of Gerard of Cremona, for it had been largely superseded at Cracow by Peurbach's *Theoricae novae planetarum.* In 1482 he composed his own treatise upon this theme, *Commentariolum super Theoricas novas planetarum,* which soon became the accepted text in theoretical astronomy at Cracow. In this, and in other lesser works, he went even further than Jan of Głogów in challenging accepted astronomical theories.[31] He attacked the Ptolemaic explanations for the motions of the moon, he pointed out that recent commentators had provided sound Aristotelian bases for accepting the earth's motion on its own axis, and he explored the possibility that an understanding that motion could be relative might overcome sense evidence of solar motion. In all these arguments, his reasoning was partly based upon observational data and partly

[29]See the discussion of these points in L. Birkenmajer, *Stromata Copernicana, Studja, poszukiwania i materiały biograficzne* (Cracow: Polska Akademia Umiejętności, 1924), pp. 125ff.; A. Birkenmajer, "Les astronomes et les astrologues," p. 464. The works of Jan of Głogów are best approached through Stefan Swieżawski, "Materiały do studiów nad Janem z. Głogowa (+1507)," *Studia Mediewistyczne,* 1961, *2:* 135-173; and Władysław Senko, "Wstęp do studium nad Janem z Głogowa," *Materiały i Studia Zakładu Historii Filozofii Starożytnej i Średniowiecznej,* 1961, *1:* 9-59 and 1964, *3:* 30-38.

[30]The biography and works of Wojciech are now best treated in Ryszard Palacz, "Wojciech Blar z Brudzewa," *Materiały i Studia Zakładu Historii Filozofii Starożytnej i Średniowiecznej,* 1961, *1*: 172-198.

[31]See particularly the introduction to L.A. Birkenmajer, ed., *Commentariolum super Theoricas novas planetarum*...(Cracow, 1900). Some of the most crucial points are further discussed by the same author in his *Stromata Copernicana,* pp. 83-103.

derived from the implications of natural philosophy at Cracow in the preceding generations.

Estimates by historians vary, but it is probable that by the time Copernicus arrived in Cracow, Wojciech had begun to lecture on other topics and that he did not have any direct influence upon the young astronomer.[32] But Wojciech nevertheless represents the level which astronomy had reached in Cracow and justifies the judgment made by both the fifteenth century German chronicler Hartmann Schedel and his contemporary, the Hungarian historian A. Bonfinius, who recognized that the Jagiellonian University was indeed an important center for astronomical study.[33] In our present concentration upon Nicholas Copernicus therefore, these astronimical and mathematical developments in the arts curriculum loom very large indeed. But they were not the only currents of the day.

5. The University and Natural Philosophy

Let us turn now briefly to a consideration of the third of the major developments referred to above. At this point, the distinction between the arts faculty and other academic disciplines blurs. During the fifteenth century important changes took place at Cracow in the philosophy of nature which was taught and discussed in the university.

Of the many late medieval philosophers of nature known at Cracow in the early part of the century, it was the physical writings of John Buridan and his followers, Nicole Oresme, Albert of Saxony, Marsilius of Inghen, and others, which proved most popular in the lectures and

[32]See A. Birkenmajer, "Copernic philosophe," in *Etudes d'histoire des sciences en Pologne,* p. 623.

[33]"Cracoviae est celebre gymnasium multis clarissimis doctissimisque viris pollens, ubi plurimae ingenuas artes recitantur. Astronomiae tamen studium maxime viget. Nec in tota Germania, ut ex multorum relatione satis mihi cognitum est, ille clarius reperitus." Hartmann Schedel, *Liber chronicarum* (Nürnberg, 1943), p. 267. "A coniectoribus et astrologis, quibus referta Cracovia est." Mentioned by S. Katona, *Historia critica regnum Hungariae*, XVII (Buda, 1793), p. 258. Each cited in A. Birkenmajer, *Etudes d'histoire des sciences en Pologne*, p. 491 and 464, n. 86. The implications in both quotations for Polish and German cultural history have been hotly disputed. See Markowski, *Burydanizm w Polsce,* pp. 242-244.

disputations there.[34] As a result many Cracow masters wrote commentaries on their views of Aristotle which are only now beginning to be studied systematically. Paul of Worczyń, for example, is the author of *Quaestiones* on a half-dozen of Aristotle's physical works, as well as an extended treatise on the *Nicomachean Ethics.*[35] Benedict Hesse's *Quaestiones super octo libros Physicorum Aristotelis* is only one of nearly a dozen treatments of this same work at Cracow in the early fifteenth century.[36] All, in one degree or another, advocate Buridan's ideas on motion.

Very slowly, later in the century, other philosophical positions began to be represented in the arts faculty. In the *Quaestiones Cracoviensis*[37] and the *Quaestiones in Physicam Aristotelis* of the aforementioned Jan of Głogów,[38] fundamental questions about thc organization of knowledge and the sciences were raised in non-Aristotelian guises. One fundamental issue was whether natural philosophy was a theoretical or practical science. (Benedict of Hesse asked it in its most common form: "Utrum philosophia naturalis est scientia speculativa?" while in another

[34]Buridan's influence in Polish philosophy was early noted by Kazimierz Michalski, "Jan Buridanus i jego wpływ na filoszofię scholastyczną w Polsce," *Sprawozdania z Czynności i Posiedzeń AU w Krakowie,* 1916, *21,* 10: 25-34. See also his "Zachodnie prądy filozoficzne w XIV wieku i stopniowy ich wpływ w sródkówej i wschodniej Europej." *Przegląd Filozoficzny,* 1928, *31:* 15-21. The question has been most recently studied *in extenso* by Markowski, *Burydanizm w Polsce.* pp. 70-81 and 200-208 which trace the arrival of Buridan's thought to Cracow.

[35]His shorter treatises are contained in MS Cracow B.J. 2073, while the commentary upon the *Ethics* in in MSS Cracow B.J. 714, 720, 741, 2000, and 3352. See Jerzy Rebeta, "Paweł z worczyna," *Materiały i Studia Zakładu Historii Filozofii Starożytnej i Średniowiecznej,* 1964, *3:* 120-156.

[36]See Markowski, "Krakowski Komentarze do "Fizyki" Arystotelesa zachowane w średniowiecznych rękopisach Biblioteki Jagiellońskiej," *Studia Mediewistyczne,* 1966, *7:* 107-124.

[37]MS Cracow B.J. 2087. See the preliminary edition by Ryszard Palacz in *Studia Mediewistyczne,* 1969, *10.* The work is also discussed by the same author in "Les Quaestiones Cracovienses—principale source pour la philosophie de la nature dans la seconde moitié du XV s. à l'Université Jagellone à Cracovie," *Mediaevalia Philosophica Polonorum,* 1969, *14:* 41-52.

[38]MS Cracow B.J. 2017. See M. Zwiercan, "Les 'Quaestiones in Physicam Aristotelis' de Jean de Głogów enfin Retrouvées?" *Med. Phil. Pol.*, 1963, *11*: 86-92.

version he asked: "Utrum philosophia naturalis, quae vocatur physica, sit scientia speculativa vel practica?"[39])

Easily answered by Aristotle in favor of the theoretical, this question was more difficult to decide in the fifteenth century, for upon an answer to it depended not only the place which one assigned to physics but also ultimately the relationship between the three criteria for truth: faith (*fides*), natural reason (*ratio naturalis*), and authority (*auctoritas*), whether it be biblical, patristic, or philosophical. The tendency throughout the century was to see natural philosophy more and more as a practical, or semi-practical, non-abstract science. As a result, although the surface of the arts curriculum in the late fifteenth century continued to appear Aristotelian, its content was increasingly designed to oppose ancient authority and to emphasize what natural reason might on its own derive from prime principles, self-evident propositions, and the mathematically demonstrable principles of the philosophy of nature.[40]

Despite the fact that a great many of these concerns are clearly expressed in the extant treatises from the masters of the fifteenth century, there remains unanswered the crucial question of the extent to which these treatises are derived from the arts faculty. Ironically, although scholars know with some precision what was taught in the arts course of the thirteenth and fourteenth century European university, our knowledge is embarrasingly thin when we turn to the fifteenth century. This is true at Cracow as well as at Paris, Oxford, and elsewhere.[41] We are able, however, to point to a fourth tendency which

[39]MSS Cracow B.J. 2376 f. 4 v., 2100, f. 2 r.

[40]A great deal of Polish scholarship in the past three decades has been devoted to the philosophical and scientific tradition at the University of Cracow in the fifteenth century. In addition to the works already cited, special mention should be made of S. Swieżawski, "Filozofia w średniowiecznym Uniwersytecie Krakowskim" in *Historia kultury średniowiecznej w Polsce,* 2 vols. (Warsaw: Polskie Towarzystwo Historyczne, 1963), I, pp. 129-159; Zdzisław Kuksewicz, ed., *Z dziejów filozofii na Uniwersytecie Krakowskim w XV wieku* (Wrocław, Warsaw, Cracow: Oss 1965); R. Palacz, ed., *Filozofia Polska XV wieku* (Warsaw, 1972); and Palacz, "Z problematyki badań nad filozofią przyrody w XV wieku," *Studia Mediewistyczne* 1970, *11:* 73-109; 1971, *13:* 3-107; 1973, *14:* 87-198.

[41]See particularly the comments made by Sven Stelling-Michaud, "L'histoire des universités au Moyen Age et à la Renaissance au cours des vingt-cinq dernieres années,"

might have had a bearing upon Copernicus at the University of Cracow. This is the revival of literary and rhetorical concerns in the arts faculty as the result of humanistic influences in Poland.

6. *The Advent of Humanism*

During the first half of the fifteenth century, the triplex influence of native Poles in contact with humanism abroad (particularly at the church councils of that period), of the direct penetration of Italian influences into the country, and of the indirect mediation of these same cultural concerns through Hungary created conditions in which the first flowering of the Polish renaissance took place.[42] Four individuals in particular have generally been associated with this: Zbigniew Cardinal Oleśnicki, Johannes Długosz, Gregory of Sanok and Jan Ostroróg. All had close contacts with the University of Cracow, and it is not therefore surprising that within the *studium* we should find humanistic influences.[43]

Beyond the tradition of the medieval *artes liberales,* a specific concern for antiquity on its own terms and for the *studia humaniora* in an Italian sense may be observed throughout the 1420's and 1430's. For example in the academic addresses given by professors recommending students in the arts for degrees and in the maiden speeches given by the new licentiate or bachelor, one finds a growing emphasis upon the humane arts versus more practical and mechanical concerns.[44] Frequently, as in the use of Ovid by Łukasz of Wielki Koźmin in 1412,[45] or

XI Congrès International des Sciences Historiques, Rapports, I (Stockholm: Almquist and Wiksell, 1960), pp. 104-107, and the bibliography in his notes 71-89.

[42]For a fuller treatment of each of these themes, see my paper "The World of the Young Copernicus: Society, Science, and the University," in *Science and Society* (Ann Arbor: University of Michigan Press, 1975), pp. 19-44.

[43]In addition to works cited in the paper mentioned in the previous note, see particularly Ignacy Zarębski, "Okres wczesnego humanizmu," in Lepszy, ed., *Dzieje Uniwersytetu Jagiellónskiego*, pp. 151-172.

[44]These speeches have, since I used the manuscripts in 1966 and 1970, now been discussed and partially printed by Kowalczyk, *Krakowskie mowy uniwersyteckie,* with whom however, I differ on the extent of tradition versus innovation reflected in the speeches.

[45]MS Cracow B.J., 2215, ff. 251-255.

in the appeals to Lucian in 1420 and to Quintillian in 1429 by Mikołaj Kozłowski,[46] the use of classical material is sharply reminiscent of the *ars oratoria* in fifteenth century Italy.

Even clearer appeals to humanistic concerns came from Rector Jan of Ludzisko in the 1440's. Although born in social obscurity about 1400, Jan attended the University of Cracow and received a bachelor's degree in 1419 and a master's three years later. He continued his studies in Italy, where he earned a doctorate in medicine at Padua in 1433. He then came profoundly under the influence of the revival of antiquity and the pedagogical ideal of Guarino Guarini, with whom he studied in Ferrara. When he returned to Cracow in 1440 to lecture in the arts faculty, his rhetorical expertise caused him to be designated as the official orator for the *studium*. For the next seven years, in this capacity and as sometime rector of the faculty, he advocated a revision of the arts curriculum which would include more antique literature, both poetry and history.[47]

There are at least eight of Jan's orations extant, all of them delivered as part of official university functions.[48] For example, in 1440 he welcomed a deputation from the Council of Basle and expressed the firm conciliar sympathies of the university faculty. In 1447 he opened the academic year as university orator, but instead of speaking perfunctorily about abstract academic goals, he issued an impassioned appeal to the faculty and King Kazimierz Jagiellończyk to ameliorate the plight of the peasantry. His most humanistic oration came in June 1440 when he described what true eloquence was: an appeal to the whole man to move him to action.[49]

Although no concrete measures for reform of the curriculum were undertaken while Jan was at the university, in 1449 a major reorganiza-

---

[46]MS Cracow B.J., 2216, ff. 105-112, 173-179.

[47]See *Pol. Słow. Biog.*, X, pp. 461-462, for biographical data. The study by B. Nadolski, "Rola Jana z Ludziska w polskim Odrodzeniu," *Pamiętnik Literacki*, 1929, *26:* 198-211, is excellent.

[48]All are contained in MS Cracow B.J., 126, and have now been published in a model initial edition by J.S. Bojarski, ed., *Johannis de Ludzisko Orationes* (Wrocław, Warsaw: Ossolineum 1971).

[49]See the analysis of his oratorical corpus by J. Stanisław Bojarski, "Jan z Ludziska i Przypisywane mu mowy uniwersyteckie" *Studia Mediewistyczne*, 1973, *14:* 3-85.

tion was implemented.[50] Designed to regularize the financial base of the *studium,* to enforce academic discipline among the professors and students, and to improve the material and physical conditions of the masters and scholars, this reform also made basic changes in the arts curriculum. It introduced into that faculty a *Collegium minus,* which consisted of two chairs, an older one founded by Stanisław of Lelow, called Nowko, and a more recent foundation in the name of Catherine Mężykowa.

The curriculum of this *Collegium* was not specified in detail, except that the former chair was to include lectures on Boethius' *De consolatione philosophiae,* Alanus de Lille's *De planctu naturae,* and the latin writers Valerius Maximus, Vergil, Ovid, Horace, Terence, Stacius, and others. The latter chair was to treat the *ars dictaminis* and Cicero's *Rhetorica ad Herennium,* as well as the *Poetria nova* of Geoffrey of Vinsauf, the *Labyrinthus* of Eberhard of Bethune, and the *Chronicon Polonorum* of Vincent Kadłubek, the twelfth century Polish historian.[51]

The intent of this reform was largely frustrated in the first years after it was introduced. Not until the polymath Jan Dąbrówka became vice-chancellor of the university in 1458 was it effectively implemented. He was able to insure that the lectures required were actually given and that the new foundation played an important role in the education of students at Cracow. Under his leadership the *Collegium minus* enrolled about 40% of the total number of students in the arts.[52] The humanistic orientation of this part of the *studium* thereafter touched all who were associated with the university.

Jan Dąbrówka was one of the most important figures in the history of humanism in the arts faculty of the fifteenth century. Born about 1405, he became Master in Arts at Cracow in 1427 and lectured for six years

[50]The details of this process for curricular reform are best traced in Morawski, *Histoire de l'Université,* II, pp. 220-236; and Zarębski, "Okres wczesnego humanizmu," in Lepszy, ed., *Dzieje Uniwersytetu,* pp. 172-175.

[51]Details on the curriculum of the Collegium Minus are found in a document, *Conclusiones Maioris Collegii,* written by Jan Dąbrówka (see below) in 1466: MS Cracow, Archiwum U.J. 63, printed in *Codex dipl. Univ. Crac.,* III, p. 47.

[52]Karbowiak, "Studya statystyczne," pp. 77-81; Zarębski, "Okres wczesnego humanizmu," pp. 176-179.

thereafter on rhetoric.[53] At this point he began the formal study of Aristotelian philosophy, scholastic theology, and canon law. By 1458 he was both *doctor decretorum* and *theologiae doctor.* In his numerous writings, on the *trivium* and *quadrivium,* law, theology, the sciences, and the historical writings of Vincent Kadłubek, he reveals a deep knowledge and appreciation of classical literature and the works of Petrarch, Pier Paolo Vergerio (whose educational treatises he adapted to the Cracow *studium*), and other Italian humanists.[54] On numerous occasions he served as rector of the university and was its vice-chancellor from 1458 until 1465. Under his leadership there was firmly established in the university the humanistic tradition of criticism—both in the narrow area of pure scholarship and in the broader area of the philosophy of man. His death in 1472 marked the end of the beginning for the early renaissance in Poland and was as important in its own way as the birth of Copernicus in the following year.

In the arts faculty, Jan Dąbrówka's presence was keenly missed, and three years after his death a minor reorganization of the curriculum took place.[55] In addition to some institutional mechanics, this reform expanded the range of antique literature required in the *Collegium minus.* Plautus, "aut alia poetica," was added to the Nowko chair, while Quintillian's *De institutione oratoria,* "et alia que ad oratoriam spectant," were included in the Mężykowa offerings.

7. *Conclusion*

By the end of the century, therefore, when Copernicus enrolled in the arts faculty at Cracow, the university held for him the three-fold

[53] Among the older treatments of Jan Dąbrówka, see Morawski, *Histoire de l'Université,* II, pp. 223-228. More recently, however, Marian Zwiercan, *Komentarz Jana z Dąbrówki do Kroniki mistrza Wincentego zwanego Kadłubkiem* (Wrocław, Warsaw: Ossolineum, 1969), has provided a thorough study of his works. See particular the comments in the introduction, pp. 5-11.

[54] W. Szelińska, "Dwa testamenty Jana Dąbrówki," *Studia i materiały z dziejów nauki polski,* Series A: V (1962), pp. 18-21, describes the text of an inventory of Jan's library and notes Vergerio's works.

[55] See Barycz, *Historja Uniwersytetu Jagiellońskiego w epoce humanizmu* (Cracow: Uniwersytet Jagielloński, 1935), pp. 21 f.

promise of mathematics and astronomy which were abreast of any developments elsewhere in Europe, of philosophical questioning which was undermining the foundations of much that had been characteristically medieval, and of a critical humanistic attitude which was transforming older cultural and educational values. It is fortunate that Copernicus, whether because of Cracow or because of independent influences, eventually realized in himself the full promise which the age offered him and gave us his *Revolutions*.

COMMENTARY: IN DEFENSE OF CONTEXT

Nicholas H. Steneck

University of Michigan

Constructing historical and intellectual context for events whose relevance to that context is tenuous or, as in the present case, virtually unknown is a difficult and often thankless task. Historians operate most effectively when they deal with tangibles. As intangibles and unknown connections become involved, the historian's craft becomes more like that of the novelist who allows imagination to supply pieces that time has ravaged.[1] Such is the problem that Professor Knoll faces in his attempt to tie Copernicus to the intellectual heritage of his first years of university training. Intuitively, as novelists seeking to fill in the missing gaps, we expect that a brilliant and questioning mind such as that of Copernicus must have been affected by the years, however brief, spent in the arts *studium* at Cracow. Reasonably, as historians, we know that there is little evidence to connect the famous pronouncement on heliocentricity to the late medieval astronomy and natural philosophy taught at the *Collegium maius.* Accordingly, any ties between Copernicus' student days and his ideas on heliocentricity must remain conjectures. Knoll's caution that "it is difficult to speak with any certainty about what specifically Copernicus gained in [his] years..." at Cracow, expresses well the constant doubt that must always accompany research of this sort.

Such doubts do not, however, invalidate the work that Knoll has done in his survey of the intellectual context present at the University of Cracow at the time Copernicus was a student there. Context provides more

[1]See the collected essays on "Methods of Historical Inquiry" in *The Nature of Historical Inquiry,* ed. Leonard Marsak (New York: Holt, Rinehart and Winston, 1970), particularly "The Historian's Purpose: History and Metahistory" by Alan Bullock.

than simply the soil from which ideas spring. Just as importantly, context provides the backdrop against which ideas can and must be judged if their significance in time is to be properly understood. It provides the one true and directly proximate measure that we have of the traditional or innovative nature of certain specific ideas that seem to move man's intellectual development in new directions. And what is for our purposes crucial, it provides a backdrop that need have no other direct connection to the subject at hand than being the "tradition that precedes" or the "prevailing philosophy" at a particular time and place. Viewed in this way, context becomes a neutral entity whose direct ties to the tradition that follows are of only secondary importance. But it also becomes an entity that must be completely understood in its own right before any further judgments can be made.

This second use of context can be better understood by briefly comparing two ways of dealing with one of the intellectual currents that Knoll associates with late fifteenth-century Cracow, namely, the natural philosophic current. At issue is one central problem: to reach a comparative appreciation of Copernicus and John Buridan, Nicole Oresme, Albert of Saxony, Marsilius of Inghen, and others who make up what is commonly referred to as the *via moderna,* an understanding of the correspondence of Copernican science to the science of late medieval Paris. More specifically, what is at issue is an answer to the following question: are or are not the achievements of Copernicus analogous to the achievements of earlier figures?

Traditionally, this question has been attacked from two directions, either beginning with Copernicus and looking back in time or beginning with the *via moderna* and attempting to relate its innovative or traditional tendencies to the innovations of the Copernican system.[2] In both instances, it is assumed that there is some justification for isolating certain select elements of the world views under consideration and for seeking to compare these elements, one with another, in order to reach an assessment of the relationship between them. Just which elements are

[2]For the former approach, see Thomas Kuhn, *The Copernican Revolution* (New York: Vintage, 1959), pp. 100-133; and Angus Armitage, *Copernicus* (New York: Barnes, 1962), pp. 37-44. For the latter, Edward Grant, "Late Medieval Thought, Copernicus, and the Scientific Revolution," *Journal of the History of Ideas,* 1962, *23:* 197-220.

isolated for study varies greatly, depending on the affection of an individual historian for specific aspects of the development of science, e.g. actual descriptions of the world (astronomical systems), metaphysical assumptions, methodological considerations, etc. But what does not vary is the assumption that method is method, that metaphysics is metaphysics, that scientific explanation is scientific explanation no matter what the period. As psychological entities, scientific thinkers are seen to change very little; as systems of thought, scientific explanations are assumed to have a degree of autonomy that provides a basis for their own definition.

The straightforward nature of this approach to the history of science would seem ideally suited to answer our basic question. Yet it has not! When there is agreement among scholars about the importance of one aspect of the science of the *via moderna,* such as the use of the hypothetical method, there is no agreement on whether this aspect prepares the way for Copernicus[3] or sharply divides him from the medieval context.[4] There is not even any consensus on whether the hypothetical method is somehow unique to the *via moderna*[5] or to the Middle Ages.[6] When there is agreement about one aspect of the Copernican achievement, such as his preoccupation with mathematical astronomy to the almost total neglect of physics, there is no agreement on whether this aspect is compatible[7] or incompatible with the medieval context. And here it seems, following Knoll's suggestions regarding Cracovian classifi-

---

[3]By making "Die ungedachten Gedanken . . . nicht mehr die unmöglichen Gedanken" (Hans Blumenberg, *Die kopernikanische Wende* [Frankfurt am Main: Suhrkamp, 1965], p. 37. See also Amos Funkenstein, "The Dialectical Preparation for Scientific Revolutions," particularly the section dealing with " 'Impossibilities' in Ancient and Medieval Astronomy," pp. 165-203.

[4]Grant, "Late Medieval Thought," p. 213; and more recently, *Physical Science in the Middle Ages* (New York: John Wiley, 1971), pp. 86-87.

[5]See, for example, the comments by Benjamin Nelson on Edward Grant, "Hypotheses in Late Medieval and Early Modern Science," *Daedalus,* Proceedings of the American Academy of Arts and Sciences, 1962, *91:* 612-16.

[6]William Wallace, *Causality and Scientific Explanation,* Vol. I (Ann Arbor: University Press, 1972), pp. 151-155, looks to Paduan science for the source of Copernicus's methodology.

[7]For example, William Donahue argues that the Aristotelian distinction between mathematical astronomy and physical reality made it possible for medievals to believe

cations of the sciences,[8] that this problem is further complicated by the fact that the thought of the *via moderna* underwent some change before becoming the context for Copernicus' thought. There is no agreement on whether the Copernican break with tradition, however cautious, represents a degree of freedom from authority that is atypical of the more conservative nature of, *or* the very essence of, the innovativeness of fourteenth-century science,[9] or whether the medieval use of experience can in any way be seen as preparing the way for the more realistic science of the Copernican era.[10] In brief, the traditional approach to this problem has led to almost as many conclusions about the relationship between Copernicus and medieval science as there are statements on the subject.

Faced with such ambiguity, let me suggest a second way of proceeding. The many conclusions that have arisen in conjunction with attempts to relate Copernicus to the medieval context stem not so much from uncertainty about the nature of the Copernican achievement—although this is certainly a contributory factor—as they do from a lack of understanding about what exactly the medieval context is. Given this

that "a wide range of opinions concerning the heavens. . . [could be] mutually compatible" (William Donahue, "The Solid Planetary Spheres in Post-Copernican Natural Philosophy," p. 245). It should be noted, as a point of clarification, that most medievals classified mathematical astronomy and the study of physical reality (natural philosophy) under the more general heading of philosophy, broadly considered, thus in theory at least, making both proper disciplines for philosophers to pursue. James A. Weisheipl, "Classification of the Sciences in Medieval Thought," *Mediaeval Studies,* 1965, *17:* 54-90; and Nicholas H. Steneck, "A Late Medieval *Arbor scientiarum,*" *Speculum,* 1975, *50:* 245-269; see also Grant, *op. cit.,* "Late Medieval Thought," p. 215.

[8]See Knoll's discussion of the question: "Utrum philosophia naturalis, quae vocatur physica, sit scientia speculativa vel practica?", in section five of his paper.

[9]The latter, more optimistic interpretation stems from the writings of Pierre Duhem and his famous thesis regarding the Condemnation of 1277. More recent interpretations, particularly by Anneliese Maier, have tended to place most medieval science into a more traditional Aristotelian framework, thus playing down its anti-authoritarian nature. For a recent restatement of the issues associated with this problem, see Grant, *Physical Science,* pp. 26-29. Grant does lean toward Duhem's interpretation; however, his views are not as facile as the very narrow-minded and unsympathetic review of his book by Noel Swerdlow would have one believe (*Speculum,* 1973, *48:* 364-365).

[10]For a recent discussion of this crucial problem, see Wallace, *Causality,* pp. 88-103, and 114-116.

fact, it seems evident that a more appropriate way to proceed is to begin by carefully and quite generally describing just what is the subject under investigation: the organizations of science, the limits and ambiguities of the scientific treatise, the motivations and preoccupations of the scientist, etc. And what is even more important, these initial descriptions must be made quite apart from any *a priori* judgments about significant questions within the overall development of science. To admit that there are certain general questions that apply to the whole of the history of science is to assume a continuity that may not in the end be justified. Moreover, the last thing we need to know when attempting to characterize a specific context, particularly the context of the late Middle Ages, is how particular persons answered particular questions. A great deal of preliminary investigation must come first.[11] Let me explain.

Before any investigation of the science of a period can be carried out, it must first be known what is to be included as "scientific" and under what conditions "scientific discussions" took place. If, for example, we assume that discussions about the general makeup of the physical world comprise *science,* then it must immediately be noted that in the fourteenth century, limiting the discussion to Parisian science, *scientific discussions* took place in at least two separate places, in the arts and in the theological faculties of universities. Within these two localities the overall orientation of the scientific discussion varied considerably. In the arts, most instruction centered on the texts of Aristotle, and so "scientific explanation" under these conditions means clarifying ambiguities and difficulties in the Aristotelian world view. Theologians, on the other hand, focused their attention on different texts, primarily

[11]For a penetrating discussion of the weaknesses of the traditional history-of-ideas approach to problems such as the one under consideration, see Michel Foucault, *The Archaeology of Knowledge,* trans. A.M. Sheridan Smith (New York: Pantheon, 1972), pp. 135-140. As should be obvious from the discussion that follows, although I agree with Foucault's basic approach to intellectual history and with his focus on continuity as opposed to discontinuity, I do not find his total preoccupation with language to the almost total neglect of context particularly persuasive. This is the reason that some of his specific interpretations, such as those contained in *The Order of Things* (New York: Pantheon, 1970), are methodologically fascinating but historically weak. Recently, I made more specific recommendations regarding the applicability of Foucault's method to the study of late medieval science in an as yet unpublished paper entitled "Digging for the History of Science: A Medieval Site."

on the Bible and on the *Sentences* of Peter Lombard, and so their lectures sought to clarify different problems and to clarify these problems in different ways. Accordingly, it is extremely important to specify just what aspect of the science of the *via moderna,* its arts or theological side, is being used to reach a particular generalization. What is true for one type of literature may not necessarily be true for the other.[12]

The fact that university classrooms were the laboratories of the medieval scientist points to yet another generalization that needs to be considered, the didactic motivation that lies behind most medieval science. In contrast to modern scientists, those medievals who discussed nature rarely had as their objective unlocking the secrets of nature. Individuals such as Albertus Magnus who were truly interested in the fine details associated with the world that surrounded them were unusual,[13] or found their calling outside the university and in other endeavors, such as medicine and alchemy. More commonly, the *artista* was engaged in clarifying Aristotle, in bringing concord to the discordant text and commentary tradition that preceded him,[14] the *theologus* in laying the physical and metaphysical foundations for theological arguments. As a result, their teachings need to be carefully interpreted before too much significance is attributed to a particular point. "Probable" and "hypothetical," for example, can have several meanings

[12]A case in point can be drawn from medieval psychology. The highly specialized discussions of cerebral psychology and species multiplication, which comprised an integral part of the *De anima* tradition in the arts (see, Nicholas H. Steneck, "The Problem of the Internal Senses in the Fourteenth Century" [unpublished PhD dissertation, The University of Wisconsin, 1970]), had no place in most theological discussions. When one discussed psychology from the standpoint of a theologian, the physical basis of cognition seems to have been of little importance. By like token, some aspects of cognition, which were of great interest to theologians (e.g. *notitia intuitiva*), had no real foundation in Aristotle's writings and so were not discussed. Quite obviously, then, it is important to recognize these limitations when generalizing about the relevance of epistemology and psychology to larger issues.

[13]For a discussion of Albert's interest in nature as illustrated by his psychological writings, see Nicholas H. Steneck, "Albert the Great on the Classification and Localization of the Internal Senses," *Isis,* 1974, *65:* 193-211.

[14]For a discussion of the didactic nature of medieval psychology, see Nicholas H. Steneck, "A Late Medieval Debate concerning the Primary Organ of Perception," *Proceedings of the XIIIth International Congress of the History of Science,* Section IV (Moscow: Editions "Naouka," 1974), pp. 198-204.

depending on context. It is one thing to mean by "probable" that given the limitations of evidence, one line of reasoning is as plausible as some other line of reasoning. It is another to mean that in the order of nature, what can be thought to exist might exist, that there just might be, to cite an example from astronomy that has born no fruit, yet another planet inside the orbit of Mercury that is not seen because of its close proximity to the sun.[15]

By raising fundamental questions such as the ones just mentioned, it is possible to pinpoint much more precisely what characterizes a particular context. Even more than this, by pursuing such research, a firm foundation is established that can then serve as a point of reference for comparative studies of other contexts. And finally, since direct causality is not being sought, the comparative analysis can proceed even if the actual connections between various areas under investigation are either unknown or remote.

To return to Copernicus, if we disregard speculations having to do with the direct connection of ideas, the process of looking for the soil from which ideas spring, important insights can still be gained by comparing his science to the general context of medieval science. Take, for example, the difficult problem of authority and tradition. Copernicus did not operate within the confines of the university method; his approach to astronomy has nothing in common with the tradition of *De caelo* and *De sphaera* commentaries that were without a doubt dominant in medieval astronomical writings. He ignored, for the most part, the physical questions that predominate in this tradition and turned instead to other questions. Still, there is no reason to assume that Copernicus' break with the didactic tradition of the late Middle Ages sets him at complete odds to that tradition. Within university science there was a great deal of give and take, of modification of some important positions held by Aristotle and others, just as within the Copernican system, Ptolemy and Aristotle were bent, but never to the breaking point. Neither tradition totally rejects the prevailing descriptions of nature, neither accepts an entire world view just because of its authority. Both

[15]This suggestion is made by Henry of Langenstein (d. 1397) in his commentary on Genesis (mss. Vienna, National-Bibliothek, 3900, fol. 37vb) and is intended to be interpreted in much the same fashion as the better known medieval discussions of diurnal rotation.

traditions actively worked within the domain of larger assumptions and attempted to smooth out the difficulties they encountered in working with these assumptions. Medievals, if viewed on their own terms, labored no more under authority than did Copernicus.[16]

Now it is true, if I may anticipate finally one objection to this approach to the history of science, that such broad comparisons will in the final analysis tell us almost nothing about how Copernicus arrived at the notion of an astronomy based on a moving earth. It is also true that research can be conducted that may eventually resolve the problem of causality to the satisfaction of most scholars. This is a distinct possibility, and I am not suggesting that such research be abandoned. Instead, what I am suggesting is that for the crucial period of transition from medieval to modern science, such research must proceed with a great deal of caution and with a constant understanding of the very limited nature of the conclusions that are reached. It is not until the intricate workings of the context of late medieval science are completely understood that we will be able to write a complete history of either the Copernican revolution or of the scientific revolution. Change can only be measured against a backdrop, and until that backdrop is fully and completely understood, our characterizations of the innovations that comprise the scientific revolution must remain more the inventions of historians of later periods that supply the evidence now buried in the hundreds of unread manuscripts from the late medieval period. By digging into these manuscripts and attempting to bring some order to the various currents that may eventually prove relevant to the history of science, Knoll has begun an important task, and for this we are gratefully in his debt.

[16]In fact, if the suggestions of Curtis Wilson are correct, Copernicus may have been even less willing to break with authority than were some medievals; see Curtis Wilson, "Rheticus, Ravetz, and the 'Necessity' of Copernicus' Innovation," *this volume, pp. 17-39.*

VI

THE DIALECTICAL PREPARATION FOR SCIENTIFIC REVOLUTIONS

On the Role of Hypothetical Reasoning in the Emergence of Copernican Astronomy and Galilean Mechanics

Amos Funkenstein

University of California, Los Angeles

> "Der Mann war noch nicht auf der Welt, der zu seinen Gläubigen hätte sagen können: Stehlt, mordet, treibt Unzucht—unsere Lehre ist so stark, dass sie aus der Jauche eurer Sünden schäumend helle Bergwässer macht; aber in der Wissenschaft kommt es alle paar Jahre vor, dass etwas, das bis dahin als Fehler galt, plötzlich alle Auschauungen umkehrt oder dass ein scheinbarer und verachteter Gedanke zum Herrscher über ein neues Gedkankenreich wird, und solche Vorkommnisse sind dort nicht bloss Umstürze, sondern führen wie eine Himmelsleiter in die Höhe. Es geht in der Wissenschaft so stark und unbekümmert und herrlich zu wie in einem Märchen."
>
> (Robert Musil, *Der Mann ohne Eigenschaften*, c. 11)

1. A Preliminary Thesis

The prehistory of new, even of "revolutionary," theories is usually subsumed under two complementary aspects. It is viewed either as a positive or as a negative preparation, or as both. We either look for "precursors" who anticipated the new theory in some way—in part, in principle, or in a sketch. Or again, we stress the growing awareness of insurmountable problems posed by a once dominant theory; an awareness which eventually leads to its replacement. Historians and theoreticians of science thus interpret the relation of old to new theories in much the same terms as Christian theologians interpreted the relation between the old and the new dispensation since antiquity. Depending on polemical or dogmatic necessities, they defined the Law given to Moses either as a growing burden or as a positive evangelical preparation (*παιδαγωγὸς εἰς Χριστον*).[1]

Both categories are vague and do not offer a genuine principle of mediation, or continuity, between successive theories. Without any claim to having hit upon a better, and more generally applicable formula, I wish to draw attention to a quite precise sense in which a dominant theory may be said at times to "anticipate" a new theory, even an adversary one. A good many examples can be gathered to illustrate the following circumstance. Well-reasoned, elaborated theories may, or may not, specify possible instances of their own falsification. The criterion of an *experimentum crucis* is, after all, relatively modern.[2]

[1]Or as both an evolutionary and revolutionary transition: e.g. Gregory of Nazianz, *Orationes Theologica V*, p. 25 ff., ed. J. Barbel (*Testimonia, Schriften der altchristischen Zeit* 2 [Editione "De Tempi et alia," Düsseldorf, 1963], p. 261); extensively used by Anselm of Havelberg, *Dialogi* I, 5 Migne, PL 188, 1147. J. de Ghellinck, *Les mouvements théologiques du XIIe-siecle* (12th ed., Bruges, Bruxelles-Paris: 1948), p. 375 ff., esp. p. 376 n. 8. The passage reads as, say, a Kuhnian description of a paradigm switch and the following period of slow adaptation. The analogy is not only metaphorical. A good many key terms of our historical reasoning are a secularized version of corresponding old terms of historical self-reflection in the Jewish or Christian tradition. Cf. my remarks in Amos Funkenstein, *Heilsplan und natürliche Entwicklung* (München: Nymphenburger Vlg., 1965), pp. 17-67 and (recently) in "Periodization and Self-Understanding in the Middle Ages," *Medievalia et Humanistica*, 1975, *5*: 3-13.

[2]K. Popper's fruitful criterion of falsification (*The Logic of Scientific Discovery* [2nd English ed., New York: Harper, 1968], pp. 40-48; 78-92) has both historical and philosophical deficiencies. Historically, the distinction between experimental and untestable hypotheses is not earlier than the 17th century; of the latter Newton claimed *hypotheses non fingo*. Antiquity and Middle Ages regarded observation as but an inclining or warning instance. Philosophically, Popper cannot overcome Goodman's paradox of induction any better than Hume or Carnap. It makes no difference at all whether we ex-

Since the beginning of consistent theoretical reasoning, however, sound theories have often specified explicitly that which, in their own terms, must be regarded as a wrong, if not impossible or absurd, position. A conceptual *revolution* consists more often than not in the deliberate adaptation of such well-defined "absurdities" (or, better yet, the absurd consequences of contradictory assumptions) as the cornerstone of a new theory. Such were the beginnings of the atomistic theory. Parmenides had proved that to ascribe any degree of reality to negation amounts to attributing being to non-being. Being suffers no differentiation or change. That which "is not" cannot be "thought of." The atomists comitted themselves consciously and deliberately to this absurdity in order to save movement and variety. Their atoms were Parmenidean "beings" embedded within the void, i.e., within a non-being endowed with reality.[3] Similarly, Aristotle's theory of motion may be said to have paved the way towards the principle of inertia more than any of its alleged forerunners, including the impetus theory. For he anticipated its conceptual content as the absurd (or impossible) consequence of a misleading assumption, the (atomistic!) assumption of movement in the void.[4] The latter example, the transition from the Aristotelian to early modern mechanics, will soon occupy a good part of our attention, but first some qualifying remarks should be advanced.

---

pect the future verification or only falsifiability of a generalization (x) (Ax Bx). In both cases, we can construe $B(x_{t \leq o}) \equiv B(x_{t \leq o}) V \sim B(x_{t > o})$, i.e. construe past "green" as "grue," (Cf. N. Goodman, *Fact, Fiction, and Forecast* [2nd ed., Indianapolis: Bobbs-Merrill, 1965], pp. 59-83).

[3]Diels-Kranz, *Fragmente der Vorsokratiker* 67 A6 = Aristotle, *Metaphysica* A4, 985b4: Λεύκιππος δὲ καὶ ὁ ἑταῖρος αὐτοῦ Δημόκριτος στοιχεῖα μὲν τὸ πλῆρες καὶ τὸ κενὸν εἶναί φασι, λέγοντες τὸ μὲν ὂν τὸ δὲ μὴ ὄν, τούτων δὲ τὸ μὲν πλῆρες καὶ στερεὸν τὸ ὄν, τὸ δὲ κενὸν τὸ μὴ ὄν. Cf. Simplicius on Aristotle's *Physics* 28, 4 ff + 67A8. ". . . was in Wahrheit bei den Eleaten vorhanden war, spricht Leukipp als seiend aus": G.W.F. Hegel, *Vorlesungen über die Geschichte der Philosophie, Werke,* ed. E. Moldenhauer u. U. M. Michel, *18* (Frankfurt: Suhrkamp, 1971), p. 355; this remained the accepted interpretation.

[4]Aristotle, *De Caelo* 301b1-4: ἔτι δ' εἰ ἔσται τι σῶμα κινούμενον μήτε κουφότητα μήτε βάρος ἔχον, ἀνάγκη τοῦτο βίᾳ κινεῖσθαι, βίᾳ δὲ κινούμενον ἄπειρον ποιεῖν τὴν κίνησιν; and even more clearly *Physica,* Δ8, 215a19-22: ἔτι οὐδεὶς ἂν ἔχοι εἰπεῖν διὰ τί κινηθὲν στήσεταί που· τί γὰρ μᾶλλον ἐνταῦθα ἢ ἐνταῦθα; ὥστε ἢ ἠρεμήσει ἢ εἰς ἄπειρον ἀνάγκη φέρεσθαι, ἐὰν μή τι ἐμποδίσῃ κρεῖττον. Cf. S. Sambursky, *Laws of Heaven and Earth* (Hebr. ed., Jerusalem: Mossad Bialik, 1954), p. 97, and H.G. Apostle, *Aristotle's Physics* (Bloomington: Indiana UP, 1969), p. 254 n. 12. Against this remark of Aristotle, cf. the passage in Plu-

The audacity to think the unthinkable is well known to historians of mathematics. All the expansions of the realm of numbers beyond the rational numbers were once considered to be such impossibilities of thought and, at the time of their conception, "amphibians between being and non-being" (Leibniz),[5] tolerated only by virtue of their performance. The history of mathematics may be read as a running commentary on the incompleteness theorem. Time and again the inability to solve problems within one field led to the construction of new fields, since "no antecedent limits can be placed on the inventiveness of mathematicians in devising new rules of proof."[6] New mathematical disciplines have often accompanied scientific revolutions. Some grew out of a conceptual revision in science (the calculus), some made revisions within a science possible (non-Euclidean geometry). Nevertheless, conceptual revolutions in the sciences or in philosophy are different from those in mathematics even where they, too, involve the assertion of the absurd. The inherited body of mathematical theorems is not proven to be wrong, or only approximately true,[7] but rather richer, by the legitimation of a mathematical entity or operation which was previously taken intuitively to be a non-number or a non-procedure. Yet physical

tarch, *De facie in orbe lunde* 923A, that the very natural movement of the moon, as every natural movement, would continue unless impeded, is compatible with the Aristotelian doctrine of celestial motions. On the medieval treatment (or lack thereof) of these passages of Aristotle see E. Grant, "Motion in the Void and the Principle of Inertia in the Middle Ages," *Isis,* 1964, *55*: 269. On the concept of impetus as a possible forerunner of the inertial principal, cf. A. Funkenstein, "Some Remarks on the Concept of Impetus and the Determination of Simple Motion," *Viator,* 1971, *2*: 329-348. Note the emphasis, in Aristotle's argument, on the lack of a sufficient reason for a body to cease moving in space. This is exactly Descartes' (and later Leibniz's) philosophical justification of the inertial principle, namely on the ground of the principle of sufficient reason. Cf. below, p. 186.

[5]G.W. Leibniz, *Mathematisch Schriften,* ed. C.I. Gerhardt, 5 (Halle, 1849-63), p. 357.

[6]E. Nagel and J.R. Newman, *Gödel's Proof* (New York, N.Y.U. Press, 1958), p. 99; less optimistic H. Weyl, *Philosophy of Mathematics and Natural Science* (New York: Athenaeum, 1963), p. 235.

[7]As, eg., Newtonian celestial mechanics compared to the theory of relativity. Only where there is a dynamical relation between "data" and "hypotheses," observations and their mathematical formulation, may one speak of "approximation"; not in the history of mathematics as such. This dynamical relation between the given datum and the imposed explanatory structure was elaborated, before Popper, by the neo-Kantian interpretation

theories are concerned not only with consistency and richness, but with truth and meaning. Where such theories introduce an absurdity in terms of a previous explanatory endeavor, the latter is destroyed, or at least proven inaccurate.

Nor should the dialectical preparation for scientific revolutions be confused with the readiness, already manifested in Greek astronomy, to entertain several explanatory models and to operate with those explanations best capable of "saving the phenomena" (*σώξειν τα φαινόμενα*), disregarding the question of their physical reality. I do not underestimate the emancipatory value of the recognition of a plurality of models in spite of, or if you wish, because of, the epistemic resignation involved in it.[8] I agree with Feyerabend that the pursuance of a plurality of alternative explanations is, at least today, imperative. At any rate, the history of astronomy is a paradigmatic case of the benefits of theoretical "anarchy."[9] But the history of astronomy in Antiquity and the Middle Ages shows also that it is one thing to look for many alternative explanations within given assumptions, and another to become conscious of such assumptions and revise them. Perhaps due to its liberality, astronomy had greater difficulties than mechanics in becoming aware of its most deeply rooted preconception, the assignment of circular, "perfect" motion to the planetary orbits.[10] But this, too, we shall discuss later.

---

of science. The neo-Kantian "Ursprungsprinzip," based on a peculiar interpretation of the so-called infinite judgments, allowed the conversion of mere "negations" into "beginnings" — a constructive violation of the principle of excluded middle. This in turn allowed not only each hypothesis to become, in due time, a "datum" — but, in a never-ending process of spontaneous clarification of scientific concepts, the very concept of passive data to be turned into a mere incentive and limiting case for ever more precise distinctions in science. I am about to discuss this matter in another place.

[8]Simplicius, *In Aristotelis quatuor libros de coelo commentaria* I.6 (ed. Heiberg), p. 32: *οὐδεν οὖν θαυμαστὸν, εἰ ἄλλοι ἐξ ἄλλον ὑποθέσεων ἐπειϱαθησαν διασῶσαι τὰ φαινόμελα.* P. Duhem, *To Save the Phenomena. An Essay on the Idea of Physical Theory from Plato to Galileo,* transl. E. Donald and C. Maschler (Chicago: Chicago University Press, 1969), p. 23.

[9]P. Feyerabend, "Consolations for the Specialist," in *Criticism and the Growth of Knowledge*, ed. I Lakatos and A. Musgrave (Cambridge: Cambridge University Press, 1970), pp. 197-229. On the whole, Feyerabend's guiding principle, "anything goes," fits best the history of mathematics, and least of all, say, zoology.

[10]Cf. below, p. 199.

We ought, then, to pay close attention to the terms in which a theory defines "improbabilities" and, still more important, "impossibilities." The more precise the argument, the likelier it is to be a candidate for future revisions. Once the impermissible assumption is spelled out with some of its consequences, it is but a matter of time and circumstances (a different climate of opinion, tensions within the old theory, developments in other fields, new factual evidence) until the truly radical alternative is reconsidered. The starting point of scientific reasoning, the Socratic curiosity (ϑαυμάξειν) consists not only in asking *why* and *how* within a given theory or as if no theory existed. It consists rather at certain critical junctures in asking *why not*?—despite a definite, enduring, argued consensus to the contrary.

In the following considerations I shall try to examine the anticipation of both the Copernican and Galilean revolutions within the systems they replaced. The historical perspective will blur our schematic contraposition considerably, and mitigate the scheme of a dialectical transition. The historian, as so often, finds himself entangled in the web of nuances after embarking from a clear cut thesis. We shall find that, although Copernicus feared that traditional astronomers would condemn his model for its obscurity, and although Descartes regarded the inertial principle to be inconceivable in the terms of Aristotelian physics, nevertheless the main tenets of neither the Copernican theory nor Galilean mechanics were formulated mainly *out* of such a contraposition with past theories, and certainly not out of literary reminiscences of passages where "absurdities" were defined. In effect, many of the "absurdities" in the terms of Aristotelian or Ptolemeian science had already become, through the medieval exercises in hypothetical reasoning, mere improbabilities. And finally, we shall try to show that, in one sense only, Galilean physics did assert the impossible—in its method rather than in its content: namely inasmuch as it employed counterfactual states (or imaginary experiments) to represent the limiting case of actual states of nature.

2. *"Impossibilities" in Aristotelian and Medieval Mechanics*

If revolutions are a conscious and "resolute attempt. . . to break with the past,"[11] philosophies of nature in the 17th century certainly *interpreted* the inertial principle and theories of motion founded on it as a revolution, as an emancipation of knowledge from childish anthropomorphisms.

> "But that when a thing is in motion," says Hobbes, "it will eternally be in motion, unless somewhat els stay it. . . is not easily assented to. For men measure, not only other men, but all other things, by themselves: and because they find themselves subject after motion to pain, and lassitude, think every-thing els growes weary of motion, and seeks repose of its own accord. . . . From hence it is, that the schooles say, Heavy bodies fall downwards, out of an appetite to rest."[12]

The conceptual content of the inertial principle lay in the abolition of the absolute distinction between movement and rest and its replacement by the absolute distinction between movement and acceleration. And we deem the method by which this principle was elicited to be no less revolutionary. The conditions under which a body will continue to move uniformly in a straight line in a given direction are understood to be unobservable, if not outright counterfactual; the product of an "imaginary experiment." Depending upon our own methodological perspective, we call such conditions "empty," "mythical," "ideal," "fictional," or simply

---

[11]Alexis de Tocqueville, *The Old Régime and the Revolution,* trans. S. Gilbert, (Garden City: Harper, 1955), p. vii. This consciousness of purposeful assent distinguishes the political usage of the term "revolution" from the more passive concept of radical change in earlier times (Cf. R. Griewank, *Der neuzeitlische Revolutionsbegriff, Entstehung und Entwicklung* (2nd ed., Frankfurt am Main: Europäische verlagsanstalt, 1969).

[12]Thomas Hobbes, *Leviathan,* I, 2, ed. C.B. Macpherson (Penguin Books, 1968), p. 87. Hobbes, however, rightly observes that the assumption that "nothing can change it selfe" is shared by both the old and the new science. Descartes, who first formulated the principle, sees likewise Aristotle's conception of motion as "naturally" wrong and as belonging to an infantile phase of human thought (Descartes, *Principia Philosophiae,* ed. Adam-Tannery, VIII.1 (Paris: Vrin, 1973), Pt. 2, 37, p. 62 f. Cf. *La logique ou l'art de penser* 1, 9 (1st ed. 1662; Engl. trans. J. Dickoff and P. James, *The Art of Thinking: Port Royalé Logic* (New York: Bobbs-Merill, 1964), p. 69. The difficulty in conceiving of the inertial principle is accentuated in Cartesian physics, by the fact that it is a mere "tendency" of bodies which is never actualized. The universe is a material continuum.

"counterfactual."[13] Whatever the evaluation of such constructs may be,[14] their function is clear: they isolate a group of phenomena and consture a counterfactual case as the limiting case of all factual cases within that group.[15] In the inability of the Aristotelian philosophy of nature to sever a phenomenon from its *context* and study it "in itself," the natural philosophy of the 17th century saw the main impediment of past science.[16]

[13]"Die Vorstellung, die himmlischen Körper würden sich für sich in gerade Linie fortbewegen, wenn sie nicht zufälligerweise in die Anziehungssphäre der Sonne kämen, ist ein leerer Gedanke" (Hegel, *Vorlesungen über die Geschichte der Philosophie* [see above, n. 4], *Werke,* XIX, p. 193); this statement despite his subtle interpretation of limiting concepts, especially of the infinitesimal as a pure relation (*Wissenschaft der Logik, Werke,* V, p. 297 ff). On the importance of "ideal experiments," see E. Cassirer, *Substance and Function* (Engl. trans. Chicago 1923), pp. 120-122, 168, 175. Useful "fictions" and imaginary experiments are treated by H. Vahinger, *Die Philosophie des Als Ob,* (2nd ed. Berlin, 1913), pp. 28-36 ("abstraktive neglektive Fiktionen"), pp. 417-423, 423-425, ("scheinatische Fiktionen") and esp. pp. 105-109, 451-471. On the counterfactual status of conditionals in thought-experiments see N. Rescher, *Hypothetical Reasoning* (Amsterdam: Brill, 1964), pp. 7-8 (p. 89 bibliography). The "utility" of "limit myths" and other "entia non grata" despite their inconvenience is again stressed by W. V. O. Quine, *Words and Objects* (Cambridge, Mass.: The M.I.T. Press, 1960), pp. 51, 248-251.

[14]The insecure standing of imaginary experiments led K. Popper, *The Logic of Scientific Discovery, op. cit.,* pp. 442-456, to interpret them as a mere auxiliary measure, permissible only as a "concession favorable to the opponent" in a *reductio ad impossibile.* This characterizes at best, as we shall see, Aristotle's use of thought-experiments, but certainly not their function in early modern physics. Popper's awkward position was, perhaps, an attempt to refute either Cassirer or Vahinger, yet neither are mentioned.

[15]An attempt to formalize idealized models was recently made by L. Nowak, "Laws of Science, Theories, Measurement," *Philosophy of Science,* 1972, *34*: 533-548. Whewell had already described what he called the "method of curves" as enabling us to obtain "data which are more true than the individual facts themselves" (W. Whewell, *On the Philosophy of Discovery* [3rd ed., London, 1860], pp. 206-7).

[16]Johann Clauberg, *Differentia inter Cartesianam et in scholis vulgo usitatam philosophiam, Opera omnia philosophica,* (Amsterdam, 1691), pp. 1217-1235: "Vulgaris philosophia non tam accurate considerat rem ut in se et sua natura est, sed potius prout se habet in respectu aliorum, quo ipso tamen interna ejus natura plerumque occulta manet." The "inner nature" of things regarded "in themselves" are the Cartesian "simple natures" (Descartes, *Regulae ad directionem ingenii* VI, Adam-Tannery X, pp. 383-84).

But neither is Aristotle's insistence that "whatever moves, moves by another,"[17] an anthropomorphic abstraction, nor can it be ascribed to his inability to construct thought-experiments or to use mathematical models. Rather, it was a well-argued position which led Aristotle to deny "indefinite" motion, the only possible motion in the void.[18] Aristotle does indeed use "thought-experiments" so as to refute the atomistic theory of motion. They may better be called "arguments from incommensurability," set forth with a basic technique which resembles Galileo's resolutive method. A finite body is imagined under a series of conditions in which one variable diminishes gradually and the relations between the variables involved is a continuous function. Yet unlike Galileo, Aristotle does not formulate a universal law valid for the factual and the counterfactual conditions alike. If Galileo regarded the counterfactual condition as a limiting case of all factual conditions, the crux of this group of arguments is, on the contrary, to reject any mediation between factual and counterfactual states. They are always incommensurable.

In this way Aristotle argues against "movement in the void" (*Physics* Δ8, 215a24-216a26) or against the "weightless body" (*De caelo* Γ2, 301a20-b16). Movement in the void and movement in the plenum cannot be thought of conjunctively. Other things (force or weight) being equal, the velocity of a body moving (presumably, by force) in the void must always be greater than the velocity of an equal body moving in a medium however rare. Velocity increases in an inverse proportion to resistance, i.e. in direct proportion to the rarity of the medium. The analogous proportion $v_1/v_2 = m_1/m_2$ becomes meaningless if $m_2 = 0$ (void), since they have no common ratio (λόγος). The movements of

---

[17]Aristotle, *Physica* VII, 1.241b34: *ἅπαν τὸ κινούμενον ὑπό τινος ἀναγκη κινεῖσθαι* (omne quod movetur ab alio movetur). This statement knows no exceptions, for even natural movements assume a previous violent removal of the object from its *οἴκεως τόπος*. A body moving by its nature has but "*a* cause of movement in itself," and is not a "self mover" (Cf. W. Wieland, *Die Aristotelische Physik, Untersuchungen über die Grundliegung der Naturwissenschaften und die Sprachlichen Bedingungen der Prinzipienforschung bei Aristoteles* [Göttingen: Vandenhoek und Ruprecht, 1962], p. 231 ff).

[18]Above n. 4.

[19]"Proportion" means here not necessarily an arithmetical commensurability (which different media may not have), but the ideal, or rather general, commensurability as put

two equal bodies in a given time through the plenum and the void are incommensurable. On the other hand, differences of weight would mean nothing outside the medium and thus bodies of different magnitudes would move up or downwards with equal velocity in the void; "but this is impossible." Just as earlier he had anticipated the inertial principle by considering the imaginary lateral movement of bodies in the void, so Aristotle anticipates here the basic presupposition of the law of free fall—as a counterfactual consequence of the atomistic universe.

The bipartite structure of Aristotle's argument[20] becomes even clearer if compared with *De caelo* Γ2, where he introduces the assumption of "weightless bodies" only to discard it in the same way. Imagine, he says, a weightless body, and compare it to a heavy or light body of the

forth in the fifth book of Euclid's *Elements*. Cf. T.L. Heath, *The Thirteen Books of Euclid's Elements* (2nd ed., New York: Dover, 1956), pp. 116-121, 131; *ibid.*, p. 120 against the arithmetical understanding of ἀνταίρεσις (Heiberg) in *Topics* 3, 158b29. Only the doctrine of μεγέθη, that is, of magnitude as such, allows Aristotle to apply the principle of analogy to many sciences without violating the "postulate of homogeneity" which forbids one, for example, to prove geometrical proportions arithmetically (An. post. A7, 75a 38 ff. and H. Scholz, "Die Axiomatik der Alten," *Mathesis universalis. Abhandlungen zur Philosophie als strenge Wissenschaft*; (2nd ed., [Basel-Stuttgart: Schwade und Co. Verlag, 1969], p. 37 and n. 25). The domain of validity of the general principle of analogy thus constitutes unities as commensurable unities in the sense of *Met.* Δ6, 6, 1016b34 (the one κατ' ἀναλογιαν). (J. Stenzel, *Zahl und Gestalt bei Plato und Aristoteles* [3rd ed., Darmstadt: Wissenschaftliche Buchgemeinschaft, 1959], p. 159). This has nothing to do with ideal numbers, which Aristotle abhors. The distinction between physics and mathematics lies not in the object of their inquiry (they deal with the same objects) but, in the words of a modern interpreter, in their "degré d'abstraction"; for pure "mathematical objects" do not exist. A. Mansion, *Introduction à la physique Aristotelienne*, (Louvain-Paris, 1946), pp. 143-195.

[20]*Physics* Δ8, 214b12 ff. and *De caelo* Γ2, 301a20 are undoubtedly parallel arguments. Here and there Aristotle construes a body in imaginary conditions to which, mutatis mutandis, a similar body under real conditions is compared. Here and there he proves his case first as to "natural," then as to "forced" motion. Had Aristotle meant, in *Physics* Δ8, to prove that any unspecified motion in the void is the same whether forced or up- and downwards, he could not argue in *De caelo* Γ2 that the behavior of weightless bodies up- or downwards is different than in forced (lateral) motion. I. Dühring, *Aristoteles, Darstellung und Interpretation seines Denkens* (Heidelberg, 1966), 320 f., misunderstood the passage to apply to all bodies in the void as having one velocity; for a similar inexactitude, cf. D. Ross, *Aristotle* (5th ed., London 1966), pp. 87-9 and H. Apostle, *Aristotle's Physics*, loc. cit. (above n. 4).

same size: the weightless body will always traverse a smaller distance up- or downwards, and a longer distance in the horizontal directions. You may then cut the heavy or light body (or augment it) until it traverses the same distance as the body without weight: "but this is impossible." The weightless body must be imagined as always faster or slower than the heavy or light body, no matter how big or small the latter. Both in the case in which $weight_1/weight_2 = distance_1/distance_2$ (the natural motion up- or downwards) and in the case in which $weight_1/weight_2 = distance_2/distance_1$ (forced lateral motion), the equation becomes meaningless if weight = 0. "Nothing" has no proportion to any finite magnitude. Note that Aristotle's proof rests on a further, tacit assumption that *some* bodies evidently move up- or downwards without constraint.[21] This we "see with our eyes." Without this assumption, all he has proven is that either every body is weightless, or none is. The counterfactual cannot be conceived of conjunctively with the factual; it belongs to another world altogether.[22]

Nowhere does the Stagyrite explicitly distinguish between logical-conceptual and physical necessities (or impossibilities).[23] As a matter of

---

[21]Such apparently are the absolute natural movements of earth and fire as against the relative natural movements of the water and the air. This, Aristotle remarks elsewhere (*De caelo* Δ4, 311b19-20), we "see with our eyes." Aristotle might have developed his awkward explanation of projectile motion in order to save the *immediacy* of absolute natural motions; cf. my remarks in *op. cit.*, "Remarks on Impetus," 331 ff.

[22]Aristotle proves elsewhere (*De caelo* A9, 277b27-279b4) that alternative "worlds," even if they were *a priori* possible (which he denies), would still be physically impossible. Our world contains all the matter that can be, and a world greater than ours (or a *pluralitas mundium*) would have an excess of vacant forms. "World" and "this world" convert perfectly.

[23]Aristotle distinguishes, however, between absolute (or simple) necessity, hypothetical necessity (*Met.* E5, 1015a20-1015b15; *Physics* B9, 199b34-200b8; *De gen.* B11, 336b33 ff.) and at times introduces the "contingent" necessity of past (as against future) contingents (*De interpretatione* 9, 18b35 ff.). Cf. Düring, *Aristoteles* 243-4; K. Jakko, T. Hintikka, "Aristotle's Different Possibilities," *Inquiry*, 1960, *3*: 18-28. G. E. M. Anscombe, "Aristotle and the Sea Battle," *Mind,* 1956, *65*: 1-15, also notes the lack of distinction between physical and logical necessity (although, we saw, it is sometimes introduced temporarily, for the sake of the argument). Cf. also S. Mansion *Le jugement d'existence chez Aristotle* (Louvain-Paris: Editions de l'Institut superieur de philosophie, 1946), pp. 68-74.

fact, he tries time and again to prove the conceptual absurdity of that which is physically impossible, or its converse. He accumulates arguments, at times begging the question. Yet the foundation for the distinction has already been laid in the arguments from incommensurability as set forth in some of his "imaginary experiments." In the course of such experiments, an assumption is shown to be necessarily impossible not on the grounds of merely logical-conceptual consideratons (at least not within this particular argument), nor because it does not immediately correspond to sense perception, but because it implies a world totally different from ours. This totally incommensurable (and hence incompatible) state stands in opposition to Galileo's imaginary, yet commensurable, limiting cases. Alternative worlds to ours are, for the Stagyrite, strictly disjunctive; and since ours exists, others are nonexistent. Our κόσμος is unique, and nothing in it could profitably be taken out of its *context* and examined under ideal—nonexistent—conditions. For these reasons Aristotle was not willing to see things "as they are in themselves" but always only "as they are in respect to each other."[24]

* * *

Aristotle's theory of motion, as with most of his doctrines, underwent serious transformations and modifications after its reception in the West in the 13th century. Inasmuch as these changes pertain to natural philosophy, they may be attributed to three distinct (though not always actually separable) motives. Their driving force was either (1) mainly interpretive, i.e., the wish to correct the *philosophus*; or (2) theological, born out of the insistence upon the contingency of all orders, no matter how necessary they might seem to the *lumen naturale*; or finally (3) mathematical-conceptual speculations, admittedly without relevance to the physical world.

(1) In the first group I include the theory of impetus as well as Ockham's conceptualistic reduction of the term "motion." Both were intended to replace Aristotle's cumbersome and deficient explanation of projectile motion without at the same time abandoning Aristotle's absolute disjunction between rest and motion (i.e., the principle *omne*

[24]Above n. 16 (Clauberg).

quod movetur ab alio movetur).[25] The theory of impetus—suggested already in antiquity and revived in the Middle Ages[26]—explained the continuation of projectile motions *cessante movente* by assuming a force imparted to the moved by the moving body. Buridan went so far as to assume this force, or quality, to be of itself indefatigable. Unless exhausted by friction in the medium, the acquired impetus, he contended, would last indefinitely.[27] It might even provide a better explanation for the (eternal) celestial motions than the "intelligences" (or movement by μιμεσις). Undoubtedly, *impetus* and the inertial principal have nothing in common: impetus is a quality, somewhat analogous to heat, and as such is the cause of the continuation of motion; the inertial principle, on the other hand, is nothing but a *methodological guide as to where one can refrain from the search for causes*[28]—namely when a movement continues uniformly. Rest thus becomes a special case of movement. The concept of "impetus" therefore is much less a "forerunner" of the inertial principle than Aristotle's characterization of the (absurd) motion in the void. And yet, almost inadvertently, Buridan's generalization of *impetus* introduced a grave methodological shift. "Simple" motion in Aristotelian terms (the natural motion of a body to—or within—its οἰκεῖως τόπος) meant, and had to mean, the natural motion in its totality—an *actually* existing motion and, for some simple bodies,[29] one immediately *discernible* through the senses. In early

[25]Cf. above p. 173 and n. 17.

[26]For the literature on the concept of impetus cf. Funkenstein, *op. cit.*, "Remarks on Impetus," p. 329, ns. 1, 5, 6.

[27]Johannes Buridan, *Questiones super libris physicorum* 8.12 (Paris, 1509), quoted by A. Maier, "Die Aristotelische Theorie und die Impetus-Hypothese," *Zwei Grundprobleme der Scholastischen Naturphilosophie* (Rome: Edizioni di Storia e Letteratura, 1951), pp. 207-214; and later *Questiones super libris quattour de caelo et mundo* ed. E. Moody (Cambridge, Mass., 1942), pp. 180-184, 240-243. As I tried to show in the article mentioned above (n. 26), Buridan was careful not to make the acceptance of the concept of impetus contingent on the acceptance of his far-reaching cosmological suggestions (such as the abolition of the intelligences).

[28]On the necessity of an analogous methodological principle for every systematic explanation of motion cf. A. Koslow, "The Law of Inertia: Some Remarks on Its Structure and Significance," *Philosophy, Science, and Method: Essays in Honor of Ernest Nagel*, ed. S. Morgenbesser and others (New York: St. Martin's Press, 1969), esp. pp. 552-4 (condition of normalcy).

[29]Fire and Earth; above p. 175, n. 21.

modern dynamics, in contrast, simple motion became an abstraction, one component in the factoral analysis of motion, a mere "tendency"—albeit a "clear and distinct" idea. Buridan's concept of impetus was a basic step towards this insight. For him, natural motion became only one aspect of actual motions. The introduction of impetus forced him to assume that at least *all* sublunar motions are mixed motions.[30]

Motion was still to be regarded, however, as a change from one state to another, not as a state in itself which requires *no* causes to account for it. Was Ockham's radical simplification of the problem perhaps an anticipatory step towards the recognition of uniform motion as a state? Ockham simply rejects the theory of impetus as well as the Aristotelian account of projectile motion by denying the existence of the problem.[31] The term "movement," as extension or quantity, is a connotative term, denoting an object and connoting a series of places which it occupies consecutively. Both movement and conservation of movement are two expressions for one and the same phenomenon. We need only one cause to explain why a body left l_1 and reached l_n through $l_2 \ldots l_{n-1}$. If it left l_1, it necessarily occupies other places. Yet Ockham, because he was preoccupied with the reduction of our concepts to those singulars and absolute qualities which they stand for *in recto* or *in obliquo,*[32] stopped exactly at the point where he might have hit upon the distinction underlying the inertial principle—namely the distinction between uniform motion (or rest) and change of motion.

(2) The second group of transformations to which the Aristotelian cosmology was subjected was theological in nature. Theological concerns led not only to the abandonment of such basic tenets as the *aeternitas mundi.* They underlie likewise the growing use of hypothetical discussions—imaginary experiments—in the Middle Ages. They aim

[30]Inasmuch as all of them are a product of impetus ± natural movement and ± friction of the medium. Buridan, however, had difficulties in distinguishing between impetus as a cause of movement and impetus as a cause of acceleration.

[31]Ockham, *Sent.* I, 26 M (Lyons 1495; reprint London: Gregg Press, 1962); A. Maier (above n. 27) 154 ff.; Ph. Böhner, *Ockham: Philosophical Writing* (Edinburgh: The Nelson Philosophical Texts, 1957), pp. 139-141; for further literature, cf. my "Remarks on Impetus," *op. cit.*, 337 n. 27.

[32]Below pp. 184-186.

to prove the contingency not only of singulars, but likewise of "natures" or orders. The doctrine of motion is no exception, although at times it is difficult to distinguish between the interpretation and refutation of Aristotle. Consider the attempts to define the nature of motion in the void (even by those who deny its existence categorically): is it instantaneous or successive? The mainstream of medieval interpreters decided, against Averroes, that it is a movement in time; and in this respect Thomas Aquinas and Duns Scotus follow Avicenna.[33] By the 14th century, those who will assert the void against the Aristotelian tradition (Nicholaus of Autrecourt) or even a continuous "rudimentary motion" of unimpeded bodies in it (Hasdai Crescas, following in part Avempace)[34] will thus look back on a chain of discussions from hypothetical reasoning—discussions in which the void, and movement in it, appear (at least temporarily) as a *logical* possibility. Where such arguments appear with critical rather than interpretative intent, we may assume mostly theological motives. The possibility of other worlds than ours or the possibility of the movement of the total universe, if God so

[33]Averroes, *Comm. in Phys. Arist.* IV, 71, *Aristotelis Opera* (Venice 1560) IV, 130D ff.; Thomas Aquinas, *Physic.* IV lect. 11-13, *Opera* ed. Marietti, 51b ff.; Duns Scotus, *Sentent* II, 2.9, *Opera* (Quarracchi) 299 ff. These and other references in K. Lasswitz, *Geschichte der Atomistik vom Mittelalter bis Newton* (Marburg-Leipzig 1890, reprint Darmstadt: Wissenschaftliche Buchgesellschaft, 1963) I, pp. 207-8; cf. A. C. Crombie, *Augustine to Galileo. Science in the Middle Ages and Early Modern Times* (London: Mercury Books, 1961) and Wolfson (n. 34 below). The most comprehensive treatment of these and related questions is to be found in E. Grant, "Motion in the Void," *op. cit.,* and *idem, A Source Book in Medieval Science* (Cambridge, Mass: Harvard University Press, 1974), pp. 334-350.

[34]Hasdai Crescas, *Or Adonai* 1, 2.1 (Vienna 1859) 16: "And since movement is not possible but in time it is necessary that movement should have a radical time (*Zman Shorshi*) if we assume movement in place." Crescas wants to prove (against Aristotle and Maimonides) the possibility of infinite motion. His "rudimentary time" is to be conceived as a constant factor in any given movement to which acceleration (by force) or deceleration (through resistance) should be added or subtracted. H. Wolfson, *Crescas' Critique of Aristotle: Problems of Aristotle's Physics in Jewish and Arabic Philosophy* (Cambridge, Mass.: Harvard University Press, 1929), pp. 183, 205, 403-409, shows the source of Crescas' discussion in the discourse of Averroes (where Avempace's is also discussed and rejected) and the contentions of Gersonides. Cf. also M. Jammer, *Das Problem des Raumes: Die Entwicklung der Raumtheorien* (Darmstadt: Wissenschaftliche Buchgesellschaft, 1960), pp. 84-87.

wished, in a straight line again becomes thematic, especially in the 14th century. The implications of action-at-a-distance or of changes in the material structure of the cosmos were reexamined: Would, for example, the sublunar elements at some future time stay in their proper places and cease to intermix?

Aristotle, as we have seen, tended to assume the coincidence of logical-conceptual and physical necessities (or impossibilities), though at times he chose to separate the logical, or purely conceptual, argument from the physical argument and to construe the latter as a "thought-experiment"—solely in order to reduce assumptions *ad impossibile*. Thought-experiments in the Middle Ages, however, had quite another function. The separation of logical from physical necessities now became inevitable and definite since cosmology was subjected to preconditions imposed upon it by the postulate of God's omnipotence, on which no definite rational structure, save the principle of contradiction,[35] could infringe.[36] It is one and the same tradition which starts with the medieval discussions on the distinction between God's absolute and ordained power, continues with Descartes' contention that even

[35]The term *principium contradictionis* stands, in the Middle Ages and until the 18th century, for both the principle of contradiction and the principle of the excluded middle. Whether the discussion of future contingents was an exception is contestable. Another possible exception was Maimonides' systematic doctrine of negative attributes. Only negative essential attributes are a legitimate mode of locution about God; in fact, Maimonides allows only, when speaking of God, for negations of negations in the form, e.g., "God is not evil." But [~~A(g)] is *not* equivalent to A(g); it is not a privation (*steresis*), but an infinite sentence. This is tantamount to the exemption of the divine attributes from the law of excluded middle—or even excluded n-th possibility. Inasmuch as non-essential attributes are concerned, Maimonides allows for so-called "attributes of action" (*te'are pe'ula*). But if the statement "God created the world" says *nothing* about God, and is equivalent only to the sentence "the world was created," Maimonides must be seen as operating with a model analogous to material implication.

[36]The problems related to the power of God were in part a heritage of Antiquity. The "necessity" of past contingents (above n. 23) Aristotle expresses in one place as the circumstance that even the gods cannot reverse the past (*Ethica Nicomachea* Z2, 1139b7-11; cf. Thomas Aquinas, *Summa Theologiae* Ia q.25, a.4 [Marietti]; sources: e.g., Plato, *Nomoi* 934a6; *Ilias* 24, 550-551; 522-4). The Patristic literature preserved several reminiscences and additions; already *Sap. Solomonis* 12:18 contains a peculiar defense of God's power *vis-à-vis* the moral order of nature: *subest enim tibi, cum volueris posse* became an often-quoted verse (e.g., Petrus Damiani, *De divina omnipotentia,* Migne, PL 145, 599 f.;

eternal truths, albeit "clear and distinct," are created,[37] and leads to Leibniz' distinction between the *necessité logique* (or *metaphysique*) which is based on the principle of contradiction, and the *necessité physique* (or *architectonique,* or again *morale*)[38] based on the principle

---

again with the discussion whether God may reverse time). Both Celsus and Porphyrius invoked the paradoxes of omnipotence or of actions *Contra naturam*—the paradox of reversing past events or of the possibility of God's self-annihilation (A. v Harnack, "Porphyrius gegen die Christen," *SB der königl. Ak. der Wiss., Phil-hist. Klasse,* I, [Berlin, 1916]; Origenes, *Contra Celsum* V, 23; Augustinus, *Contra Faustum* 26, Migne, PL XL//, 480). By the 12th century, even before the reception of Aristotle, it was fairly well agreed that God's omnipotence is subject to the law of contradiction, and that He cannot reverse time. Another tradition of the problem, the insistence of certain schools of the Moslem Kalām on the utter dependence of every thing and event on the direct divine causation, was transmitted through Maimonides' refutation into the 13th century. (But see W.J. Courtenay, "The Critique of Natural Causality in the Mutakallimun and Nominalism," *Harvard Theological Review,* 1973, *66*: 77-94.) On the history of the medieval discussion cf. H. Gronziel, *Die Entwicklung der Unterscheidung zwischen der potentia Dei absoluta und der potentia Dei ordinata* (Kath.-theol. Diss.; Bresslau, 1926); E. Borchert, *Der Einfluss der Nominalismus auf die Christologie der Spätscholastik* (BGPFM XXXV, 4-5, Münster, 1940), pp. 46-74; J. Miethke, *Ockhams Weg zur Sozialphilosophie* (Berlin: De Gruyter, 1969).

[37]Cf. below p. 185 n. 47.

[38]G.W. Leibniz, *Essai de Théodicée,* in *Die philosophischen Schriften* ed. C.J. Gerhardt, (Berlin, 1885; reprinted Hildesheim: Olms, 1965): G. VI, p. 50 (*necessité géometrique* against *necessité physique* or *morale*); *ibid,* G. VI, p. 321 (n. *absolue* — n. *morale*); *Tentamen anagogicum* G. VII, p. 278 (*determinations Géometriques,* founded on absolute necessity, against *determinations Architectoniques,* founded only on a *necessité de choix*; the negation of the first implies a contradiction, only of the second imperfection); *De rerum originatione radicali* G. VII, p. 303 (physical necessity as hypothetical necessity which always relies on *another* reason, e.g., causality; as against absolute or metaphysical necessity, *cuius ratio reddi non potest*). *Ibid.,* p. 304 (logical absurdity against moral absurdity). *Principes de la Nature et de la Grace,* G. VII, p. 603 (*principe de necessité — principe de la convenance*) cf. *Théodicée* G. VII, p. 44 (impossible—inconveniens; the latter term is of scholastic origin, where it described, at times, the *potentia dei ordinata*). In a letter to Arnauld, G. II, p. 62, Leibniz speaks of two kinds of a priori truths, those founded on the principle of contradiction and those founded on the principle of sufficient reason. The Cartesian origin of the reference to moral necessity is clearest where Leibniz distinguishes, e.g., *Noveaux Essais* G. IV, p. 6, between *certitude* (*morale s'entend ou physique*) and necessité (*ou certitude métaphysique*). With this principle of sufficient reason, which "inclines without necessity," Leibniz believed that he had mediated between the postulate of the complete rationality of the structure of our universe and the equally cogent postulate of its utter contingency: "Atque haec est radix contingentiae, nescio an

of sufficient reason.[39] The relative necessity of existing orders could still be maintained either by holding that a world altogether different than ours entails logical-conceptual contradictions (e.g., "void" as an existent non-being), or because an arbitrary abrogation of the order of creation does not "fit" ("is not convenient with") the notion of divine providence.[40]

The "necessity" attributed to nature could be redefined in roughly three different modes, and the Middle Ages tried all of them successively. (a) The rational structure of the world could be attributed, if not to an absolute necessity in the sense in which God is an *ens necessarium,* at least to a necessity relative to God's *existence.* (b) "Physical" possibilities and impossibilities could be conceived as different from both logical possibilities on the one hand and the possibility of singular events on the other. Such was Thomas Aquinas' subtle application of modal categories to the just established contraposition of the *potentia dei absoluta* and the *potêntia dei ordinata.* If viewed from the vantage point of God's

hactenus explicata a quoquam" (G. VII, p. 200; cf. *Théodicée* G. VII, p. 126). The following discussion will show that he was the last rather than the first philosopher of rank who tried to justify the intermediate status of physical necessities in terms borrowed from reflections on the domain of God's omnipotence.

[39]Whether the principle of sufficient reason (POSR) has to be split into two distinct principles, and whether other "principles" (continuity, identity of indiscernibles, perfection) are independent of the POSR is a major dividing line between Leibniz-exegetes (Russell, Parkinson, Reisher against Martin, e.a.). Leibniz himself not only holds to the unity of the POSR, but explicitly identifies it with the others (e.g., G. VII, p. 199: POSR identified with the "predicate—in-notion" principle as the way in which the notion "*sui subjecti . . . implicite esse*"). I hope to show elsewhere that, in effect, although the POSR has but one positive formulation, it refers to two different forms of negation: the endless negation, which allows Leibniz to prefer being (or an attribute of a being) to its indefinite negation, and definite negation, which occurs if a thing has as many "reasons" to be as not to be; in which case it will not be, sînce *nulla in rebus est indifferentia.* The importance of this dual structure for Leibniz' methodology of science is, I hope to show, considerable.

[40]In essence, this is Anselm of Canterbury's position. *Proslogion,* 7, *Opera Omnia* ed. F.S. Schmitt (Edingurgh, 1956), I, 105 ff., which is likewise an answer to Damiani. God's objects of volition have to remain constant, and thus the order imposed on the world. Not very different is Maimonides' position, except that he shows that the very concept of "order" or "Law" cannot determine things *omnimodo*; a residue of contingency is part of every law, whether terrestrial or celestial, inasmuch as it informs matter. From this residue of contingency, miracles as well as special providence draw their place in the order of things.

absolute power, everything is possible which does not entail self-contradiction. *Potentia dei ordinata*, on the other hand, is not only (as some of his interpreters suggest) the actual order of things as they are. Thomas not only takes the assumption "all things are possible to God except that which involves contradiction" as literally as Ockham will take it later,[41] but emphasizes as well that God "often" acts against natural norms (*frequenter faciat contra consuetum cursum naturae*) in the case of singular events. And finally he believes, as much as any 14th century thinker, that "whatever God produces by means of secondary causes, God can produce and conserve immediately without their aid," a maxim inherited from Tempier's condemnation list.[42] What distinguishes Thomas from the later, nominalistic insistence on the utter independence of singulars from each other is the emphasis on the necessity of *some* order of created things—not necessarily the existing order, but always (*de potentia dei ordinata*) a perfect order.[43] There are infinitely many perfect orders from the vantage point of God's ordained power—each of them, however, incommensurable with the other. (c) The legitimacy of assuming a rational-physical order, e.g., some forms of necessary interdependence of created things could be doubted altogether. This was the thrust of the Terministic criticism since the 14th century: it left only the

[41]Thomas Aquinas, *Questiones disputatae,* I (*De potentia Dei*), 9 Ia 3 (Turin-Rome: Marietti, 1927), pp. 7-11; *Summa Theologiae,* Ia 25.3. Thomas' starting point is the Aristotelian distinction between absolute and hypothetical necessity. Ph. Boehner, (*Ockham: Philosophical Writings* [Edingurgh: The Nelson Philosophical Texts, 1957], pp. xix-xxi) has tried to deduce the main guiding maxims of Ockham's thought from this maxim. In the following remarks, I shall try to show (1) that the differences between Ockham and Thomas are often exaggerated and (2) that Ockham operated with a multiplicity of principles which, although compatible with each other, are not necessarily derivable one from another—as the principle of immediacy and the abolition of the *principium individuationis.*

[42]*De potentia* 9.3a.7, Marietti edition, p. 62; for the principle in Ockham cf. E. Hochstädter, *Studien für Metaphysik und Erkenntnislehre Wilhelms von Ockham* (Berlin-Leipzig: De Gruyter, 1927), pp. 12-26, esp. p. 17 f.

[43]*Summa Theologiae* Ia 25a 5, esp. *respon. ad tertium; ibid.*, a.6, *ad tertium; De potentia* 9.6a 1 ad XII: *ars divina non totam se ipsam explicat in creaturarum productione; et ideo secundum autem suam potest alio modo aliquid operari quam habeat cursum naturae. . . nam et homo artifex potest aliud artificium facere per suam artem contrario modo quam prius fecit.*

categories of logical and factual necessities. In the name of what may be termed the *principle of immediacy,* Ockham denies that a "nature" (a law) is anything else but the singular instances of its concretization, and maintains that the universe could exist—even *de potentia dei ordinata*—as a conglomerate of totally unrelated things, "one" only in number.[44] The principle of immediacy, as we saw, is not new: new is its nominalistic interpretation, as epitomized in the *method of annihilation.* "Real" things, singulars, are those capable of being created independently; in other words, the test for a false attribution of reality to a thing is the impossibility of thinking its existence while all other things in the universe are "annihilated": *omnis res absoluta, distincta loco et subjecto ab alia re absoluta, potest per divinam potentiam existere alia re absoluta destructa.*[45] Aided with this principle, Ockham argues that no mediation between singulars (through universals) is necessary, and that no mediation between one's cognition and singulars as its objects (e.g. intelligible species) is necessary either. Existential judgments depend only upon *notitia intuitiva* caused immediately by the objects. No *principium individuationis* is necessary to account for individual things—not even the Scotistic *haecceitas.*[46]

[44]This again does not mean that Ockham assumes such a disconnected conglomerate. Atomism is not a necessary consequence of the Terministic criticism; it has not even become—in spite of Nicholaus of Autrecourt's contentions—more probable. Nor does this criticism necessitate a higher expectation for cosmic changes to come. Ockham's principle of economy, albeit a guide for reason and not necessarily for creation (Boehner, p. xxi; Miethke, pp. 238-244), forbids him to anticipate changes in the empirically elicited *cursus naturae* (except those revealed). To anticipate another order means likewise to impose a plurality without necessity.

[45]*Quodlibeta,* VI, p. 6 (Boehner, *Philosophical Writings,* p. 26); *Sent. prol.,* q 1 HH; Hochstetter, *Studien,* pp. 56-7. Miethke, *Ockhams Weg,* mistakes this criterion to allow a positive definition of things, but since only the possibility is thus guaranteed, it remains a negative, i.e. critical, principle.

[46]Ockham, it seems, only exchanged one necessity for another. What is it that guarantees our intuitive notions? Not their logical independence (secured by the method of annihilation), for it may at best assure us what a thing is *not* rather than what it is, and thus serves only as a clue. Nor indeed any *adequatio rei an intellectum,* for it would require mediation through *species.* Only a strict causal dependence remains. An intuitive notion of a singular, as already shown by Hochstetter, *loc. cit.,* is immediately caused by the singular and again causes on its side the emergence of genuine abstract notions. But if this is so, how could God cause a *notitia intuitiva de rebus non existentibus*? Ockham

It may be doubted whether Ockham's version of immediacy guarantees God's omnipotence any better than the assumption that some reference structure to other objects (*ordo ad invicem*) is a constitutive element of every "thing." Ockham, however, believed the following to be the case: namely that his criterion of isolation through imaginary destruction was necessary to save the utter contingency of the world. And here, so it may seem, we find at last a real anticipation of imaginary experiments underlying the *metodo risolotivo e compositivo.* Was not the Galilean analysis founded upon the very same principle of the extrapolation of a phenomenon from its context? Indeed, the career of the "method of annihilation" may have its roots in the Terministic analysis of our concepts; but the differences between the use made of this method in the 14th century, and the place which it occupies later in science (Galileo) or philosophy (Descartes,[47] Hobbes[48]) are considerable.

never arrived at a satisfactory answer; but his insistence, at first, that God could at best impose the image of non-existent things, not an "evidence" of this existence, — the position which A. Maier, ("Das Problem der Evidenz in der Philosophie des 14 Jahrhunderts," *Ausgehendes Mittelalter* [Rome: Edizioni di Storia e Letteratura, 1967], pp. 367-522, esp. 373) found enigmatic, — is understandable on this background only — hard as it was to defend. Another, related epistemological difficulty arises from this close causal link between objects and our intuitive cognition of them. How could there be an intuitive negative cognition, the notion of "x not being here?" For certainly such a notion is an immediate existential judgment. Here, again, Hochstetter shows that Ockham never reached a satisfactory solution; nor could he have.

[47]Because the "substances" which Descartes establishes through the method of annihilation are abstractions rather than an infinity of singulars — the *sum res cogitans* and extension (matter) — the method could serve him as a positive, rather than merely as a critical, tool. Descartes, like the Nominalists, reduces immediate evidence to substances and their absolute qualities; the difference between his intuition (or *ideae clarae et distinctae*) and theirs is that he knows but three such kinds of substances — God, matter, and "souls." This "nominalistic" reading of Descartes may solve some exegetical problems. When he argued (letter to Mersenne, 15 April 1630; or *Meditationes de prima philosophia,* resp. VI, ed. Adam-Tannery, *Oeuvres de Descartes,* VII [Paris, 1964], pp. 432; 436) that God could invalidate even the most basic mathematical operations, or create mountains without valleys, did he mean, as most interpreters believe (A. Koyré, *Descartes und die Scholastik* [Bonn, 1893, reprint Darmstadt: Wissenschaftliche Buchgesellschaft, 1971], pp. 21-26, 85-6) to exempt God, *de potentia eius absoluta,* from the laws of contradiction? What does it mean that eternal truths are "created?" It can be shown that the key to understanding the ontic status of "eternal truths" is the doctrine of substances. Only *substances* exist — one matter, souls, and God. "Eternal truths" do not exist, but are valid in reference

The *Terministae* used the heuristic *topos* (isolation through annihilation) with a negating, that is, critical, intent only—much as earlier Thomas used the distinction between logical and contextual necessity. In a general sense, they wished to establish what the Aristotelian cosmos —which neither Ockham nor his followers wanted to destroy—is *not*: namely, in any sense "necessary." In particular, they wished to establish what *things* are not: connotative notions such as extension, motion, time should not be hypostatized nor have any claim to an ontic status. It is significant that one of these abstractions (extension), rather than the individual *res extra animam*, became the "substances" of Descartes, guaranteed by a *cognitio intuitiva* (as were Ockham's *res*): the resolutive method demanded the extrapolation of relations rather than of things, and with a constructive rather than critical intent.

In summary: alternative, counterfactual worlds, or conditions for a particular universal phenomenon, were both incommensurable and impossible for Aristotle; still incommensurable, though not impossible, for Thomas; irrelevant to our world, but indeed possible, for the conceptualists of the 14th century; but certainly not the limiting case, of all factual instances, as they became for the 17th century. Such a

---

to substances. To say that God could abstain from creating mathematics is as much as to say that He could abstain from creating matter. To say that 2 + 2 could equal 5 is to say that, had He wanted to, God could add to four parts of extension, *e nihilo,* a fifth whenever we add 2 + 2. "Clear and distinct ideas" do not refer to logical necessities, but to *physical* ones; being "inconceivable," the criterion of physical impossibility is not the same as a logical contradiction. I hope to enlarge upon this interpretation at another occasion, for it is this intermediate status of physical necessities which will occupy the methodologies of science in the 17th century, time and again. If our interpretation is correct, then Descartes saw the necessity of geometry as being something short of logical necessity—a position challenged by Leibniz, and maintained again, with a richer argumentative apparatus, by Kant.

[48]Hobbes commenced his analysis of phenomena by imagining the whole outside world destroyed. All that remains, he argues against Descartes, is not only the self, but likewise its memories, from which the concept of space as a phantasma of things outside us may be reconstructed: *verum ipse factum.* Hobbes' analysis of the *status naturalis* of society without a sovereign is likewise a repetition of the method of annihilation. Thomas Hobbes, *De Corpore,* 7.1-2, *Opera,* ed. W. Molesworth (London, 1839-45). The importance of Hobbes' method of annihilation for both his methodology of science and his political theory was recognized by Högingswald, *Hobbes und die Staatsphilosophie* (München: Ernst Reinhardt Verlag, 1924), p. 123.

constructive mediation (instead of a critical confrontation) underlies the methodological revolution of the 17th century. But was it anticipated as an impossibility?

(3) There exists, finally, a third group of medieval hypothetical considerations concerning the nature of movement. While the Ockhamistic arguments from divine omnipotence tended to reduce our conceptual apparatus to a minimum of connotative notions, i.e., to eliminate a plurality of coextensive terms, the trademark of the *Calculatores* was, on the contrary, the introduction of mathematical distinctions without relation or foundation in the actual course of nature. Interestingly enough, however, they also defended their procedures with reference to divine omnipotence. The efforts of the Merton school to quantify the relation between uniform motion and uniformly difform motion (uniform acceleration or deceleration) led, as is well known, to a rigorous analysis of rates of change of interdependent variables (the method known as *latitudo formarum*) and in particular to the formulation of the rule of mean speed.[49] Important as these mathematical exercises were, they were likewise understood (and attacked) as mere abstractions, without bearing upon reality; valid only *secundum imaginationem.*[50] For these reasons we look in vain in the Middle Ages for an effective, elaborate mathematical formalization of physical entities proper, such as the *impetus.* Dominicus de Soto was in fact the first to link the theory of *impetus* with the method of formal latitudes, in general remarks, and only a short time prior to Galileo.[51]

---

[49]First formulated by Heytesbury in his treatise *De motu*: A. Maier, "Die Mathematik der Formalattitüden," *An der Grenze von Scholastik und Naturwissenschaft* (2nd ed., Rome: Edizioni di Storia e Letteratura, 1952), p. 287 n. 58; cf. also C. Wilson, *William Heytesbury, Medieval Logic and the Rise of Mathematical Physics* (Madison: Wisconsin University Press, 1960), p. 18 ff., 115-147; Clagett, *Mechanics, op. cit.,* pp. 255-329.

[50]Cf., for example, Nicole Oresme, *Tractatus de configurationibus qualitatum et motuum,* in ed. M. Clagett, *Nicole Oresme and the Medieval Geometry of Qualities and Motions* (Madison: Wisconsin University Press, 1968), p. 158 (Prooem.): *cum ymaginationem de uniformitate et difformitate intensionum ordinare coepissem. . . ; ibid.,* 164: *Omnis res mensurabilis exceptis numeris imaginatur ad modum quantitatis continue.* Cf. also Clagett, *ibid.,* introduction, p. 12 f.

[51]Perhaps Oresme's return to the position generally held before Buridan (except for the then unknown "energy" of Philoponus), namely that the impetus is a defatigable quality (not unlike heat), may be interpreted not as a regress, but rather as an introduction of the

With some justice, Galileo's analysis of free fall can be seen as a continuation of the work of the *Calculatores*, although his use of their method was firmer, his mathematical intuition better. The mathematical apparatus at his disposal was essentially theirs as well. Only as it developed did the new mechanics create for itself the new mathematical tools which it needed. His starting point was nevertheless radically different—methodologically as well as in the very concept of movement which he employed. He assumed uniform motion to continue without additional causes, and accepted mathematical abstractions as a true model, or picture, of physical reality, a valid though not an exhaustive description. Both his method and concept of motion are essentially interdependent, and *because of this interdependency* had already been formulated as absurdities by Aristotle. Aristotle formulated the conditions under which movement in an abstract, mathematical space, without a medium, may be conceived. Such a movement will be uniform and continue without further causes, but will also be incommensurable with any movement which takes place in an actual context, that is, in the medium. For the latter will always be either faster (if natural) or slower (if violent) than the motion outside a medium. The Middle Ages did not change this assessment, even though it emphasized the ultimate contingency of physical impossibilities and even though it took increasing delight in lending mathematical precision to such abstract, and in their eyes physically irrelevant, distinctions as that between uniform and difform motion, or the meaning of initial velocity. When Galileo, in his earlier treatments of mechanical problems, introduced the notion of a (circular) motion which is neither natural nor violent, but indifferent, he embarked on a direction which would lead to the formulation of the principle of inertia (which he himself never clearly conceived). But

---

distinction between movement and acceleration into the impetus theory. Buridan's theory was ambiguous. He viewed the impetus as cause (and result) of both movement and acceleration. For Oresme it accounts for acceleration only. What, then, accounts for uniform movement? Or, which is a reformulation of the same question, whence does the initial acceleration (which causes impetus, and the impetus again a lesser acceleration) come from except from the contiguous *movens* (A. Maier, *Zwei Grundprobleme der scholas-tischen Naturphilosophie* [Rome: Edizioni di Storia e Letteratura, 1951], p. 254 ff.; Clagett, *Mechanics,* p. 552)? Oresme, I believe, would answer that no *actual* motion is ever uniform, or uniformly difform.

whether we regard Galileo's "indifferent" motion—or later his circular *impetus*—as a decisive step or not, its conception was by no means "original."[52] As it was already a well-argued absurdity within Aristotelian mechanics, Galileo's originality consisted in the assertion of the absurd, or the allegedly incommensurable, and by showing its commensurability.[53] But this depended utterly on his willingness to affirm another absurdity (a methodological one), namely the priority of mathematical models even if they leave a necessary unresolved residue of phenomenal variables, that is to say even if they are valid strictly speaking only under counterfactual conditions. Galileo had a fair notion of the methodological meaning or difficulties of his procedure. He even comes close to speaking of abstract models (or counterfactual conditions) as limiting cases.[54] At any rate, he was well aware of the absurdity of the new method from the point of view of the old one, as when he ridiculed Simplicio for his unwillingness to assume the case of that which, in reality, could not possibly be the case. Simplicio, after protesting once again against the exaggerated use of mathematical models to explain physical phenomena, is taught by Salviati that the *filosofo geometra,* when he wishes "to recognize in the concrete the effects which he had proven in the abstract, must deduct the material hindrances."[55] We may add: no matter how subtle the model, it will

[52]As argued by S. Drake, *Galileo Studies* (Ann Arbor: University of Michigan Press, 1970), pp. 240-256; id., *Galileo Galilei on Motion and Mechanics,* (with I.E. Drabkin; Madison: University of Wisconsin Press, 1960), p. 170 n. 25.

[53]Cf. also C.B. Boyer, "Galileo's Place in the History of Mathematics," *Galileo, Man of Science,* ed. E. McMullin (New York: Basic Books, 1967), p. 239 (on the commensurability of the circle and straight line).

[54]Galileo, *Two New Sciences,* trans. S. Drake (Madison: Wisconsin University Press, 1974), pp. 222-224, (*Opere* VIII, pp. 274-275), ". . . Archimedes and others imagined themselves, in their theorizing, to be situated at infinite distance from the center," etc.

[55]*Dialogo. . . sopra i due massimi sistemi del mondo Tolemaico, e Copernicano, Le Opere di Galileo Galilei,* VII (Firenze: Edizioni Nationale 1898), p. 242 (Engl. trans. S. Drake, Berkeley-Los Angeles: California University Press, 1963, p. 207). Note that Galileo allows Simplicio to introduce his objection at the worst opportunity. In the context of the argument (whether the rotation of the earth should have visible effects on the mechanical behavior of bodies in its proximity) the diurnal motion of the earth becomes even better grounded if we accept the objection that a material sphere must touch a tangential plane on more than one point. Salviatti goes on to explain that of course such a concession should not be made. A material sphere of the sort that would be to Simplicio's liking

never fully recapture *all* the "material hindrances" once removed.[56] Aristotle may not have known all that; but he knew enough of it to argue against all attempts to sever the phenomenon from its context.

3. *"Impossibilities" in Ancient and Medieval Astronomy*

The role of hypothetical arguments in the preparation for the Copernican revolution is easier to assess—simply because it did not involve, as did the Galilean revolution, a change in the mode of hypothetical reasoning itself. Copernicus, it is now agreed in nearly every discussion of his achievements, did not introduce any conceptual revisions into the existing body of astronomical science, nor were his observational techniques any better.[57] From the outset his theory could be discussed on the same level as the Ptolemaic model and with the same vocabulary—whether or not we accept Copernicus' claim of greater parsimony.

would be a contradiction in terms. Does Galileo, when he speaks of disregarding the material hindrances, address himself only to the necessity of accepting the inexactitude of actual measurements? So he was understood to mean by A. Koyré, "Galileo and Plato," *Metaphysics and Measurement: Essays in the Scientific Revolution* (Cambridge, Mass.: Harvard University Press, 1968), p. 37. This would mean, in our terminology, only the isolation of the "quasi-factual" from the actual, not of the counterfactual from the sequence of the quasi-factual conditions; but Galileo probably meant both steps in the *Dialogi,* without distinguishing them methodically. Cf. also E. Cassirer, *Das Erkenntnisproblem,* I (Berlin: Bruno Cassirer), p. 383. Later, in the *Two New Sciences,* Galileo seems to aim at such a distinction (cf. n. 54).

[56]Cf. also M. Bunge, *Causality: The Place of the Causal Principle in Modern Science* (Cambridge, Mass.: Harvard University Press, 1959), pp. 125-132 ("Paradoxes of isolation").

[57]Both Thomas Kuhn, *The Copernican Revolution* (New York: Vintage Books, 1957) and Hans Blumenberg, *Die Kopernikanische Wende* (Frankfurt am Main: Suhrkamp, 1956)—the most influential contributions to the recent discussion on the Copernican theory—appeared independently of each other, and with the same initial question: How the Copernican theory, with no theoretical or practical edge over the Ptolemaic, could become a "paradigm" (Kuhn) or a "metaphor" (Blumenberg). The question was not new; but the sociological (Kuhn) and cultural-philosophical approaches (Blumenberg) were. As to the theoretical merits of the Copernican system cf. also D.J. De Solla Price, "Contra-Copernicus: A Critical Re-Estimation of the Mathematical Planetary Theory of Ptolemy, Copernicus and Kepler," in *Critical problems in the History of Science,* ed. M. Clagett (Madison: Wisconsin University Press, 1969), pp. 197-218.

Moreover, in the astronomical tradition which he challenged, the movement of the earth was not regarded in all respects as impossible. At least the assumption of a diurnal rotation was always held to be capable of "saving the phaenomena" as a geometrical model. Ptolemy himself goes so far as to concede that it is the "simpler" hypothesis; only physical arguments make it "ridiculous."[58] Yet inasmuch as these physical arguments are concerned, the Ptolemaic models themselves were deficient. The epicycles violated the demand that celestial, perfect motions should be "simple," i.e., not a composite of several motions. The equant was regarded by many as an arbitrary ad-hoc device.[59] From the vantage point of Aristotelian physics, Copernicus only replaced a physical impossibility by a greater one. But the Aristotelian refutation of the Earth's movement had already been considerably mitigated in the Middle Ages; and instead of an impossibility, a movement of the Earth became, at best, an improbability.

Again it was in the classical age of hypothetical reasoning, the 14th century, that some physical arguments were vigorously reconsidered. Of the accepted refutations of a diurnal rotation Buridan accepted only one as cogent: that an object, if thrown upwards, would never return to its starting point, either in a straight line, or otherwise.[60] The other physical arguments were discarded as inconclusive. Nicole Oresme denied the absolute validity even of this argument: the descent of such an object, he argues, which can be seen as a compound movement of the

---

[58]Claudius Ptolemy, *Syntaxis Mathematica*, I, 7. The (*prima facie*) simplicity was stressed later also by Buridan in his reconsiderations of the arguments; see below n. 63.

[59]Proclus, *Hypothesis Astronomicarum Positionum* VII, pp. 50-58; cf. Duhem, *op. cit.* (above n. 8). Copernicus, as is well known, took great pride in creating an *Astronomia sine equante*; and this achievement promoted consideration of his theory (cf. R.S. Westman, "The Melanchthon Circle, Rheticus and the Wittenberg Interpretation of the Copernican Theory," *Isis*, 1975, *66*: 165-193). But the inaccuracies of the Copernican theory soon demanded the usage of every thinkable device; the equant was reintroduced; its arbitrary character was removed with Kepler's second law. Kepler achieved, from the viewpoint of the astronomical tradition, a *general theory of equants*.

[60]Johannes Buridan, *Questiones super libros Aristotelis de coelo et mundo,* ed. E.A. Moody (Cambridge, Mass., 1942), pp. 225-232; further commentary and translation, Clagett, *op. cit., Mechanics,* pp. 594-599.

air and the inclination downwards, is not impossible.[61] We remember that one of the gains of the mature theory of impetus was the insight into the complex nature of all movements, whether "natural" or forced.[62]

The key methodological term in both Buridan's analysis and its revision by Oresme is *persuasio*. Both of them do not seek absolute demonstrations, but rather sufficient certitude. The discussion of the nature of scientific evidence is indeed one of the most interesting discussions initiated by the nominalists.[63]They introduced most stringent criteria of demonstrative endeavors. Logical truths, self-awareness and immediate sense perception alone could claim to have evidence, whereby the last was open to doubt in view of the charge that God, *de potentia eius absoluta,* could produce in us the immediate notion of non-existing things (*notitia intuitiva de rebus non existentibus*).[64] Descartes' paradigmatic use of the *cogito* is not only reminiscent of Augustine's *De libero arbitrio*, but likewise of the scholastic discussions; and his *spiritus malignus* is a late offshoot of the question whether God can deceive our intuition,[65] with the sole, yet important, difference that for Ockham our intuitions establish a multiplicity of concrete, individual "things," while for Descartes they establish but two abstract "substances."[66] Already in the 14th century Ockham's position gave birth to a skeptical abolition of all scientific certitude: Nicolaus of Autrecourt allowed merely for

[61]Nicole Oresme, *Le livre du ciel et du monde,* ed. A.D. Memit and A.J. Denomy, *Medieval Studies,* 1941-43, *3-5*, esp. 1942, *4*: 270-79, transl. and commented by Clagett, *ibid.*, pp. 600-609.

[62]Above p. 187. n. 51.

[63]The following after A. Maier, "Das Problem der Evidenz," (above n. 46).

[64]For literature, see above, no. 46.

[65]Cf. Maier, p. 22 n. 47. The history of the *cogito* was written by L. Blanchet, *Les Antécédents historiques du "Je pense donc je suits"* (Paris, 1920). It is curious that R. Popkin, *The History of Scepticism from Erasmus to Descartes* (2nd ed., New York: Harper, 1964) traces the roots of early modern scepticism to antiquity, and interprets Descartes' use of sceptical arguments against the background of the new scepticism only—as if the rich tradition of discussions on the epistemological and critical implications of God's omnipotence in the later Middle Ages were negligible.

[66]Above, no. 47. While Gilson looked for Descartes' Thomistic sources, and while Koyré interpreted him largely as a Scotist, I suggest here that he may be interpreted as a Nominalist.

degrees of confirmation (*gradus probabilitatis*),[67] none of which are demonstrative.

Buridan developed his theory of evidence against Autrecourt. Probability, he calls to our attention, should not be confused with certainty; objective certainty again cannot be limited to those positions which evade skeptical reduction, since all statements including plain logical truths can be *subjectively* put to doubt. When he asked an old maid whether God in His omnipotence could violate the principle of contradiction she did not hesitate to answer in the affirmative. Objective certainty should not be confused with subjective credulity, nor is it confined to the truths of logic. Its criterion is the lack of an alternative reasonable argument, as against arguments from probability only.

It is worth while to view Buridan's arguments for and against a diurnal rotation of the earth in the light of his concept of demonstrative evidence. All but one argument turn out to be arguments from probability alone, and their opposites therefore become possible. Once Oresme had reduced even the argument from falling bodies to the level of probability alone, the preference for the geostatic point of view rested not on any particular argument, but on the overall "persuasion" of all arguments taken together. At least from Oresme onwards, the impossibility of an axial rotation of the earth became a mere improbability.

As against the continuous speculations on the physical possibility of an axial rotation, the heliocentric hypothesis was barely mentioned in the Middle Ages. Although likewise capable of "saving the phaenomena," it was seldom dignified even with hypothetical considerations—partly because the Aristarchan theory, even if considered on its mathematical merits alone, may have seemed to violate the principle of parsimony beyond tolerance,[68] and partly because the Aristarchan model simply was no articulated model at all, but only a suggestion. Now despite this silence, the physical arguments against the heliocentric

[67] J. Weinberg, *Nicolaus of Autrecourt* (Princeton: Princeton University Press, 1948).

[68] Inasmuch as it introduced an immense (if not, as Archimedes proves, infinite or incommensurable) distance between the Earth's orbit and the fixed stars. Cf. Archimedes, *The Sandreckoner,* ed. and trans. by T.L. Heath, *The Works of Archimedes* (New York: Dover, 1953), p. 221 f. No Greek author actually accuses Aristarchus of violating the principle of parsimony; but it is easy to imagine such arguments in analogy, say, to Aristotle's refutation of the "infinity of elements" (Anaxagoras) with the same principle.

system were discussed indeed under different headings. In part, the physical contentions against an orbital motion of the earth will coincide with contentions against the axial rotation. There existed a well-developed tradition of answers to the latter. In part, specific arguments were needed to keep the earth at the gravitational center of the universe, whether rotating or not. If the physical possibility of a heliocentric universe was not raised, however, the much more comprehensive possibility of a plurality of worlds (*pluralitas mundium*)—leaving our universe without an absolute center—was raised in the 14th century as a critical theological argument.

Whether the world is one or many is a question which Ockham,[69] among others, answered with reference to the divine omnipotence. A plurality of worlds is not a logical impossibility; and since the uniqueness of our universe is not a logical necessity, it cannot be a physical necessity either.[70] Aristotle's argument from the amount of matter[71] does not hold in view of God's power to increase the amount of matter *ex nihilo*. Matter, it should be remembered, is (against Aristotle) actual even as prime matter, though not necessarily quantified (i.e., extended) matter, and is always the matter of this or that singular: "materia prima mea non est materia prima tua."[72] Nor is proper place an absolute determination. Aristotle contended that, given two or more separate "worlds," a body inhabiting any one of them would not have one, but rather two natural places towards which to move. If it moves "naturally" towards one center of gravity 1_1 (or any other natural place), it evidently moves *eo ipso* away from the other, 1_2. Ockham meets this argument by pointing out that one could argue in the same way as an orthodox Aristotelian; namely that, within our world (even if assumed to be

[69]William of Ockham, *Sentent.* I dist. XLV, q.1 E-H. *Opera plurima* (Lyon, 1494-96; reprint London: Gregg Press, 1962), III. A brief review of the history and importance of the problem is given by H. Blumenberg, *Die Legitimität der Neuzeit* (Frankfurt am Main: Suhrkamp, 1966), pp. 113-124.

[70]Cf. above, pp. 179-185.

[71]*De caelo* A9, 277b27-279b4. The argument may be a later insertion of Aristotle; cf. L. Elders, *Aristotle's Cosmology: A Commentary on De caelo* (Assen, 1966), pp. 137-49.

[72]S. Moser, *Grundbegriffe der Naturphilosophie bei Wilhelm von Ockham* (Innsbruck: Verlag Felizian Rauch, 1932), p. 44 (*Summulae in libros physicorum* [Bononiae, 1494] I, p. 14).

unique), a body moving "naturally" upwards towards a point 1_1 within its natural environment (a periphery) can be said to move away from the opposite points therein. It is not even necessary to conceive a natural place—taken not as a reference to a point, but as a generic concept—as a continuum. "Natural places" cannot be understood as definite places, but are relative to begin with; if this is true in respect to the circumference, this *could* be true also in respect to the center. Natural places are relative (connotative) concepts referring to singular bodies and their material contiguous environment. Gravity could be understood as *actio in distans,* even though Ockham did not think of it in analogy to magnetic attraction.[73]

Neither the hypothetical considerations of the Terminists nor the speculative considerations of Cusanus could have served Copernicus in their entirety, since he postulated an absolute center as strongly as did the pre-Terministic natural philosophy. But once the decentralization of the universe, for Copernicus still an impossibility, was reconsidered, the concept of gravity as a force rather than a passive cause[74] could be reintroduced with better defenses than given by Averroes.

But was the motion of the earth, if no longer a physical impossibility to the medieval scientific understanding, an exegetical impossibility? For the sun in Gibeon stood still, and many more direct and indirect references to the immobility of the earth are scattered throughout the Scriptures. Yet the exegetical problems posed by the Copernican theory could be easily met (*historia ipsa docet*), as some of the problems posed by the Aristotelian cosmology before it, with an old exegetical and historiosophical device, the principle of *accommodation.* The ancient polemical question why Christ came so late (*quare non ante venit Christus*); the necessity of a "burdensome" law, good but for a time (*bonum in suo tempore*) to be abrogated with the new dispensation; the necessity of sacrifices and other anthropomorphic usages and images; the very fact that the sacred texts do not speak a philosophical language

[73]A. Maier, "Die Aristotelische Theorie," p. 157; *id.,* "Das Problem der Gravitation," *An der Grenze von Scholastik und Naturwissenschaft,* p. 166 f.

[74]Thomas Aquinas, *In libros Aristotelis de caelo et mundo expositio,* 3.Z lectio 7, *Opera* (Roma, 1952) III, p. 252 f.; against Averroes, *Physica* 8 summa 3.2 text 28, *Opera* V, pp. 91vb-92va; A. Maier, "Die Aristotelische Theories," 149 ff. and my remarks, *Viator,* 1973, 2: 337 f.

—these and other difficulties were explained as instances of the divine accommodation[75] to the degree of comprehension of the masses, or of the average persons (*simplices*) at the original time of revelation. Now as to discrepancies between the Bible and the ancient or medieval scientific cosmologies, the principle of accommodation allowed two different approaches. Although adapted to the comprehension of the *simplices*, the Scriptures could nevertheless contain, in a deeper sense, all the wisdom there is, which the allegorical interpretation could decipher. This maximal claim was naturally the more prevalent. At times, however, we find the admission that the Scriptures neither include nor imply the whole body of scientific knowledge. Ibn Ezra (d. ca. 1150) maintains in his explanation of *Gen.* 1, that "the heaven" and "the earth," the "wind" and the "light" are but the observable four elements, not the entire universe; and that the whole story of creation is the narrative of the creation of the objects immediately perceived[76] in proportion

[75]One of the best formulations of this doctrine is the famous passage in Augustine, *Epistulae* 138, zi, p. 5, ed. Goldbacher, *Corpus Scriptorum Ecclesiasticorum Latinorum* 44, 130: "Aptum fuit primis temporis sacrificium, quod praeceperat deus, nunc vero non ita est, aliud enim praecepit, quod huic tempori aptum esset, qui multo magis homo novit, quid cuique tempori accommodate adhibeatur, quid quanto impertiat, addat, detrahat, augeat minuatve immutabilis mutabilium sicut creator ita moderator, donec universi seaculi pulchritudo, cuius particulae sunt, quae suis quibusque temporibus aptae sunt, velut magnum carmen. . . excurrat." Cf. also *Adv. Judaeos* III, 4 Migne, PL 42, 53: "ut rerum signa suis quaeque temporibus conveniant." On this whole theme cf. the literature above p. 1, n. 1.

[76]To *Gen.* 1:1: "The heavens": with a definite article to indicate that he speaks of those [heavens] seen; "heaven" and "earth" are later interpreted as sublunar elements. To *Gen.* 1:2: "And void" (*vabohu*): ". . . For Moses did not speak about the world of coelestial bodies (*ha'olam haba*; this otherwise eschatological term is used here in an astronomical-literal sense), which is the world of angels (Intelligences), but only about the world of generation and corruption (*'olam hahavaja vehahashata*)." This interpretation he held throughout. We do not mean that he shied away from references to hidden meanings; but only that he did not understand *Gen.* 1, either literally or allegorically, to contain a comprehensive cosmology. It is easy to see, indeed, why his influence on Spinoza was so decisive: Spinoza's aim was likewise to find an immanent criterion for the permissible limits of allegorization (*Tractatus Theologico-politicus* VII, ed. Vloten [Haag: Martin Nijhoff, 1882] I, pp. 171-175). Throughout his commentary, ibn Ezra gives (or preserves even when disagreeing) enough examples of text-criticism to serve as a starting point for the exegetical revolution of the 17th century. Richard Simon praised him highly.

to the way in which they are perceived. That the moon is called a great luminary, while the bigger planets (except Mercury and Venus) are only called stars—this mode of speech corresponds merely to our point of view.[77] Further hermeneutical devices would not be needed to reconcile the Copernican theory with the Scriptures.

The medieval exegete thus enjoyed considerable freedom. And already Oresme had discarded the exegetical arguments against the motion of the earth almost *en passant* long before Copernicus or Galileo: they were the least disturbing parts of the hypothesis. He refers to the principle of accommodation, which allows the Scriptures "to speak the language of men."[78] How, then, can we account for the later hostility of the Church to the heliocentric universe? It had little to do with the Scriptures, but rather with the embarrassment at having allied Christian theology for so long with so wrong a theory—a most disturbing thought precisely because the heliocentric system was not any more difficult to comprehend than the Ptolemaic. "Accommodation" could hardly account for the cosmological errors of the complex and sophisticated theological literature of the Middle Ages.[79]

Was, then, the Copernican theory "the assertion of the absurd"? This was precisely Copernicus' claim, and the reason he gave for his long silence. If it was the common imagination he feared to insult, he was

[77]Ibid. to *Gen.* 1:16: "And should one ask: did not astronomers teach that all planets excepting Mercury and Venus are bigger than the moon, and how could it be written [in the Scriptures] "the big ones"? the answer is that the meaning of "big" is not in respect to bodily size but in respect to their light, and the moon's light is many times [stronger] because of its proximity to the earth."

[78]The hermeneutical principle used by Oresme (and later invoked by Galileo and even Bellarmine), "Scriptura humane loquitur," goes back to the rabbinic rule *dibra tora kileshon bne 'adam.* Originally, it belonged only to the field of jurisprudence; against R. Akiba's tendency to derive legal implications and inferences from every iota of the law, R. Jishmael insisted on legal formulae to be rhetorical figures only. A Saduceen origin of the principle is assumed (without definitive proof) by J.Z. Lauterbach, "The Saducees and Pharisees," *Rabbinic Essays* (Cincinnati: HUC Press, 1951), p. 31 f. and n. 11. Only later, in Jewish and Christian medieval exegesis, did the principle acquire its broad allegorical-philosophical use, as a comprehensive principle of accommodation. On Maimonides, e.g., cf. A. Funkenstein, "Gesetz und Geschichte: Zur Historisierenden Hermeneutik Bei Moses Maimonides und Thomas von Aquin," *Viator,* 1970, *1*: 147-163.

[79]Cf., on the whole problem, K. Scholder, *Ursprünge und Probleme der Bibelkritik im 17. Jahrhundert* (München: Chs. Kaiser, 1966), pp. 56-78 (Kopernikus und die Folgen).

certainly right. The immediate *picture* of the world certainly was affected by the Copernican revolution more than, in due time, by Galileo. Yet in the scientific terms of his time, he did not assert the impossible, but at best the improbable.

4. Reservations and Final Remarks

A serious objection to our insistence on the importance of absurdities in the transformation of theories can now be met. Are we not, in fact, merely saying what nobody doubts: that a new, revolutionary theory challenges the presuppositions of the theory it wishes to replace; of what heuristic use is such a *dictum de omni et de nullo*?

We ought to distinguish between conscious, well-argued presuppositions and hardly-conscious preconceptions. The latter may be exposed by a revolutionary theory; the former lead to it. The impossibility of an inertial motion or the impossibility of the Earth's movement belong to the first category: and so, to an extent, does the resolutive method. And the abundance of *well-defined* assumptions in both Greek mechanics and astronomy is a testimony to their maturity. Such assumptions were well argued: that is, their alternatives were considered up to a point, and rejected. This dialectical-argumentative character of Greek science, particularly of Aristotle, explains much of its fascination; and also its capacity to point beyond itself.

Yet both Aristotelian mechanics and Ptolomaic astronomy also contained assumptions of which they were hardly aware, or, if they were aware of them, did not think the alternative worthy of further argument. Such, for example, was the (never clearly articulated) assumption of the inability of matter by itself to resist motion. Bodies move faster or slower either because the force acting on them is stronger or weaker, or (if they are of the same size) because the media they are placed in are of different densities. How, then, should the medium resist motion in any different way than bodies moving through it? Unless we wish to enter into an infinite regress, this resistance should be attributed to some body *per se*. Aristotle never asked this question, because he took the factor of medium resistance to be immediately given to perception, a datum from which science proceeds. But it caused other difficulties. The ambiguity of the "medium" is nowhere clearer than where it both impedes and facilitates motion—at the same time!—as is the case of projectile movement.[80] Moreover: suppose we accept, formalizing Aristotle's dynamics, that $V_1 = F_1/R_1$. No matter how small F is, we always obtain V = 0; a flea could move a mountain. The Mertonians (Bradwardine), as is well known, found a mathematical solution to the problem:[81] $V_2 = \log_{(F_1/R_1)}(F_2/R_2)$. They did remove the overt absurdity, but not the ambiguity of the relations between force, medium and resistance. Contrary to the very concept of motion, where Aristotle envisaged, as an (impossible) alternative, the concept of a continuous movement in space (that is, without the resistance of the medium), he never seriously asked the corresponding question, what force is needed for such a hypothetical motion to begin. Or at times he indicates that bodies moved by different forces will move, in the imaginary space, with the same velocity. Not before the concepts of inertia, gravity and mass were clearly established could ambiguities be removed, and the Aristotelian dynamical concepts of matter be discussed as vague misconceptions. But this happened only as a result of the new mechanics, not in its beginnings.[82]

A similar case was the "obsession with circularity" (Koyré). Plato only articulated a common consensus, which was never seriously questioned during the career of ancient or medieval astronomy: that the perfect, regular motion of the planets can be imagined only as circular. Copernicus, far from reversing this assumption, believed rather to have strengthened it with his system. The only hint of a challenge came from

---

[80]Aristotle, *Physics* Δ8, 215a14-17; above p. 175. Cf. also M. Jammer, *Das Problem der Masse in der Physik* (Darmstadt: Wissenschaftliche Buchgesellschaft, 1964), pp. 16-30. Similar difficulties were present in the Aristotelian concept of gravity: Was it to be attributed to density or to size? Not even the medieval introduction of a *quantitas materiae* helped to mediate between both sources of gravity.

[81]A. Maier, "Der Funktionsbegriff in der Physik des 14. Jh." *Die Vorläufer Galileos im 14. Jh.* (2nd ed., Rome: Edizioni di Storia e Letteratura, 1966), pp. 87; 94-110 (difficulties and different solutions). Clagett, *op. cit., Mechanics,* p. 421 ff.; E.J. Dijksterjuis, *The Mechanization of the World Picture* (Oxford: Oxford University Press, 1961), pp. 188-193.

[82]On Galileo's difficulties with the correct estimation of resistance, see R.S. Westfall, *Force in Newton's Physics: The Science of Dynamics in the Seventeenth Century* (London: Macdonald, 1971), pp. 1-55, esp. 33 ff., 39 ff.

atomistic quarters, and its language reminds us that the abandonment of the circle as a pattern of perfect motion was, in a sense, the abandonment of an aesthetic ideal.[83] But these are almost less than hints of opposition. Inasmuch as the axiom of circularity was a precise one (unlike the precursors of the concept of mass), it could be reversed in one concentrated, even if most laborious, effort; inasmuch as it was never challenged before Kepler, and did not even enjoy the benefits of a hypothetical challenge (*de potentia dei absoluta*), it was the last to be reversed of the chief tenets of ancient astronomy.

I do hope that my remarks on the role of the assertion of the impossible in the transformation of theories will not be misconstrued as a strictly "dialectical" scheme of development. Were the history of scientific theories a fundamentally immanent, autonomous affair, the construction of such a scheme would have considerable use. Even Hegel's *Wissenschaft der Logik* has recently been interpreted as a methodology of the sciences of some merit.[84] But the history of science does not progress merely through acts of self-reflection towards greater differentiation and exactitude. And although I did not mention many of the external circumstances leading towards the conception and reception of either the Copernican or the Galilean theories, I assume their role to be

[83]"Admirabor eorum tarditatem," Cicero lets his Epicurean dismiss the *anima mundi,* "qui animantem immortalem et eundum beatum rotundum esse velint, quod ea forma neget ullam esse pulchriorem Plato: at mihi vel cylindri vel quadrati vel coni vel pyramidis vidatur esse formosior." Cicero, *De natura deorum* (ed. Plassberg) I, p. 10; cf. II, p. 67: "conum tibi ais et cylindrum et pyramidem pulchriorem quam sphaera videri. Novum etiam oculorum iudicium (!) habetis. . . ;" astronomical arguments follow. Another source of vague objections to the axiom of circularity was the contention of Nicolaus Cusanus that no actual body has the perfection of a geometrical figure, but at best approximates it. This contention is of interest in view of Kepler's explanation later for the elliptic orbits, with a similar argument.

[84]D. Henrich, *Hegel im Kontext* (Frankfurt am Main: Suhrkamp, 1972), esp. pp. 95-156 (Hegels Logik der Reflexion). Less exact, R. Bubner, *Dialektik und Wissenschaft* (Frankfurt am Main: Suhrkamp, 1973), esp. 129-174, esp. 139 ff. Except for my insistence on external, non-immanent causes in the formation of theories, there is a fundamental difference between Bubner's (or similar) usage of a dialectical logic of discovery, which emphasizes the immanent elaboration of contradictions within a theory, and my basic question as to how clearly can a theory formulate, in its conceptual horizon, its opposite theory. That is, I contend that the function of dynamical negation is historically a limited one.

of no less importance than the capability of original theories to define the cornerstone of the theory which later succeeds them.

Such external circumstances—either of the new astronomy or of the new mechanics, or of both—were the fierce humanistic critique of the scholastic administration of knowledge; speculative philosophies of nature in the Renaissance, both with its sensualistic epistemology (which long before Galileo emphasized the distinction between primary and secondary qualities) and its sun-oriented, reductionist cosmology; the growing practical needs for exact methods of measurement, not least in ballistics; the growing awareness, already in the Middle Ages, of the insufficiency of the Aristotelian doctrine, and the growing distrust, since the 14th century, in universal generalizations without basis in "experience." Even shifts in esthetic ideals played a modest role. I see some correspondence between the foundation of universal harmony on elliptical orbits and the predilection for elliptic forms in Baroque architecture. In both cases harmony, still defined as "unity within multiplicity" ceased to be static, that is, expressed in *one* ideal geometrical form, and became, so to speak, dynamical. What was previously regarded as a deviation from the ideal form was comprehended for the first time as an integral part of the form or of a form of forms which are capable of endless transitions one into another; the actual varieties becoming "aspects" of one principle, its "points of view."[85]

For a century and a half, a rapid succession of "new sciences," "new methods," "new observation," whether alleged or genuine, was the principal trademark of intellectual endeavors. The age between the end of the sixteenth and the beginning of the eighteenth century saw itself as utterly innovative and "critical," and painted a picture of Aristotelian and Scholastic science to serve as a dark background against its emanci-

[85]H. Wölflin, *Renaissance und Barock* (7th ed., Basel-Stuttgart: Schwabe et Co., 1968), esp. pp. 45-52. More precisely, both the aesthetic and scientific upgrading of the ellipse may be linked to the emergence of projective geometry (out of the occupation with perspective). Mathematically, it taught (a) that the circle is but a special case of the ellipse and (b) that the ellipse can be generated from the circle and vice versa through shifts in "points of view"; aesthetically, the "harmony" became a dynamic harmony, in which one "form" leads to and generates another within a work of art; scientifically, the universe became also a continuum of forms (rather than an assembly of discrete forms).

patory achievements.[86] The courage to assert the impossible became a fashion. But historians of science should avoid ever-renewed paraphrases of the *querelle des anciens et de modernes.* The radical contraposition of the "old cosmos" with the "new universe" in the style of Cassirer or Koyre is as exaggerated as Duhem's hunt for the medieval precursors of Galileo (preferably in Paris), while Annelise Maier was seldom tempted to draw the outline of a characteristic pattern from her cautious, rich, differentiated picture of specific medieval problems and achievements.[87] Our discussion has attempted to draw attention to *one* such characteristic pattern only: the fact that Greek science, and even more so, its scholastic reformulation, defined with growing clarity the starting point and the method for its own overthrow.

Whether or not a mediation of this kind between successive theories may be seen in the recent history of scientific theories, I do not know. For it is not sufficient to say that the postulation of a constant velocity of light, irrespective of the velocity of its source; or the relativization of the absolute distinction between uniform movement and acceleration; or the abolition of the *lex continui,* or again the assertion of an absolute limit on the determination of events through exact measurement—that these and other unexpected new beginnings were impossibilities in the terms of classical mechanics. An a-historical consideration of classical mechanics will certainly establish its *capability* to point out, with preci-

[86]On the role of this contraposition cf. also S. Toulmin, *Foresight and Understanding. An Enquiry into the Aims of Science* (London: Hutchinson, 1961), Ch. VI. Only at the end of a scientific transition can all the preconceptions of the replaced system be formulated with clarity.

[87]But see her summary, "An der Schwelle der exakten Naturwissenschaft," *Metaphysische Hintergründe der scholastischen Naturphilosophie* (Rome: Edizioni di Storia e Letteratura, 1955), pp. 339-402, esp. p. 398 ff. The deepest reason why later scholasticism could not immerse itself into the context of problems of an exact natural science is, she argues, its belief "dass ein wirklich genaues Messen, auch schon in den einfachsten Fällen, prinzipiell unmöglich ist." And she concludes: "So sind sie an der Schwelle einer eigentlichen, messenden Physik stehengeblieben, ohne sie zu überschreiten—letzten Endes, weil sie sich nicht zu dem Verzicht auf Exaktheit entschliessen konnten, der allein die exakte Naturwissenschaft möglich macht." But this denial of a possibility of measurement has deeper roots. Measurement entails theoretical isolation and the acceptance of limiting cases, and with it, as we have tried to argue, the constructive rather than critical function of "possible worlds" other than ours.

sion, what it must have regarded as impossibilities; just as it was capable of formulating instances of falsification, that is, to construe the tools for the observation of "facts" disagreeing with the theory. A historian, however, has to prove more: namely that such capabilities were, at times, actualized, either in disciplined hypothetical speculations, or in axiological inquiries. For "that which is not" is not easily asserted, or even "thought of."

COMMENTARY: DIALECTICAL ASPECTS OF THE COPERNICAN REVOLUTION: CONCEPTUAL ELUCIDATIONS AND HISTORIOGRAPHICAL PROBLEMS

Maurice A. Finocchiaro

University of Nevada

Two of the most crucial ideas in the Copernican revolution were the motion of the earth and inertial motion. Copernicus himself primarily elaborated the former idea, which may be conceived here as referring to the earth's daily axial rotation and to its annual orbital revolution around the sun. On the other hand, it was Galileo who began to elaborate the idea of inertial motion, which may be conceived here as motion which continues uniformly in the absence of external forces or impediments.

Professor Funkenstein is concerned with tracing what may be called the "dialectical prehistory" of each of these ideas. That is, both ideas underwent a great deal of development prior to Copernicus and Galileo but did so as absurdities; hence, a great deal of the merit and originality of these two thinkers lay in their intellectual and moral courage to assert the supposedly absurd. In other words, since Antiquity those ideas had been the subject of a great amount of systematic hypothetical reasoning which resulted in their achieving the status of more or less well-refuted options.

It seems to me that for the case of the earth's motion, although the main factual features of its dialectical prehistory have been rather well-known, to conceive those facts as such a dialectical prehistory sheds new light on them. On the other hand, for the idea of inertial motion, the factual claims as well as the dialectical interpretation of its prehistory represent, in my opinion, novel insights.

According to Professor Funkenstein, the motion of the earth, which was an impossibility in ancient astronomy, came to be regarded as an improbability during the Middle Ages, and such it was at Copernicus's time. He is careful to add that this was true as far as the scientific intellect was concerned, whereas from the point of view of the scientific imagination the motion of the earth remained as much of an impossibility as ever. On the other hand, from the point of view of the religious sensibility and Biblical exegesis, the situation was more complex. Professor Funkenstein claims that, although the earth's motion was typically regarded as an exegetical impossibility, it really was not since an "accommodation" would have been theoretically possible.

The dialectical prehistory of the idea of inertial motion is more complicated. To begin with, one has to realize that there is a very important difference between it and the idea of the earth's motion. If we moderns have accepted both ideas, we accept them in different ways. Although both phenomena are unobservable, the motion of the earth *could* be observed from other vantage points in the solar system, for example from the sun, or by a God looking at the solar system as a whole. But inertial motion, which the law of inertia tells us is the natural motion of bodies, could not be observed even by a God for the simple reason that it is not actualized anywhere, or worse, given the law of universal gravitation, it cannot in principle be actualized anywhere. What a God could observe would be better approximations to inertial motion than we can, but only more or less good approximations to inertial motion can be observed. In other words, though both terrestrial and inertial motion are facts, they are different kinds of facts: the earth's motion is a fact in the world; inertial motion a fact in the mind of the scientist, so to speak; the former is a real fact, the latter a "counterfactual fact"; inertial motion is a limiting but ideal case of actual motions subject to ever diminishing impediments.

This counterfactual character of inertial motion is very important in Professor Funkenstein's account because his dialectical prehistory of the idea involves an examination of the history of its impossibility, refutations, and rejections. The importance—one might say the paradox—is that inertial motion, even nowadays when it is "accepted," is regarded as a physical impossibility, strictly speaking.

Professor Funkenstein thinks he can find an articulated rejection of inertial motion in Aristotle's argument against motion in the void. Presumably Aristotle in some sense saw the counterfactual character of inertial motion or motion in the void. But for him this meant that such motion was "incommensurable and hence incompatible" with actual motion and hence impossible, in a mixed physical-conceptual sense. On the other hand, in the Middle Ages theological sophistication led to the distinction between what God actually created and what He might have created; hence they could distinguish clearly between logical and physical impossibilities. And so, although we still find them regarding inertial motion (or motion in the void) as incommensurable with actual motion, Thomas Aquinas regards it as not impossible, while in the 14th century it was regarded as irrelevant but indeed possible. In any case, the result was that for all of them a principle of such motion had no place in a science of motion and in that sense was to be rejected. And this makes clear why a philosophical-methodological revolution was required for Galileo; he had to show that what was previously regarded as a reason for the unacceptability of inertial motion, could be regarded, and should be regarded, as a reason for adopting the idea.

I believe that such an account can be carried further and its relevance to the Copernican revolution increased by examining the connection between the above-mentioned two elements of the Copernican revolution: the motion of the earth and inertial motion. In fact, as long as we treat the two separately, as Professor Funkenstein does, I do not see how the two prehistories could have had any role in the emergence of Copernican astronomy and Galilean mechanics. For what he examines is two supposedly well-refuted options which are then adopted and prevail. It would not suffice to say that a study of the biography of Copernicus and Galileo would reveal, at least in part, how and why they chose the rejected options, and that a study of more nearly external types of factors would reveal how and why they became generally accepted. This would not suffice because what I am concerned with here is the relevance of the dialectical prehistory of our two ideas; that is I am trying to determine in what sense if any the Copernican revolution has a dialectical origin, namely in what sense its origin could lie with the history of the rejections of those two ideas.

Let me tackle this problem from a different angle. Suppose Professor Funkenstein were saying that the Copernican revolution emerged partly because (a) the motion of the earth came to be regarded as less and less of an absurdity, and party because (b) inertial motion came to be regarded as less and less impossible. Then most of his account could be interpreted as an attempt to justify (a) and (b). At the same time one could say that such a gradual reduction of the absurdity of those ideas involved all along the disciplined use of hypothetical reasoning. But hypothetical reasoning is just another name for what others call "thought-experiments." Now it is well agreed that thought-experiments are an important element of the method of modern science. Therefore, we would be entitled to assert the existence of a methodological continuity between modern science and medieval natural philosophy and even Aristotle.

If Professor Funkenstein were saying that, it might very well be undeniable, but it would not be very exciting. At least not to me, and at least by comparison with a more "dialectical" thesis. That is, if he is taking more seriously the possibility of the emergence of an idea from its opposite, or at least if we want to explore this possibility, then we would have to entertain the following considerations.

My idea is that impossibilities and absurdities may turn into believable alternatives when they are mutually supporting. One may be perfectly satisfied with rejecting two absurdities if there is no connection between them. But if two alleged absurdities cohere with one another, in the sense that one requires the other, then both together become more believable. In other words, in combining absurdities, which is not the same as merely piling them up, the combination is likely to be less absurd than each separately and may point toward a genuinely workable alternative.

The main argument on the strength of which the motion of the earth had been regarded as impossible or improbable was based on the behavior of falling bodies. It was said that if the earth moved (and here we may think primarily of its axial rotation), then an object thrown upwards would not fall back to the place of ejection but some distance west of it, since while the object is in mid-air the earth's surface underneath would be moving from west to east, thus leaving the object behind. Another way of stating the argument was the following. When bodies are let fall freely they fall vertically, a fact that can be easily verified by going on top of a tower, dropping a rock, and seeing it fall along the edge of the tower; and this vertical fall could not occur if the earth moved because if the earth moved a rock dropped from the top of a tower would fall not at the foot of the tower, but some distance to the west, due to the eastward motion of the tower carried by the axial rotation of the earth.

Such arguments, as Galileo perceived, presuppose that the natural state of a body is one of rest and that whatever moves requires a cause to merely keep moving. This presupposition is, of course, a basic principle of the Aristotelian theory of motion. This principle, in turn, is a denial of inertial motion. And that is how the alleged impossibility of the earth's motion is based on the alleged impossibility of inertial motion. But such a connection between these two impossibilities amounts to a connection between two possibilities. In other words, if the denial of inertial motion implies the denial of the earth's motion, then the acceptance of earth's motion would necessitate acceptance of inertial motion. Thus a world in which the earth moved would also be a world in which natural motion was inertial. If it turns out that one has to accept the void if one accepts inertial motion, then one accepts that also. And in this fashion he proceeds to build a new world, or at least a new conception of the world, which may be said to have emerged dialectically from the old conception.

So far I have elaborated what I regard as Professor Funkenstein's most interesting philosophical or interpretative idea. Now I wish to examine what I regard as his most striking factual thesis; namely that Aristotle anticipated the content of the principle of inertia in his argument about the absurdity of the void. This means that in rejecting motion in the void Aristotle was rejecting inertial motion.

The first problem here relates to the question whether this thesis has any historical content. The question is, did the historical agents who adopted the principle of inertia see themselves as accepting something that Aristotle and his medieval followers had argumentatively rejected? I believe it is clear, even from Professor Funkenstein's own evidence, that such is not the case. It is he who states that in the view of 17th century natural philosophers the inertial principle constituted a resolute attempt to break with the past. Certainly, the situation here contrasts sharply with the attitude toward the Copernican idea of a moving earth.

The historicity or nonhistoricity of such claims is important in two ways. First, if the thesis is nonhistorical, that means the anticipation was not perceived by the historical agents, and if so then we are unlikely to have those peculiar historical events called "revolutions." In fact, we may share Professor Funkenstein's agreement with Tocqueville's view of revolutions as a conscious and resolute attempt to break with the past. The same conclusion is supported by the revolutionary fervor which one finds in Galileo but which Copernicus lacked.

Second, if the Aristotelian anticipation does not have historical content, that is going to create difficulties for Professor Funkenstein's dialectical prehistory. For the history of the rejections of motion in the void can have no effect on the emergence of the idea of inertial motion if the people who first formulated the latter did not see that in essence it had been considered and rejected by those who argued against motion in the void.

I might mention here that a very relevant piece of evidence is the passage in Galileo's *Two New Sciences*[1] where he examines Aristotle's argument. Galileo himself was aware of the importance of his view of that argument, for in the book's index entitled "Table of the More Notable Topics" he has an entry which reads: "Aristotle's argument against the vacuum is *ad hominem,* p. ---."[2] Unfortunately, his view supports my contention that he was unable to read into Aristotle's argument any anticipation of inertial motion. By an *ad hominem* argument Galileo and his contemporary did not mean the *ad hominem* fallacy of contemporary logicians but rather an argument in which one derives a conclusion unacceptable to an opponent (but usually accepted by the arguer) from premises which are accepted by the opponent (but not necessarily by the arguer). Galileo's view is that Aristotle's argument is *ad hominem* against those who thought the vacuum a prerequisite for motion, presumably the atomists. This means that for Galileo, what Aristotle is doing in that argument is speaking to the atomists and saying: Since there is such a thing as motion, given your view that motion without a vacuum is impossible, since it can be shown that

[1]Galileo Galilei, *Dialogues Concerning Two New Sciences,* trans. H. Crew and A. De Salvio (New York: Dover, n.d.), pp. 61-62; and G. Galilei, *Opere* (Ed. A. Favaro, 20 vols.; Florence: Barbera, 1890-1909), vol. 8, pp. 105-106.

[2]Opere, vol. 8, p. 314.

motion *in* a vacuum is impossible (because it would have to be instantaneous), we may conclude that a vacuum is an impossibility.

Hence, if Professor Funkenstein's anticipation claim is to have any validity, it is in the theoretical domain. As a logician, I would find most intriguing and interesting the examination of whether Aristotle's rejection of motion in the void can be interpreted to be theoretically equivalent to a rejection of inertial motion. However, the relevance of such a theoretical equivalence to the historical origin of the Copernican revolution seems slight. Hence in the present context I feel little need to pursue the question of the extent of that equivalence.

The same problem, among others, may arise with another one of Professor Funkenstein's claims. At the beginning of his discussion dealing with " 'impossibilities' in Aristotelian and Medieval mechanics," he points out that the revolutionary character of the content of the inertial principle consisted in "the abolition of the absolute distinction between movement and rest for the sake of the absolute distinction between movement and acceleration" (p. 171). After this statement, he adds: "And we deem the method by which this principle was elicited to be no less revolutionary (p. 171)." Then he points out the counterfactual character of the inertial principle.

In distinguishing between the content and the method of the inertial principle, Professor Funkenstein seems to be making a valid distinction. However, it is not clear what distinction he is exactly making. He could be distinguishing between what the principle says and how it was discovered; or he could be distinguishing between the content and the logical form of the principle. His use of the term "elicited," when he speaks of "the method by which it is elicited," compounds the confusion. In fact, the method of elicitation could refer either to the manner of stating the principle, or to the procedure for justifying it, or to the way in which it was discovered. Another ambiguity—that between historical content and theoretical content—is apparently avoided since Professor Funkenstein says that it is *we* who deem the method to be no less revolutionary than the content and since he substantiates his claim by references to Hegel, Cassirer, Vahinger, Rescher, and Quine. But I am afraid that this apparent explicitness generates an actual ambiguity because in our context one would have expected the claim to be that it was the historical agents who thought that the method was revolutionary as well.

This last ambiguity—between the historical and the theoretical—can perhaps be resolved as follows. To say that we the historians deem revolutionary the method of eliciting the inertial principle is to say that the historical agents were aware that this method represented a resolute break with the past. But then what becomes questionable is the propriety of Professor Funkenstein's sketch of a theoretical analysis of that method and references to Hegel, Cassirer, etc. What one would have to give instead is historical evidence.

It may be said at this point that the existence of such historical evidence about a methodological break with the past is well-known or at least easily forthcoming; therefore there may not be any real problem with the impropriety of Professor Funkenstein's evidence.

I believe that the problem would remain for the following reasons. It is indeed true that the inertial principle involved a conscious methodological break with the past. But it is something different to say that the conscious methodological break consisted in the counterfactual character of the principle. To say the latter may very well be both false and inconsistent with Professor Funkenstein's general approach. Inconsistent because he emphasized the methodological continuity in the Copernican revolution and does so precisely as regards the technique of hypothetical reasoning. The inconsistency could of course be avoided by allowing the counterfactual method continuity from the theoretical point of view and disclaiming it at the historical level. But then the contextual relevance of the continuity would become questionable.

At any rate I am doubtful that the formulation of the inertial principle was perceived to be methodologically revolutionary *with respect to its counterfactual character*. My evidence would be taken from the way that Galileo arrived at the inertial principle.[3] He did so by considering that undisturbed and downward motion undergoes acceleration whereas undisturbed upward motion undergoes deceleration, and concluding that undisturbed horizontal motion, being neither upward nor

[3]G. Galilei, *Dialogue Concerning the Two Chief World Systems,* trans. S. Drake (Berkeley: University of California Press, 1953 and later reprints) pp. 145-48.

downward, should undergo neither acceleration nor deceleration, which is to say should continue uniformly. Such evidence could of course be fashionably criticized by claiming that Galileo had the wrong principle of inertia and this is why he was not particularly aware of its counterfactual character. But to take seriously this fashion, which I do not, would only make the evidence from Galileo irrelevant to the present discussion and thus would still create a problem for Professor Funkenstein's claim.

I wish to end with a more positive note. I found quite enlightening his discussion of the theologically oriented medieval interpretation of Aristotelian mechanics. After reading Professor Funkenstein it is difficult not to appreciate the value of those speculations.

VII

ANDREAS OSIANDER'S CONTRIBUTION TO THE COPERNICAN ACHIEVEMENT

Bruce Wrightsman

Luther College, Decorah, Iowa

1. Introduction

Historically, one of the most interesting problems concerning the achievement of Copernicus is created by the appearance with the printed work, *De revolutionibus,* of a letter written by the editor, Andreas Osiander, entitled, "Ad lectorem de hypothesibus huius operis."[1] The prefixing of this letter which expresses views on the aim and nature of scientific theories at variance with Copernicus' claims for his own theory, has aroused considerable controversy during the four hundred and fifty years since its appearance. Questions have been raised not only about the propriety of and motivations for Osiander's

[1]The term "preface," widely employed to designate Osiander's letter, is a confusing misnomer, since the word "praefatio" nowhere occurs in it. Also it confuses Osiander's letter with Copernicus' own preface, which is the Dedicatory letter to Pope Paul. The continued use of this term by historians perpetuates the error that Osiander *intended* to deceive the reader into believing that it was written by Copernicus, disclaiming his belief in his own theory. But Copernicus calls his Dedicatory letter "Praefatio authoris," and this legend is carefully preserved on each page of the printed work on which it appears (by Osiander's supervision). This makes it demonstrably clear that Osiander differentiated between his writing and that of Copernicus. One might adopt Pierre Gassendi's term "Praefatiunculam" to designate Osiander's letter. But this preserves the suggestion that Osiander *intended* his letter to be taken as some kind of preface. Accordingly, I shall hereafter refer to Osiander's writing as "Ad Lectorem," and to Copernicus's "praefatio" as "the Dedicatory letter."

actions but about his views concerning the purpose and status of astronomical theories.

In many recent publications, historians and philosophers of science have dealt extensively with the latter issue, as it relates to the achievement of Copernicus. But the contribution of the author of "Ad Lectorem" to that issue has been passed over as insignificant. As is most often the case, Osiander has been dismissed as an obstructionist theologian and an "enemy of science" whose actions were unethical, whose opinions were unsound and whose contributions to Copernicus (and to science) were negative.

Osiander cannot thus be said to have suffered from a lack of attention; indeed, it can safely be asserted that no other editor in scientific history has drawn more emotional lightning and less critical analysis from historians and philosophers.[2]

With few exceptions, his views and actions have been judged reprehensible and condemned, while his personality and character have been maligned and his motivations distorted. But he has rarely been studied with the seriousness he deserves so that his possible real and positive contributions to the growth of science and to the achievement of Copernicus can become known and appreciated.

Considering the historical importance of his role as editor of *De revolutionibus* and the philosophical importance of the issue he raises in "Ad Lectorem" — the enduring problem of the aim and status of scientific theories — it is strange that his career and writings have remained so long neglected by scientific historians.[3] My own studies of Osiander admittedly show that he made no direct technical contribution to the formation, correction or validation of the Copernican theory itself. His

[2]I have dealt extensively with the irrational character of this criticism in the historiographical tradition in my unpublished doctoral dissertation: *Andreas Osiander and Lutheran Contributions to the Scientific Revolution,* University of Wisconsin, Madison (Copyright, Ann Arbor, Michigan, 1970), and in a forthcoming paper: "Andreas Osiander: Friend or Foe of Copernicus?" This work shows that the absence of critical examination is responsible for the ignorance, but not the defamation. In the absence of knowledge, however, prejudice flourishes.

[3]That this issue continues to dominate philosophical discussion in science may be seen in recent works devoted to the problem. For recent discussion, see S. Morgenbesser, "The Realist-Instrumentalist Controversy" in, *Philosophy, Science and Method,* ed. S. Morgen-

material contribution is represented only by his editorship of *De revolutionibus* and his authorship of "Ad Lectorem." Despite the fact that Osiander was not a scientist as such nor, to our knowledge, the author of a scientific work, he nevertheless merits attention. When the question is raised about the effect of "Ad Lectorem" on the reception of Copernicus' work during the sixteenth century, and when appropriate consideration is given to the high esteem in which he was held by the ranking scientists of the time, and moreover, when we consider the character and importance of the issue he raises, it is clear that Osiander not only demands serious scholarly attention but that he deserves to be regarded as something more than an obstructionist.

Considerations such as these prompt a series of specific questions that invite our consideration: If not just an obsequious, meddling churchman, then who was he? What was his interest in scientific activity? What did he do to promote it? Why did he write "Ad Lectorem" and insert it in *De revolutionibus*? What was the basis for the views he expresses in this letter? What was the actual *effect* of the letter on the reception and use of Copernicus' works? Perhaps, by answering some of these questions we may come to a better understanding of Osiander's actual, rather than alleged, contributions to Copernicus and to the growth of science. Accordingly, this paper will briefly present some of the relevant facts concerning Osiander's life, career and thought derived from his own writings and other sixteenth-century sources, to establish a proper basis for assessing the views contained in "Ad Lectorem," and his motivations for presenting them as he did. Finally, the probable effect of this action on the reception of the Copernican work will be examined.

2. *The Life and Career of Osiander (1498-1552)*

Osiander's life and his career as a Lutheran reformer and preacher spanned the crucial years during the reformation of Nürnberg, where he lived and worked for twenty-nine years. Until recently, very little was known about this period of his life because church historians concen-

besser, et. al., (New York: St. Martin's Press, 1969) pp. 200-218. That Osiander's writing is relevant to this discussion is seen from the fact that Kepler specifically cites it while attempting to demonstrate the reality and validity of Copernicus' theory (*Joannis Kepleri astronomi opera,* Christian Frisch ed., [Frankfurt a. M. and Erlangen: 1858]) p. 245.

trated on the last three years before his death when he lived in Königsberg, East Prussia and where he became embroiled in a fierce theological controversy.[4] Thus, partly due to his subsequently being declared a heretic by his opponents (who defined the church's theological position in a manner inimical to Osiander for over three hundred years), church historians have tended to neglect the long span of his activities in Nürnberg, where he made important contributions to the reformation of Germany.

During the nineteenth century, three critical studies of Osiander's life and work were published by church historians who uncovered some of the sources concerning Osiander's career, but did little to remedy the defect in earlier scholarly neglect.[5] In the twentieth century, two full-length studies have appeared dealing extensively with his life, thought and writings. The first of these, Emmanuel Hirsch's, *Die Theologie des Andreas Osiander und ihre geschichtlichen Veraussetzungen*[6] is the only study to date to systematically analyze the structure and intellectual sources of Osiander's thought. More recently, Gottfried Seebass has published his, *Das reformatorische Werk des Andreas Osiander* (Nürnberg: Einzelarbeiten aus der Kirchengeschichte Bayerns 1967, *44*), a critical study of the sources for Osiandrian research which does much to remedy the neglect by earlier historians of his Nürnberg career. Neither of these modern works, however, gives sufficient attention to Osiander's non-theological, scholarly and scientific interests, dealing with them only very briefly and treating them as "hobbies." Hence, even the best modern studies continue the pattern of neglecting Osiander's scientific

[4]The Osiandrian position has never been very popular in the Lutheran churches. For a discussion of the controversy see, A. Ritschl, *A Critical History of the Christian Doctrine of Justification and Reconciliation,* trans. J.S. Black (Edinburgh: 1872), p. 215 ff; H.J. Grimm, *The Reformation Era, 1500-1650* (New York: Macmillan, 1954) and G. Aulen, *Christus Victor* (New York: Macmillan, 1960).

[5]C.L. Lehnerdt, *De Andrea Osiandri* (Edinburg: 1837); C.H. Wilken, *Andreas Osianders Leben, Lehre und Schriften* (Stralsund: 1844); and Wilhelm Moller, *Andreas Osiander, Leben und ausgewählte Schriften der Vater und Begründer der Lutherischen Kirche* (Elberfeld: 1870). The source most historians of science rely on for their meager references to Osiander's career is the brief summary by Möller in Vol. 24 of the *Allgemeine Deutsche Biographie* (*ADB*) (Leipzig: Verlag Von Trinder and Humbolt, 1887), pp. 473-483.

[6]Göttingen: Vanderhoek and Ruprecht, 1919.

activities and fail to improve our understanding of his contributions to science. This paper is a modest attempt to correct that neglect.[7]

On the basis of the sources listed in Seebass and elsewhere, very little can be known of Osiander's childhood and early education before he matriculated at the University of Ingolstadt in 1515. There, he rejected the traditional scholastic curriculum and studied biblical languages and mathematics until 1519-20 under humanist scholars such as Johannes Boschenstein (1472-1542) and Johannes Reuchlin (1455-1522).[8] By them, he was not only trained in humanistic exegetical techniques but became introduced to other non-biblical Jewish literature and to the writings of Renaissance philosophers, all of which had an enormous influence in shaping his philosophical and theological views. Also during this period, which followed the publication of Luther's 95 Theses and his famous Leipzig debate with Johannes Eck, the Rector of Ingolstadt, Osiander became converted from humanistic studies to the study of theology.

In 1519, as the Reformation began to spread rapidly throughout Germany, the young Osiander was called to be a teacher of Hebrew at the Augustinian cloister in Nürnberg, around which had formed early in the Reformation an important circle of liberal, reform-minded intellectuals, patricians and political leaders led by Lazarus Spengler, the secretary of the City Council.[9] After the publication of Luther's celebrated theses in 1517 and the subsequent Edict of Worms, the attention of this circle shifted from literary to religious and political concerns, and they became active in printing and circulating Luther's early, revolutionary treatises all over Europe. Growing reformist attitudes in

---

[7]In the absence of critical studies, later commentaries rely on earlier ones and the same few primary sources (including Kepler's report of portions of Osiander's correspondence with Copernicus and Rheticus on April 20, 1541), as well as on masses of secondary literature emanating from non-critical sources.

[8]On Boschenstein, see *ADB, 3*: 184.

[9]On the activities of this circle, see G. Strauss, *Nürnberg in the 16th Century* (New York: John Wiley and Sons, 1966) and A. Englehardt, *Die Reformation in Nürnberg* (Nürnberg: Mitteilungen des Vereins für Geschichte der Stadt Nürnberg, 1932), pp. 25-31. During this period, his humanistic training also bore fruit in the publication of a Latin revision of the Vulgate version of the Old Testament, corrected by the Hebrew text (1522).

Nürnberg led the City Council to staff the city's major churches and religious houses with reform-minded scholars like Osiander. His arrival in Nürnberg thus placed him in close contact with those leaders who would decide the religious and political future of the city in the years to come. In that process, Osiander played a major part. Because of his uncompromising anti-papalist position, he quickly attained fame and influence. In 1522, he was called to fill the pulpit in one of the city's two most prominent churches (St. Lorenz), from which strongly pro-reform and increasingly anti-Roman messages thundered for twenty-seven years. Thereafter, his dominant concern was religion. And for all his wide and diverse intellectual interests, everything became pressed into the service of reform.

During the years when Germany's religious and political situation was being decided, Osiander played a prominent part in shaping the city's future along Reformation lines. And since the city was the seat of several important Imperial Diets during the crucial years following Worms, when the Holy Roman Emperor, Charles V, attempted to cope with the growing religious revolt, Osiander exerted considerable influence on the deliberations of the Electors and Counsellors through sermons and polemical writings and during political negotiations at Diets and religious conferences. He was a frequent participant in the important early meetings called to resolve the religious issue, notably at the Marburg Colloquy (1529) and at the Diet of Augsburg (1530), when confessional articles were formulated setting out the reformation position.

In the process of reforming Nürnberg, Osiander quickly gained such notoriety as an articulate, zealous reformer and a militant anti-Romanist that he won the approval and friendship of other political leaders such as Albrecht, Margrave of Brandenburg-Ansbach (later Duke of Prussia) who resided in Nürnberg during the 1522-23 Diets.[10] Also he became a close friend of Thomas Cranmer, who was soon to be elevated by Henry VIII of England to be Archbishop of Canterbury,

[10]Duke Albrecht of Prussia's connections with Copernicus and with Rheticus are better known than his relationship with Osiander, whom he called his "Spiritual Father." On their relationship, see W. Hubatsch, *Albrecht von Brandenburg-Ansbach, Geschischte Prussiens,* 1960, *8*.

and who lived with Osiander during a protracted visit to Europe to consult theologians (and Osiander) about the advancement of the King's divorce suit. While living there, he encouraged Osiander in his scholarly studies and married Osiander's niece.[11]

But while acquiring important contacts and friends, Osiander also aroused, by his combativeness, a formidable array of powerful and hostile theological and political enemies, ranging from Catholics like Johannes Eck, the papal Legate Compeggio, and the Archduke Ferdinand to the more radical left-wing reformers surrounding Zwingli, Bucer, Carlstadt and Munser.[12] Later, the followers of John Calvin joined the ranks of his enemies and still later, even the followers of Melanchthon. The latter were provoked to hostility partly because of Osiander's temperament; he was widely known in Protestant circles for his theological and intellectual independence and for his intransigence on doctrinal matters, so that even the reformers around Luther and Melanchthon began to avoid him and to exclude him from more delicate political and religious negotiations.[13]

During the years following the Peace of Nürnberg (1532) (which Charles was forced to accept because of political diversions elsewhere), conflict between Osiander and his opponents on all sides not only

[11]On Thomas Cranmer, see N.S. Tjernagel, *Henry VIII and the Lutherans: A Study of Anglo-Lutheran Relations from 1521-1547* (St. Louis: Concordia, 1965) and Cranmer's biography, *Thomas Cranmer,* J. Ridley (Oxford: Clarendon Press, 1962). Cranmer encouraged Osiander's scholarship, and, when he completed his *Harmony of the Gospels* (1538), Osiander dedicated the work to Cranmer. His connection with the English Reformation can be seen in the number of Osiander's works that were translated and printed in England by Reformers, such as Miles Coverdale, who were sympathetic to the continental reformation movement.

[12]Eck was Luther's opponent at the famous Leipzig debate and the one who procured the Bull of excommunication against Luther and his followers from the Pope. He also wrote an anonymous attack on Osiander in 1540. On the radical reformers in Nürnberg, see A.P. Evans, *An Episode in the Struggle for Religious Freedom: The Sectaries of Nürnberg, 1524-1528* (New York: Columbia Univ. Press, 1924).

[13]It was these disciples of Melanchthon who were Osiander's fiercest opponents during his last three years of life. Despite the fact that Luther was often exasperated by his independence, and even the normally irenic and conciliatory Melanchthon was often moved to anger, they still continued to correspond with him in friendly fashion as late as 1546 and 1551 respectively.

intensified but extended to include fellow "Lutherans" and friends on the Nürnberg City Council. The counsellors became increasingly embarrassed by his stubbornness and angered by his intransigence on the question of relating civic and religious authority, which was in open conflict with their avowed policy of keeping religious matters firmly in their own hands. These quarrels intensified sharply, resulting in Osiander's further exclusion from consultation, when Protestant political fortunes declined in the beginning of the next decade. At the conferences held at Hagenau (1540) and Worms (1541) in which Protestant and Catholic theologians sought vainly for a resolution of the religious conflict, Osiander's divergence from the Wittenberg position and that of the Nürnberg authorities was so infuriating that he was sent home and thereafter excluded from deliberations. During the next few years, Osiander occupied himself once more with scholarly work, producing a critical work on the New Testament, and his celebrated *Conjectures*. It is also to this period that his editorial work with scientists and his involvement with Copernicus is to be found, to which we shall return later.

During the latter part of the decade, Charles renounced the Peace of Nürnberg, marched his troops into Germany and in a series of military campaigns, defeated the mutual defense league of Protestant princes (Schmalkaldic League). The settlement he imposed upon the electoral territories and Imperial cities (like Nürnberg) – the Augsburg Interim – not only re-established Imperial control but Papal authority on a pre-Reformation basis. Some cities, such as Nürnberg, procrastinated with regard to implementing the provisions of the Interim and sought to negotiate a compromise in order to retain political and economic independence. Other individuals, either on principle or because of an uncompromising Reformation stance, urged resistance to the Interim. Some of these found their position increasingly intolerable and were forced to flee. One of these was Osiander.

The years from 1548 to 1549 when the debate over the Interim took place within the city are the occasion for several of Osiander's most bitter and outspoken writings.[14] Osiander's stubborn public resistance

[14]The circumstances and discussions concerning the introduction of the Interim into the city are discussed in B. Klaus, *Veit Dietrich, Leben und Werk* (Nürnberg: Selbstverlag des Vereins für Bayerische Kirchengeschichte, 1958).

to the Interim was so upsetting the city that, as threats of Imperial action against the city grew more intense, efforts were made to silence him. The situation grew so precarious for Osiander that, fearing arrest either by the Council or by Imperial agents, he fled for safety to Breslau and later, to Königsberg, to the protection of his old friend, Duke Albrecht. There, in gratitude for his past services, the Duke gave him a university professorship and a prominent pulpit, from which he continued his struggle against the Interim. However, he was far removed from further effective influence over events in Germany. Instead, he became quickly embroiled in a fierce theological controversy with the disciples of Melanchthon who, after Luther's death (1546), began to consolidate the confessional position of Wittenberg-oriented territories rigidly along Melanchthonian lines, making Melanchthon's theological position the norm of "Lutheran orthodoxy."[15] In the midst of this controversy, Osiander died. Thus he passed into the record books of the sixteenth century as a heretic and an enemy of the gospel. Whether he should remain similarly regarded with respect to science and his contribution to Copernicus is what we shall now examine.

3. *Literary and Scholarly Activity*

A study of the literary sources and his extant, printed works, shows that while Osiander wrote no known scientific works nor ever wrote *about* science on more than two occasions, he was a prolific writer on theological subjects and a serious scholar of Hebrew literature. We will deal with his activities as a scientific editor in a moment; for the present, we shall briefly examine the nature of his theological and exegetical writings in order to elaborate the shape of his epistemological position as it related to his interest in science and to his involvement with Copernicus.

Except for his sermons and pastoral writings, most of Osiander's extant theological and exegetical writings are highly polemical and speculative in nature, rather than straightforward textual exegesis of

[15]See the modern critical edition of the *Formula of Concord,* translated and edited by T.G. Tappert (Philadelphia: Muhlenberg, 1959), pp. 4, 473, 539, 547-550.

Scripture. His most important work, *De conjectura* (1544) is a speculative interpretation of apocalyptic passages in the Bible (especially the books of Daniel and Revelation) and the Talmud, combined with an eschatological interpretation of western history oriented along anti-Papalist lines.[16] The work consists of a series of speculations, or "conjectures" about the future. It is not primarily concerned with predicting the time of the "end" of history (though he does compute it to be 1688) but with interpreting contemporary events as "signs of the times," heralding and anticipating the "end." Thus, the struggle of his time against Rome and the forces of the Holy Roman Emperor is viewed in light of a greater cosmic struggle taking place throughout history between the Kingdom of Christ and that of Anti-Christ, with the Papacy being regarded as the latest manifestation of the latter.[17]

It is interesting to note that Osiander never uses astrological techniques or doctrines to make his conjectures about the end. His rare references to celestial events in the *De conjectura* are for the purpose of showing that nature participates in history by warning man of the impending judgement. Hence, celestial events are "signs" that *attend* significant historical events, but do not predict them or determine them as effective agents. This reveals his preference for reading "signs" in historical, rather than cosmic terms. Indeed, with one exception, in all his writings, the only references to nature are moral and religious, not scientific.[18]

The four conjectures in the work have a single aim: to show that the character of the present religious struggle is strong evidence that history is fast approaching its denouement. The first conjecture is based on the

[16] *De conjectura* (Nürnberg: J. Petrejus, 1544) was dedicated to Duke Albrecht. In the dedication he states that the purpose of the work was to expose the Papacy as the Anti-Christ and to help evangelical rulers like Duke Albrecht recognize its character from its behavior and thus, overthrow its bondage.

[17] This is a style of early Christian historical interpretation popularized by St. Augustine in his *City of God.* Given the context in which his work was written, it was clear that Osiander was interpreting the "signs of the times" differently from the more pragmatic City Council, whose position of expediency Osiander regarded as showing a dangerous indifference to religious principles.

[18] The only scientific comment anywhere in his writings is found in the *De conjectura* and deals with the difficulty of reconciling the solar and lunar calendars, but there is no technical discussion of their relationship.

prophecy of Elias contained in the Talmud in which history is viewed as encompassing 6000 years, divided into three, 2000-year periods.[19] The last period, which is the epoch of Christ's reign and struggle with the forces of Anti-Christ, becomes the subject for the second conjecture and is based on certain correspondences between events in Christ's life and the rise and decline of Imperial and Papal Rome. The third and fourth conjectures make similar comparisons between the visions contained in the biblical books of Daniel and Revelation and the history of Rome.

His discussion of ancient and recent history in the *De conjectura* shows a wide familiarity with the political events of the Holy Roman empire (and of England), a good grasp of the medieval religious and philosophical tradition and the scholastic tradition from Alexander of Hales to Aquinas, as well as knowledge of the Florentine neo-Platonic tradition from Ficino to Pico. This gives us some clues as to the kinds of materials he read (though not always the specific works) and enables us to detect some of the connections between the major strands of his thought and their intellectual antecedents.

4. Osiander's Thought

Emmanuel Hirsch has partially traced and carefully documented Osiander's intellectual heritage from the dominant motifs discernible in his writings. The most formative influences he detected were Johannes Reuchlin, Pico della Mirandola, Martin Luther (and through him, the Augustinian tradition), the Talmud, the Cabala and Scripture.[20]

[19]It is interesting that Rheticus, in his brief astrological reference found in the *Narratio Prima,* Rosen, (2nd ed. New York: Dover, 1959), pp. 121-123, uses identical values for the millenial epoch as found in the *De conjectura.* Here Rheticus relates significant events in world history to the position of the sun's epicycle, a relationship which he uses to predict the fall of the empire and the return of Christ. Rheticus even cites Elias as the source, indicating that he probably got it from Osiander, who was one of the few Hebrew scholars in Europe with familiarity in the Talmud. We *know* that Melanchthon got it from Osiander and uses the prophecy in his oration on astronomy, (*De Orione,* 1553). But Osiander does not refer to the astrological relationship to supplement the prophecy, as Rheticus does, since he does not believe that heavenly motions are effective in producing historical changes.

[20]Hirsch made two errors, completely misjudging his interest in astrology and overlooking entirely the influence of Cusa. On the matter of "dependency," there are very few

This task of determining Osiander's intellectual genealogy is rendered very difficult by the fact that he rarely cites other writers as sources of ideas and when he does refer to them, he never employs them as "authorities" to support his arguments. Despite the demonstrable influence of Luther, Reuchlin and Pico, for example, he hardly ever refers to them. Partly this was in reaction to the scholastic reliance upon "authorities" and partly because Osiander was proud of the fact that he never became the disciple of anyone or of any "school." In this respect, his intellectual model was Pico, from whom he learned to extract and synthesize insights from many sources without becoming the slavish follower of any of them.[21]

Despite his avowed intellectual independence, Johannes Reuchlin became Osiander's initial intellectual mentor.[22] Reuchlin (1455-1522) was the foremost Hebrew scholar in Europe during the sixteenth-century and achieved fame in his well-known struggles against the Dominicans.[23] Under Reuchlin's tutelage, Osiander became trained in the Hebrew language and in the use of careful philological techniques for scriptural exegesis. But, through the study of Reuchlin's major works, *De verbo mirifico* (1494) and *De arte cabalistica* (1517), he was also introduced to other Hebrew literature, especially the Talmud (in which he later became widely regarded as an expert) and, more importantly, to the mystical Jewish literature which Reuchlin had popularized all over Europe — the Cabala.[24]

---

references to Luther, to whom he is demonstrably indebted. He refers only once to Cusa by name and not at all to Reuchlin!

[21]On Pico, see Ernst Cassirer, "Giovanni Pico della Mirandola," in *Journal of the History of Ideas,* 1942, *3*: 123-144; 319-354 which includes his translation of Pico's "Oration on the Dignity of Man." See also, F.A. Yates, *Giordano Bruno and the Hermetic Tradition,* (Chicago: University of Chicago Press, 1964) and P.O. Kristeller, *Eight Philosophers of the Italian Renaissance,* (Stanford: Stanford University Press, 1964), and the passages in Hirsch showing Osiander's reliance (pp. 131-135, 154-157, 165-170).

[22]On Reuchlin, see Lewis Spitz, *The Religious Renaissance of the German Humanists* (Cambridge, Mass: Harvard University Press, 1963); J. Blau, *The Christian Interpretation of the Cabala in the Renaissance* (New York: Columbia University Press, 1944).

[23]For a graphic account of Reuchlin's struggles with the Dominicans, see S. Graetz, *History of the Jews,* III (London: D. Nutt, 1891-92), pp. 444-465. Hirsch has fully documented those places where Osiander relies on Reuchlin.

[24]For the use of the Cabala during the Renaissance, see Max Dimont, *Jews, God and History* (New York: Signet Books, 1962); C.D. Ginsburg, *The Kabbalah, Its Doctrines,*

Osiander turned enthusiastically to such mystical-symbolic thinking, like many others during the Renaissance, mostly in reaction to the sterility of Scholastic logic. Since Cabalistic techniques abandoned the ordinary meanings of words and assigned values and metaphysical properties to letters and numbers which could then be manipulated and combined in ways to make words yield new meanings, it became an effective tool for generating ideas from apocalyptic literature and a useful, effective weapon for anti-papal polemics. So, for example, Osiander would manipulate the Hebrew tetragrammaton, YHWH, to yield the name "Jesus," and thus pressed Cabala into evangelistic service to give non-Christian literature a Christological meaning.[25]

But the Cabala furnished him with more than hermeneutical techniques; it also came equipped with a set of cosmological-religious doctrines, similar to those of Gnosticism and neo-Platonism, which sought to solve the perennial philosophical problem of unifying conceptual contraries: the One and the many, the transcendent and the immanent, God and the world.[26] Despite his employment of Cabalistic techniques and ideas, however, the cosmological element remains in the background while the theological (or Christological) is in the foreground. This lack of concern for speculation about the physical world shows that his primary concern was to find a structure of mediation between the divine and human, coming to focus in the "divine-man"—Jesus the Christ—who is the single point of reconciliation between heaven and

Development and Literature (London: 1865); J. Blau, *op. cit.*; G.G. Scholem *Major Trends in Jewish Mysticism,* (Jerusalem: Schocken, 1941); C. Roth, *The Jews in the Renaissance* (Philadelphia: Jewish Publication Society of America, 1959).

[25]"To the name which one could neither read nor pronounce (YHWH), one needs only add '*s*' to get Jesus." His argument goes like this: Since God's name is ineffable, God is incomprehensible. But he became man in Jesus (whose name is announced to Mary by the Angel Gabriel). So now we can pronounce his name, IHSVH, and thus, know Him.

[26]Reuchlin stated that his aim was to "show Pythagoras, reborn through me" (Spitz, *op. cit.*, p. 67). Indeed, it was popularly believed during the Renaissance that Pythagoras had received his wisdom from a more primitive source, the Hebrews, through the Cabala which was thought to have been handed down in secret alongside the Talmud. Part of that "wisdom" which Osiander appropriated from Cabala through Reuchlin was the Danielic concept of a heavenly mediator between God and the world, a "divine man" who would reconcile heaven and earth. This is the concept Osiander employs for his Christology in *An Filius,* as well as in earlier works.

earth. More importantly, his subordination of the cosmological to the theological reveals his continuity with the Patristic-Medieval theological tradition and his direct reliance upon another important theological figure standing behind both Reuchlin and Pico—Nicholas of Cusa.[27]

Osiander's reliance upon and departure from Pico della Mirandola (1401-64) is revealed precisely at those places where they employ and modify the ideas of Cusa. Cusa's revolutionary insights concerning the reconciliation of contraries (expressed in the phrase, "coincidentia oppositorum") and his famous relativistic concept are essentially theological insights expressed in mathematical analogies. Thus, while both Pico and Osiander follow Cusa in abjuring cosmological speculation, they apply his ideas in different ways. Pico's thought, in Renaissance humanistic fashion, centers on man as a microcosm of the universe and the place where the opposition of One and many are unified. Osiander, on the other hand, insists that the One and the many converge uniquely and perfectly not in humanity, but only in the "divine-man," Jesus, who is the exemplar of humanity and the incarnate locus of the reconciliation of God and world, divine and human. This fundamental departure from Pico reflects Osiander's theological commitments to the Reformation; for, Pico sought to produce a Christian-oriented universal theism synthesized out of insights drawn from many apparently incompatible sources; his cosmos is *intellectual.* That is, unity is achieved for Pico by constructing an anthropocentric universe in which man is the measure of all things and essentially, its *creator,* who orders cosmos out of the chaos of opinions. Osiander, on the other hand, was primarily a theologian and thus, followed Cusa more closely in affirming a theocentric (that is, *Christocentric*) universe and insisting that the vision of God (or

---

[27]On Cusa, see his *De docta ignorantia* (1440), Eng. trans. by Fr. G. Heron, *On Learned Ignorance* (New Haven: Yale University Press, 1954); F. Copleston, *A History of Philosophy,* III (New York: Image Books, 1963) Part 2, Ch. 15 and E. Cassirer, *Individual and Cosmos in Renaissance Philosophy* (Oxford: Oxford University Press, 1963). Osiander not only adopted Cusa's relativistic principle and the concept of "coincidentia oppositorum," but the title of Osiander's major work is borrowed from a similar type of work by Cusa with nearly an identical title: *De conjecturis* (1440), which contains his theory of knowledge. Osiander's only citation of Cusa, however, is from his exposé of the fraudulent Donation of Constatine, on which the papacy had historically justified its temporal claims.

Oneness) is achieved by inspiration of the Holy Spirit, not by intellectual organization.[28]

Two further examples of Osiander's departure from Pico and reliance upon Cusa are helpful for illuminating his philosophical position relevant to assertions made in "Ad Lectorem." First, as one steeped in the Hebrew tradition, Osiander could never adopt Pico's view that, at the locus of convergence, "the temporal collapses into timelessness."[29] For Osiander, the application of Cusa's relativistic principle to a Christocentric universe meant that there was at least *one* point in time and history that does have absolute value as the sole reference by which all time and history are measured and by which they acquire meaning, direction and goal. That point is the incarnation of the Logos of God in Jesus of Nazareth. As with Cusa, Osiander insisted that God is fundamentally unlocatable in space and time; but he argues that while he is "everywhere and nowhere" in any spatial sense, his appearance in history nevertheless established a *temporal center* from which all history is subsequently measured B.C. and A.D.[30] Thus his application of Cusa's concept of "coincidentia oppositorum" is historical and Christological and his concern is for eschatological-historical interpretation rather than spatial-cosmological speculation.

Further, Osiander could never have agreed with Pico's use of the relativistic epistemological principle contained in Cusa's *De docta*

[28]Compare Cusa's famous comment, "Nihil certu habemus ni nostra scientia, nisi nostra mathematica" (*De conjecturis,* I:11 and II:16) with Osiander's statement in "Ad Lectorem."

[29]Cassirer also saw this as an inconsistency in Pico since, if time is relative, this would only mean that there is no point in time with absolute precedence, not that time itself is abolished.

[30]At the Marburg Colloquy (1529), Osiander similarly argued with Zwingli over the manner of Christ's presence in the Eucharist: "With such passages one cannot prove anything except that Christ at certain times was in specific places (the crib, the temple, the cross, the tomb, the right hand of God); that, however, he is eternally and always in a specific place or location, yes, and never exists without a place or could not be in many places in natural and supernatural manner, as you claim, can never be proved on the basis of these passages." (W. Koehler, ed., *Das Marburger Religionsgesprach, 1529; Versuch einer Rekonstruction* (Leipzig: Schriften des Vereins für Reformations Geschichte, 1882) pp. 420-431. Thus his conceptual insight concerning time has easily been extended to a concept of space. But, other than these deliberations on the presence of Christ in the Eucharist, Osiander never makes that extension in a way translatable into science.

ignorantia to assert that there are no given truths or infallible doctrines or no *center* of intellectual reference from which men's opinions may be judged. While he may have agreed with Pico's argument that there are "no heretics of the intellect" in the realm of philosophical speculation or scientific hypothesizing, there was for him, nevertheless, one fixed point of truth, relative to which all truth-claims must ultimately be measured —God's revelation in Holy Scripture. In his last major work, *An Filius*, he wrote: "The Holy Scripture is the only errorless source of truth." This was his consistent, lifelong conviction and the controlling motif of all his thought, on science as well as theology, and the ultimate source of the epistemological scepticism he expresses in "Ad Lectorem."[31] Before turning to that writing and to the epistemological position he articulates in his letters to Copernicus and to Rheticus, we will examine the nature of his scientific activities.

5. *Osiander's Scientific Activities; Copernicus and Cardan*

As far as the sources permit us to know, Osiander addressed himself to scientific matters on only two occasions: during 1542-43 he was occupied with editing and preparing the work of Copernicus for publication at the press of Johann Petrejus; during 1544-45 (when he published his *De conjectura*), he performed a similar editorial service for Hieronymous Cardan, preparing the manuscript of his *Ars magna* for publication, also at the press of Petrejus.

What motivated this interest in science and his involvement in promoting scientific publications? Hirsch believed that it was Osiander's interest in astrology that led to his concern with the renovation of mathematical sciences (especially astronomy). But Hirsch (who was himself a notorious astrologer) provides no hard evidence to justify this

[31]Osiander, *An Filius* (Monteregio Prussiae: Johannus Lufft, 1550) p. A4b. Here again, Osiander follows Cusa explicitly in his insistence that the Spirit of God alone can reveal the laws of the universe so that only through divine inspiration can man attain truth or certain knowledge. This belief, and his conviction that the only worthy study for a Christian scholar is the study of the scripture, was the compelling reason for abandoning formal theological study under scholastic teachers at Ingolstadt and turning to biblical language study under humanist scholars.

claim.[32] Osiander, to our knowledge, never wrote an astrological work nor cast a horoscope and, although he was frequently asked for his advice on such matters, either he never replied or his replies are lost. But his attitude toward astrological practices is known well enough through references we have in his other writings, especially in his sermons. We know therefore that he was generally scornful of astrology and specifically rejects the influences of celestial movements on such human events as plagues.[33] Hirsch has made the common mistake of confusing astrological prognostication with apocalyptic interpretation. The best example of the difference between them, illustrating Osiander's rejection of the former in the interests of the latter, is from a letter he wrote to Justus Jonas (Luther's chaplain) following the appearance of a comet in 1538: "I do not wish to tell Germany's future on the basis of stars; but on the basis of theology, I announce to Germany the wrath of God."[34]

Thus, the source of Osiander's interest in astronomical studies is theological, and can best be assigned to his concern for a more accurate calendar, improving the chronology of historical events and thus providing more accurate apocalyptic interpretations of the Bible and of western history, including the making of accurate "conjectures" about the future. In the context of Germany during the 1540's, he would not only share the general concern for interpreting the significance of the momentous events then occurring, but also the general awareness that the calendar was not in agreement with astronomical movements and

[32]See Hirsch, *op. cit.*, p. 122 for his admission of this. He mainly bases his belief on Osiander's relationship with Cardan, who wrote many astrological works (several of which were printed by Petrejus) as well as a commentary on Ptolemy's *Tetrabiblos* (*De astrologorum iudiciis*). The best known of these astrological works, *De exempli certu genitum,* contains the author's praise of Osiander and several predictions about him. Cardan himself tells us in his autobiography, *De libris propriis* (Basel: H. Petri, 1552), p. 18, how he and Osiander were brought together: a letter from Osiander and Petrejus asking to be entrusted with the publication of his next scientific work, the *Ars Magna,* which states nothing about astrology in it.

[33]Here he follows Pico's rejection of astrology. Pico's famous debate with Melanchthon on the subject of astrology is discussed in C.I. Manschreck, *Melanchthon, the Quiet Reformer* (New York: Abingdon, 1958) and in Melanchthon's Reply to Pico della Mirandolla" (1558) trans. by Q. Breen in *Journal of the History of Ideas*, 1952, *13*: 413-426.

[34]Möller, *op. cit.*, p. 559.

therefore, needed to be corrected by devising better models on which to base calculations. For this purpose then, it would matter little to him whether the model of Ptolemy or that of Copernicus were "true" in the physical sense, especially when, given his theological and epistemological views, no such theoretical construct could claim to be true in the first place.[35]

Despite his lack of formal scientific education,[36] he acquired advanced knowledge of mathematical and astronomical studies and enjoyed such a significant scientific reputation among his contemporaries that he was not only consulted by a number of well-known astronomers and mathematicians of that period, but was regarded by no less a figure than Kepler as "most expert on these matters."[37] Those scientists whom we know to have been in significant contact with Osiander on scientific matters are: Johannes Schöner, Rheticus's teacher, whom Osiander recommended for his post at the Nürnberg Gymnasium; Peter Apian of Ingolstadt University; Hieronymous Schreiber, a friend of Rheticus who lived in Nürnberg from 1521-25 and on whose copy of *De revolutionibus*, Kepler found an annotation ascribing authorship of "Ad Lectorem" to Osiander;[38] Joachim Camerarius, who sent his work on eclipses to Osiander in Königsberg;[39] Erasmus Reinhold, who sent Osiander a copy of his famous *Prutenic Tables* via Duke Albrecht, just

[35]The same general judgment could be made about his interest in mathematical studies, represented by his involvement in Cardan's work. Like the Italian Platonists whom he came to know through Pico, he had little scientific interest in mathematics. What interest in numbers he had, usually led him to Cabala and theology, not to science. Generally then, his interest in science is ancillary to theology, a position which is in essential continuity with the traditional Patristic and Medieval understanding of their relationship. The Medieval notion of science as the "handmaiden" of theology is also reflected in Cusa, for whom cosmology was subordinate to theology. Despite this essentially teleological principle (which, during scholastic times came to be understood as a principle of authority) for Osiander, the question of authority was separate from that of purpose.

[36]His teacher Boschenstein wrote a textbook on elementary arithmetic, with which Osiander was probably familiar. But how he acquired advanced mathematical knowledge sufficient to understand and correct works like those of Copernicus and Cardan, is unknown.

[37]Kepler's comments about Osiander are found in his *Apologia* (1609), Frisch edition, I, p. 245.

[38]Kepler, *op. cit.*

[39]*De Ecclipsi* (1550).

before Osiander's death;[40] Joachim Rheticus, who (needless to repeat), entrusted Osiander with the responsibility of editing Copernicus' work when he left Nürnberg, and finally, Hieronymous Cardan.

On the basis of Cardan's request, Osiander was entrusted with the task of editing his major mathematical work, *Ars magna*, for publication. Cardan's attitude toward Osiander and his opinion of his competence is reflected not only in the fact that he dedicated the work to Osiander, but wrote a long, flattering preface, in which he not only credits Osiander with sufficient knowledge to understand the work, but to edit it, correct it intelligently and to "authoritatively commend it to others." The preface reads:

> Dr. Hieronymus Cardanus sends warmest greetings to be most learned Andreas Osiander. Nothing have I ever pondered so much in my mind, most learned Andreas, than that I might commend to posterity the names of those who confer a benefit on good literature. Then, indeed, a certain special consideration I added if such men, along with learning, had joined human compassion. For this reason, and since I know that you have knowledge, which is by no means mediocre, not only of Hebrew, Greek, and Latin literature but also an intelligent grasp of mathematics and because I have also found by experience that you were most kind, there seemed to be no other man than you by whom it can be both amended (if my hand, in passing, deceived the command of my mind) and can be read with pleasure and understood and indeed even authoritatively commended [to others]. This example, unless I am mistaken, will also be followed by others who will not dedicate

[40]Reinhold's letter to Duke Albrecht reveals the judgment of experts about Osiander's skill in astronomy. Reinhold asked the Duke to present the accompanying copy of his *Prutenic Tables* to Osiander for his criticism, saying that he had intended sending the work to him in Nürnberg before he fled the city. He then describes Osiander as one, ". . . who in our time has a greater reputation because of his scholarship in mathematics." Albrecht replied: "We have not neglected to show Andreas Osiander your work about which he said, after examining it hurriedly, that it appeared to be very well done and pleased him much." (J. Voigt, *Briefwechsel der berühmtesten Gelehrten des Zeitalters der Reformation mit Herzog Albrecht von Preussen,* (Königsberg, 1841), pp. 541-543; quoted in Hans Blumenberg, *Die Kopernikanische Wende* [Frankfurt a.M.: Suhrkamp, 1965], p. 96.)

> their works to learned men unless it is in that art which they practice. Receive, therefore, the everlasting witness of my love toward you and of your service toward me, and my perpetual witness at the same time of your outstanding learning. And although you are such a man, nevertheless, though Alexander and Caesar, well-known for their deeds, still desired to be inscribed in the written records of others and since Plato, who wrote all these marvelous things by himself, nevertheless desired to be praised by the writings of others, I hope that my service to you, such as it is, will also not be unpleasing to you because even in these things a certain fortune has dominion and often the better things perish while the worst things are preserved. And no matter what your judgment may be about this means, it is certain to me that I ought to ask your pardon for my service and hope that it might come to pass that I would be able to show my attitude toward all men with a more illustrous example, who have that brilliance of mind which I always recognize that you had toward the most studious men of our age. But perhaps there will be granted a better opportunity, and even if it should not be granted, I would be unwilling that this opportunity, of whatever sort it is, should have been lost to me. Farewell, five days before the Ides of March 1545.[41]

A number of things about this preface are interesting, but one is significant to this discussion concerning Osiander's contribution to Copernicus: Cardan's praise shows the attitude of an important learned scholar of the sixteenth century toward Osiander and indicates that the selection of Osiander as a scientific editor was technically justifiable. His involvement with Copernicus was thus not an isolated case, nor a case where an obstructionist, "enemy of science" was acting repressively or unscrupulously. If that judgement can finally be laid to rest, we can

[41]Cardan, *Ars magna,* (Nürnberg: Johann Petrijus, 1545). Acknowledging the tendency of Renaissance authors to praise their benefactors in flattering terms, this was not all flattery since Cardan later dedicates another work to Osiander, *De subtilitate* (Nürnberg: J. Petrejus, 1550) in which he speaks warmly of him as "a learned man" and his friend.

then go on to rationally assess his actions and motivations with regard to his handling of *De revolutionibus.*

6. Osiander's "Ad Lectorem"

We will now turn to the "Ad Lectorem" itself and here an immediate question presses for attention: why did Osiander leave the writing anonymous? Osiander has received much criticism from scholars for "unscrupulously" affixing this writing without (as far as we know) Copernicus' permission and, by leaving it unsigned, "deceiving" the reader into thinking that the views contained in it were those of Copernicus. But when we consider the matter in light of the historical context in which it was written and during which the work was printed, the probable reason is readily apparent: it was done to *protect* the work from criticisms that, no doubt, would have been forthcoming had Osiander's name appeared on it. Osiander, it must be remembered, was such a notorious reformer whose name was well-known and infamous among Catholics, that to place it on such a work, written by a loyal Catholic canon and scholar, would have almost guaranteed closer scrutiny of it by the church and increased the chances of adverse theological reaction. Moreover, it would have cast suspicion on Copernicus himself, who had quite enough trouble with his own Bishop at that moment. To sign the writing would then have incurred the risk of compromising the security of the work as well as its acceptance, and lost the scientific hearing both he and Copernicus desired for it.[42]

Again, it must be kept in mind that the printing of *De revolutionibus* took place during a very tense political period when the situation of the Protestants (and the independence of Nürnberg) was being openly threatened by Catholic authorities. There was every reason to believe that the forces of the Holy Roman Emperor would soon re-establish control. And if the books of hostile theologians could be burned (as indeed

[42]It was Copernicus' own fears that his work would be scrutinized and criticized by the "peripatetics and theologians" that he communicated to Osiander (and expressed in his Dedicatory letter) that scholars incorrectly impute to Osiander, thus misconstruing his motivations for writing "Ad lectorem." See, for example, Zinner's allegations (*op. cit.,* pp. 246, 255-56, 271).

very many were on both sides), why not scientific works with the names of hated theologians affixed to them? Anonymity was the only feasible way to protect the work in such a precarious situation. For just such a reason, I suspect, Copernicus shrewdly declined to name his Lutheran disciple, Rheticus, in his letter of dedication to the Pope, as one of those whose assistance and encouragement persuaded him to have the work published. What other possible reason could there be for such a significant omission?[43]

Turning to the writing of "Ad Lectorem" itself, its phrasing clearly indicates that it was written by someone other than Copernicus. Osiander refers to "the author" three separate times, leaving no doubt to the careful reader that the author of "Ad Lectorem" was not Copernicus. Further, the Dedicatory Letter to Pope Paul is clearly labelled "Praefatio Autoris" at the top of each page on which it is printed, *but not at the head of the page on which "Ad Lectorem" is printed.* And this was done under Osiander's supervision! This makes it demonstrably clear that Osiander was not trying to pass off his writing as the work of Copernicus, or to deceive the reader into thinking that Copernicus did not take his own theory seriously. That common allegation should also be laid to rest. Everything we know now about these two cases where he was involved in editing and promoting significant scientific works, compels us to believe that Osiander's motivation for writing his letter was not obstructive but rather (as he states both in his letter to Rheticus and in "Ad Lectorem" itself) that he valued scientific research, strongly desired to promote scientific investigation and that

[43]Oystein Ore, in his forward to T. Richard Witmer's recent translation of Cardan's *Ars magna* (*The Great Art or the Rules of Algebra* by Giralamo Cardan, [Cambridge, Mass., 1968]), refers to Cardan's arrest, inquisition and imprisonment by the Church in 1570 and insists that there is a connection between his fate and Osiander's involvement with the publication of his work. He writes: "However, this was the time of the Counter-Reformation and one of the items specified during the process may have been Cardano's cordial dedication to Andreas Osiander. Not only had Osiander been among the original leaders of the Reformation, but he was also known to have written the Preface to Copernicus' *De revolutionibus orbium coelestium*, a work which was under strong suspicion of being heretical." Leaving aside the question as to how well Osiander's authorship of "Ad Lectorem" was known in 1570 and the dubious judgment that the work was already under strong suspicion, this does raise an interesting conjecture about the potential impact his name might have had if placed openly on Copernicus' work.

editing such works and recommending them was his habitual method of furthering the growth of science.

Third, Osiander's assertion that astronomy has an inherent hypothetical element resident in it was a commonly accepted view during the sixteenth-century. His views do not therefore warrant criticism on the grounds of contemporary historical usage. Before evaluating his opinions one must first ask, what were the methodological standards of the time and the criteria for judging acceptable astronomical explanations? Most mathematical astronomers, like Osiander, took an "instrumentalist" position with regard to such theories. Philosophical purists like the Averroists were the few who demanded physical consistency and thus sought for realist models.[44]

Osiander could see from a simple inspection of *De revolutionibus* that Copernicus, too, used conventional, hypothetical devices like epicycles to account for motions geometrically, just as all astronomers had done since antiquity. How could Copernicus claim for these devices a "reality" greater than what they obviously were – hypothetical constructs solely designed to "save the phenomena" and aid computation? That they enabled him to do so more simply than Ptolemy's system (i.e., by accounting for retrograde motion) was not sufficient to alter their status. Hence, his position that "absurdity" is encountered by reifying hypotheses. Accordingly, he urges the reader to remember that the practice of hypothesizing is an acceptable, not a "blameworthy" technique, and that such hypotheses *need not* be submitted to the tests of physical consistency. And note that he does *not* say (contrary to the claims of Zinner and Prowe) that such hypotheses as that of Copernicus are not, or may not be, true, but that they cannot be *shown* to be

[44]It was this fundamental incompatibility between Ptolemaic astronomy and Aristotelian physics, and the need to preserve both, that prompted most thinkers to insist that astronomical hypotheses be interpreted instrumentally, a view which prevailed with few exceptions. Copernicus' work brought the problem to the surface by its assertion of realism in defiance of Aristotelian physical principles (and of scripture). Pierre Duhem (whose sympathies lie with the instrumentalist position) has documented this tradition in his, *To Save the Phenomena.* Also see L. Laudan, "Theories of Scientific Method from Plato to Mach," in *History of Science,* 1968, 7: 1-63.

true.[45] Osiander specifically refers to the epicycle of Venus as an example of the absurdity involved in claiming reality for astronomical hypotheses. He was here giving voice to a common objection to the Ptolemaic theory wherein Venus had the largest epicycle of any planet (about three-fourths the diameter of its deferent) due to the fact that Venus was restricted by observations to a maximum elongation of about 45 degrees on either side of the sun. If its epicycle were to be regarded as more than a geometrical device, Venus should then reflect an enormous variation in brightness and distance from the earth, which it fails to do. (Osiander estimates this variation by a factor of sixteen for brightness and four for size). In spite of this defect in Ptolemy's theory, Copernicus' hypothesis predicts approximately the same variations. On these grounds, Osiander could well justify his claim that there was little technical or physical truth-gain in replacing an old hypothesis with one possessing exactly the same deficiency. Thus, he concluded:

> . . . it is quite clear that the cause of the apparent unequal motions are completely and simply unknown to this art. And if any causes are devised by the imagination, as indeed very many are, they are not put forward to convince anyone that they are true, but merely to provide a correct basis for calculation.

These words refer specifically to the interchangeability of the hypotheses of epicycles and eccentric circles. The question of "truth" could therefore not enter into the decision between these two alternatives.

[45]Ernst Zinner, *Geschichte der Sternkunde* (Berlin: J. Springer, 1931) p. 460 states: "It was. . . most regrettable that in the first printing in which Copernicus finally acquiesced after long hesitation, the Protestant divine Osiander, without the author's knowledge, smuggled in a preface in which the Copernican motions, particularly the hypothesis of a stationary sun and a moving earth, were designated as merely serving the purpose of representing the motion, with divine revelation as the sole means of attaining truth, which not only served to weaken the work, but placed ready weapons in the hands of the opposition." But Osiander never stated this about *Copernicus'* theory; his comments are generalizations about *all* astronomical theorizing. And he does not say that this or any theory is *not* true but that, for such technical purposes as hypotheses serve, they *"need not be true, or even probable,"* in order to usefully serve that purpose, which is quite correct. It is only when different purposes are entertained that the question of truth or probability arises.

Rather, "...the astronomer would accept...the one which is easiest to grasp."

Last, Osiander's insistence that astronomical hypotheses (or kinematic models) had no validity outside of astronomy, in physics or theology, was generally conceded to be true in the sixteenth century. But it is worth restating in view of the fact that it was Copernicus' expressed fears that his theory would meet objection on those grounds that prompted Osiander's "Ad Lectorem."[46] Osiander was not the one to create the religious issue; he was reflecting Copernicus' own apprehensions, as well as the generally accepted view that certainty, or truth is a function of divine revelation. What Osiander is clearly affirming is that religion provides no criteria for validating astronomical hypotheses.

7. *Osiander's Contributions to the Acceptance of Copernicus' Theory*

Finally, let us consider the central question with which this article has been concerned: what was Osiander's actual contribution to Copernicus and his achievement? Beyond the obvious but significant contribution of editing his major work, did his writing make any further contribution to its use and its acceptance? Inasmuch as it has become clear that he was not trying to depreciate and suppress, but to protect and promote the work, what effects did "Ad Lectorem" have in furthering these aims?

If "achievement" is construed more broadly than an empirical validation of the theory's truth, then the question of Osiander's contribution depends not only on an assessment of Copernicus' own achievement but also on an estimation of the probable effect of placing "Ad Lectorem" at the beginning of the work. With respect to the first, the literature dealing with the nature of Copernicus' achievement is vast and varied, as may be seen from the papers read at this symposium; but

[46]See the letters from Osiander to Copernicus and to Rheticus in Rosen, *op. cit.,* p. 25. These letters and the pleas in Copernicus' letter to the Pope, are the basis for Osiander's tactic, not his own religious scruples or alleged fears of Luther's or Melanchthon's disapproval. In the light of all we know about Osiander's character and personality, especially his fanatical independence from either Wittenberg or Rome, the imputation of such motives is at best incredible.

however "achievement" is construed, the survival, distribution and study of the printed work are crucial. What did "Ad Lectorem" contribute to that?

Curiously, historians of science have generally concluded that the effect of "Ad Lectorem" was to hinder, not to help the acceptance and use of the work. Indeed, Adolf Müller, S.J., makes Osiander's writing responsible for the *condemnation* of *De revolutionibus* by the Catholic Church, claiming that it "transferred it to the realm of religious scruples!"[47] A. R. Hall argues,

> If the celebrated note by the Lutheran minister Osiander... had any effect in warding off official censure (which seems doubtful) it certainly did nothing to weaken the general sentiment that speculation about the earth's movement was foolish and futile.

And Ernst Zinner, considering the effect of "Ad Lectorem" on the sale of the work, insisted that it "would have been better without the prologue," without documenting his vague judgement in any way.[48]

Even Edward Rosen argues that, "The resultant question of authorship, so long as it remained unsolved, exerted a powerful influence upon the reception of Copernicanism and upon the interpretation of astronomical theory and scientific method,"[49] without telling us how this influence was exerted.

These allegations are typical of the attitude of most historians toward the effect of "Ad Lectorem" on the reception of Copernicus' work. But when we examine the *evidence* of its reception in the expressed views of those few who actually talk about it, we learn that "Ad Lectorem" demonstrably did not prevent the use of the work among astronomers at that time or later.[50] Even those who expressed reservations about its

[47]Müller, *op. cit.*, p. 367 and Harig, "Kepler und das Vorwart von Osiander zu dem Hauptwerk von Kopernikus" in *Zeitschrift für Geschichte der Naturwissenschaften, Technik und Medizin,* 1960, No. 2, Jahrgang 1: 17.

[48]A.R. Hall, *The Scientific Revolution 1500-1800* (London: Longmans, Green and Co., 1954), p. 55 and Zinner, *op. cit.*, pp. 256-257.

[49]Edward Rosen, "The Ramus-Rheticus Correspondence" in *Roots of Scientific Thought,* eds. P.P. Weiner and A. Noland (New York: Basic Books, 1957), pp. 287-292.

[50]Besides the studies of Zinner and Duhem, see especially the recent study of Robert S. Westman, presented to the Fifth International Symposium of the Smithsonian Institution

central hypothesis had no hesitations at all about making liberal use of its contents. And even before its publication, the initial response to early reports of his theory was either positive, indifferent or superficially critical, without the assistance or hindrance of "Ad Lectorem" (hence, the contrasting reactions of Cardinal Schoenberg, printed with Copernicus' Dedicatory letter, and Luther's oft-quoted remark on Copernicus). Other than the reaction of Tiedemann Giese (written in ignorance of Osiander's authorship), there is only one reference to the effect of "Ad Lectorem" until the time of Kepler. According to A. Birkenmajer, Jan Broscius wrote:[51]

> Ptolemy's hypothesis is that the earth rests. Copernicus' hypothesis is that the earth is in motion. Can either, therefore, be true? . . . . Indeed, Osiander deceives much with that preface of his. . . . Hence, someone may well ask: How is one to know which hypothesis is truer, the Ptolemaic or the Copernican? . . .

In nearly every case where Copernicus or his theory is mentioned at all, and discussed, analyzed, criticized or praised, nowhere do commentators cite "Ad Lectorem" as their reason for doing so.[52] But typically, both those who criticized him and those who praised him were happy to use the tables he provided (or those which Reinhold produced from them). As for speculation on the mobility of the earth, they commonly leave the question aside or view it as Osiander did, not because he suggested it but because this was the way they were accustomed to treat such hypotheses. Indeed, Duhem demonstrates that this was the very attitude that had permitted technical astronomy to function since

---

and the National Academy of Sciences in April 1973: "The Wittenberg Interpretation of the Copernican Theory," in *The Nature of Scientific Discovery,* ed. Owen Gingerich (Washington, D.C.: Smithsonian Institution Press, 1974), pp. 393-429 and in a longer, modified version as "The Melanchthon Circle, Rheticus and the Wittenberg Interpretation of the Copernican Theory," *Isis,* 1975, *66*: 165-193.

[51]L.A. Birkenmajer, *Mikolaj Kopernik* (Cracow: Akademya Umiejetnósci, 1900), p. 655.

[52]D. Stimson, *The Gradual Acceptance of the Copernican World Picture* (New York: 1957), pp. 42ff and E. Rosen, "Galileo's Mis-Statements about Copernicus," *Isis,* 1958, *49*: 319-330.

antiquity, despite its inconsistencies with the principles of physics and the philosophical objections of Averroists.[53]

This situation remained until the beginning of the seventeenth-century when, for various reasons, such as the pressures of growing confessional conformity, the issue of truth was raised, along with the question of its consistency with physics and with scripture. It was only then that arguments dealing with its reality began to emerge and certain churchmen began to argue in opposition to the theory, the very thing Osiander tried to prevent. Whether this tactic was "more expedient to the Art than to Copernicus" is a matter of judgement. But as a matter of fact, Zinner's studies, and those of Duhem, show that individuals used the work, exploited its tables and constructions, criticized it and praised it, but were generally indifferent to its implications until Bruno, Kepler and Galileo began to press publicly for recognition of its philosophical and religious validity. But there is no evidence to support the views of those cited above that "Ad Lectorem" adversely affected the reception of the work, caused its eventual condemnation, prevented its use, weakened speculation or "damaged the Copernican reputation" in any way. On the other hand, it is much more plausible to claim that, for over a century, "Ad Lectorem" *protected* the work from this kind of scrutiny during an extremely tense period of ideological and political conflict and thus, actually *permitted* the work to be used and pondered during that period by those with such scruples, by advocating the way it was, in fact, being regarded and used.[54] Thus, Angus Armitage has been almost alone in regarding "Ad lectorem" as ". . . a well-meaning effort to disarm criticism and to ensure a favorable reception. . . and it seems to have succeeded in its purpose for nearly a century.[55]

[53]See Duhem, *op. cit.,* pp. 98-100.

[54]That this was due to Osiander's writing cannot be known, since no one credits it as a justification. Giorgio de Santillana is wrong in claiming that Cardinal Bellarmine uses "Ad Lectorem" as authority for the views he presents in his famous cautionary letter to Galileo: *Crime of Galileo* (Chicago: University of Chicago, 1955). Bellarmine never refers to Osiander's letter. De Santillana makes other erroneous claims in his discussion of "Ad Lectorem," such as saying that no one at that time knew that the letter was *not* by Copernicus, and his ascription of Copernicus' opponents as "fundamentalistic," a term that is hardly applicable to either Lutherans or Catholics in the Sixteenth century, and certainly not to Osiander.

[55]A. Armitage, *Copernicus, the Founder of Modern Astronomy* (London: G. Allen and

Finally, Osiander's last contribution goes to the heart of the epistemelogical issue he raises in his letter: what is the proper aim and status of scientific theories? The fact that Osiander's views continue to be criticized in modern literature indicates that his contribution to contemporary philosophical discussion is not as irrelevant as first seems to be the case. Without simply taking Duhem's position that "...logic sides with Osiander...", one can insist that his general position is philosophically defensible even today, as we see from the vast literature dealing with the methodology of science and the Realist-Instrumentalist problem.[56]

In light of the actual fate of the Copernican theory and the fact that no one any longer shares Copernicus' belief in uniform circular motion, his notion that rotation is "natural to a sphere," or his belief in the existence of substantial celestial spheres bearing epicycles, all of which are tied to his central hypothesis, Osiander's reservations were justified. In this respect, Kepler's arguments against Osiander (raised in the context of his arguments against Ursus) to the effect that astronomical arguments must be subjected to physical tests, does not invalidate Osiander's point. It only transfers the argument to the domain of physics where Osiander's argument still applies and, by extension, to *all* rational scientific theory. Even Kepler, despite his erroneous claim that Copernicus "proved" his theory to be true, was forced to admit that his own *physical* demonstrations in support of it were only "probable," which is exactly as Osiander claimed.[57]

Unwin, 1938), p. 94. Even those who attribute his action to cowardice and treachery are forced to concede that, as a tactic, it seemed to work and served its purpose in forestalling immediate conflict. See, e.g., E.R. Trattner, *Architect of Ideas* (New York: Currick and Evans, 1938), p. 32.

[56]See note 3.

[57]Kepler's admission that conclusions in physics are only probable, not certain, is found in his *Astronomia Nova* (1609), in *Johannes Kepler Gesammelte Werke,* III (Munich: C.H. Beck, 1937), p. 19. Osiander was definitely not arguing Ursus' point that the choice between astronomical hypotheses was purely arbitrary, even though there was little technical difference between them, and the decision of truth was incapable of resolution on contemporary physical principles. Again, the question was not even deemed relevant for over half a century, and, when it did become an issue, there was still no physical way to resolve it.

The fact is, as we know today from the researches of modern physicists from Mach to Einstein, that there is no "all-purpose system of reference" in the universe, no system of absolute space and time within which to make certain judgements about physical events. Such concepts as "absolute space and time" which are necessary for realist theories, have no operational meaning today. And even where they are pragmatically employed (as in Newton's system), Mach has conclusively demonstrated that such a system cannot be derived from a logically consistent set of mathematical and physical principles.

Nevertheless, where Osiander was wrong was in not recognizing the *strategic* necessity for the methodological assumption of just such an absolute referential framework (e.g. Copernicus' "sphere of the fixed stars"; Newton's "absolute space and time," Einstein's speed of light) and conferring reality on such concepts, even where the means of validating their reality is absent. But the assertion of such reality, in the presence of an established physical system (such as Aristotle's) must not only be made in defiance of such principles, but must be justified on the basis of metaphysical principles, as Paul Feyerabend has argued so persuasively.[58] But that is the subject for another discussion. Here, I only want to argue that Osiander makes an often unrecognized contribution to that ongoing question in the philosophy of science.[59] The issues in the contemporary discussion of that problem may be more refined, subtle and technically exact, but the fundamental issue is the same: whether scientific propositions (or any universal propositions about the physical world) can be known with certainty to be true or whether all such propositions remain invariably hypothetical and probable with respect to the degree of confirmation obtained for them.

[58]Paul Feyerabend has recently argued that belief in the truth of a new theory which contradicts established physical principles will be justified by metaphysical, not physical or methodological principles. See "Realism and Instrumentalism: Comments on the Logic of Factual Support," *The Critical Approach to Science and Philosophy,* ed. M. Bunge (New York: Free Press of Glencoe, 1964), pp. 280-308. E. Burtt, *The Metaphysical Foundations of Modern Science* (London: Kegan Paul and Co., 1957), pp. 297-298 and Alexandre Koyré, "Science as a Social and Historical Phenomenon," *The Validation of Scientific Theories* (New York: Beacon Press, 1957), pp. 182-183, both argue that such justification is ultimately theological in nature.

[59]As Laudan (*op. cit.*) argues, such problems are recurring ones raised by all theorizing, not simply by particular theories.

So while Osiander's views were methodologically wrong, his tactics were probably politically correct for that time and ultimately correct for ours. The contemporary claim that there is no objective or absolute point outside the universe (or outside the community of scientific discourse) by which to absolutely validate truth claims made within it is precisely what Osiander meant when he said: that one cannot ". . . state anything certain, unless it has been divinely revealed to him." And if we strip Osiander's statements of its theological claims, is this not the same point as the contemporary physical recognition that all absolutistic claims about the universe are fundamentally indeterminate? If there were some operational benchmark from which absolutistic, objective claims could be made, some "God's-eye" view of the universe, how could it be known apart from some divine revelation? Where Osiander was wrong was in his failure to recognize that this was exactly what Copernicus was assuming and doing and that it is only by the very act of positing such a view that assertions of reality can be made and some progress toward greater generality in scientific theorizing, be achieved.

Thus, we are forced to give some validity to Osiander's position and credit him with a contribution to the larger achievement of Copernicus, which resides partly in the perennial issue of the aim and status of scientific theories, especially at times when such theories center absolutely on one system of principles and claim exclusive privilege as the sole interpretation of the universe.

In summary, Osiander, as one primarily interested in religion and in political and ecclesiastical reform, played only a *supporting* role in contributing to the furtherance of science. But if he does not merit an award for that, at least he deserves honorable mention. His contribution may not seem impressive; a sequoia seedling is not very impressive either, but it is a sequoia, nevertheless. And if his action was simply a tactical expedience and the reflex of his theology, so be it. It worked. And in the scale of values by which success is measured in the modern world, that is all that matters.

VIII

THE SOLID PLANETARY SPHERES IN POST-COPERNICAN NATURAL PHILOSOPHY

William H. Donahue

St. John's College, Santa Fe

The inordinate amount of time between the first publication of the Copernican system and its widespread acceptance has frequently provided cause for remark. This slow response is usually ascribed to the combined effect of a number of causes such as the technical nature of Copernicus' work and the "instrumentalist" preface written by Andreas Osiander. It is also recognized that the intellectual climate of the mid-seventeenth century differed greatly from that of the mid-sixteenth century, and that this helped determine the fate of the Copernican planetary system. Yet, partly because historians have tended to neglect theories of the heavens as given in sixteenth century works on natural philosophy, very little had been said of the nature and effects of this change of climate. Moreover, one influential work on this topic, Pierre Duhem's otherwise admirable essay translated under the title, *To Save The Appearances,*[2] seriously misrepresents the structure of learning in the sixteenth century in treating astronomy as a branch of natural

[1]This essay is based upon, and in part extracted from, the author's doctoral dissertation, entitled *The Dissolution of the Celestial Spheres, 1595-1650.* Copies are available from the University Library, Cambridge, England.

[2]Pierre Duhem, *To Save The Appearances,* trans. E. Doland and C. Maschler, with an introduction by S.L. Jaki, (Chicago: University of Chicago Press, 1969). First published in 1908.

philosophy. The present essay is intended to complement and correct Duhem's work. Although it is far from comprehensive (the field is too extensive for that) the viewpoint which it presents is novel and may suggest further and more fruitful investigations.

Of obviously crucial importance to the acceptance or rejection of any planetary theory is the theorist's opinion of the status of the devices required or implied by the theory. If the planetary circles are thought to be imaginary then they may be multiplied without inconvenience; however, if they are considered real then they must conform to the requirements of physics, whatever those might be. Not infrequently, practical astronomers believed in devices which contemporary philosophers could not accept: thus, according to recent studies by Bernard Goldstein,[3] a large majority of medieval astronomers believed in their often complicated inventions, while philosophers tended to follow Aristotle. According to the Aristotelian division of disciplines, astronomy is a branch of mathematics while physical reality is the province of philosophers alone.[4] As a result of this distinction, a wide range of opinions concerning the heavens came to be accepted as mutually compatible. A writer could make apparently contradictory statements about the heavens provided that he made them in different contexts.[5] In this instance, the two disciplines were not considered to have any responsibilities to one another, although the idea of separation does not entail this. Indeed, it is possible for mathematicians to feel constrained by physics, though in its strictest interpretation the Aristotelian epistemology does not allow physics to be determined by astronomical theories.

It was the former, more complete separation which generally prevailed when Copernicus was a student about the turn of the sixteenth

---

[3]Postgraduate Seminar in the History of Science, the University of Cambridge, Spring, 1972.

[4]Aristotle, *Metaphysics,* Book XIII, Ch. 2; *Physics,* Book II, Ch. 2. This division was adhered to by nearly all academic natural philosophers; however, an important exception must be noted. Gregor Reisch, in his often reprinted textbook *Margarita philosophica,* (Freiburg, 1503), presents mathematics as belonging to *"philosophia realis speculativa,"* along with metaphysics and physics (*op. cit.,* fol. [ ] 4^{r}). Here, he appears to be following Nicholas of Cusa; cf. Reisch, *op. cit.,* Book IV, Tractate 1, Ch. 1.

[5]Compare Dante, *Paradiso,* (London: J.M. Dent, 1900), Cantos I, II, with *Il Convivio,* ed. M. Simonelli, (Bologna, 1966), II, 3.

century. This was not to last much longer, however. The greatest intellectual efforts of the first half of the century were directed towards recapturing the spirit of the ancients and understanding the full implications of formerly isolated ideas. Vesalius and Copernicus emulated Galen and Ptolemy, respectively, but surpassed them in their concern for consistency and harmony. At the same time, Fracastoro and Amici attempted to revive the homocentric astronomy developed by Eudoxus and Callippus and approved by Aristotle.[6] Their arguments, which received much attention, reaffirmed the supremacy of Aristotelian physics and tried to develop an astronomical system in accord with physical principles.[7] After the middle of the sixteenth century, most astronomers, faced with the Copernican and homocentric arguments, retreated to the ancient though mildly disgraceful position that astronomical hypotheses should not be regarded as physical truths, as they are only intended to save the appearances.[8]

So far, this brief account follows Duhem's quite closely. At this point, however, his argument begins to go seriously astray. He proceeds immediately to consider the realist arguments of Junctinus and Clavius, showing how these astronomical writers differed sharply with their contemporaries. He thus practically ignores the distinction between mathematics and physics, and fails to notice extremely important developments in natural philosophy, without which Clavius' position would be merely another of the countless curious aberrations which the period produced. To correct this error will require a brief excursion into contemporary matter-theory, to show the gradual but drastic change then taking place in men's opinions of what the heavens are made of.

The theory of the celestial substance most widely accepted in the middle ages supposed a radical distinction between the aether and the elements. Thomas Aquinas argued thus, probably following the pseudo-Dionysius,[9] in his insistence that the heavens are made of a *matter*

[6]Aristotle, *Metaphysics,* Book XII, Ch. 8.

[7]Duhem, *To Save The Appearances, op. cit.,* pp. 49-52.

[8]*Ibid.,* pp. 69-91; cf. Jacques Besson, *L'art et science de trouver les eaux et fontaines cachées soubs terre,* (Orleans, 1569), Book I, Ch. 2; Ariel Bicard (*alias* Hartmann Beyer), *Quaestiones novae in libellum de sphaera,* (Paris, 1552), Book IV, Ch. 1, fol. 70^{V}.

[9]Thomas Litt, *Les corps celestes dans l'univers de Saint Thomas d'Aquin,* (Centre de Wulf-Mansion. Philosophes Médiévaux, 7), (Louvain: Publications universitaires, 1963), p. 37.

wholly different from that of the elements.[10] He describes this matter as intermediate between the eternal and the temporal,[11] and strengthens the Aristotelian analogy between the heavens and the animal soul[12] by likening them to the human intellect.[13] Dante's views as expressed in the *Paradiso* are much like St. Thomas's.[14]

A similar view is clearly expressed in many philosophical works of the sixteenth century. This is especially so at Padua, where the opinions of Averroes elicited much support,[15] but it is also to be seen elsewhere in Italy,[16] in Germany,[17] and doubtless in other countries as well. There was not, however, universal agreement that the heavens and the earth should be so radically separated: a weakening of the distinction is found in a number of fairly orthodox works of the later sixteenth century.[18] This marks the beginning of a trend whose nature is not entirely clear, but of which several components may be discerned. One of these, already mentioned above, was the classical revival, which was characterized by an enlightened obedience to ancient authority whereby it was the ancients' methods, rather than their opinions, that were respected.[19] Another component was the intrusion of certain elements of the Stoic natural philosophy, as expounded by no less an authority than Galen,

[10]*Ibid.*, p. 59 (citing 37 passages).

[11]Thomas Aquinas, *De divinis nominibus,* Book III, Ch. 10, text 875.

[12]Aristotle, *On the Generation of Animals,* Book II, Ch. 3, 736 b 35.

[13]Thomas Aquinas, *In De Coelo,* Book II, Ch. 10, text 394.

[14]Dante, *Paradiso, op. cit.,* Canto II, 133-141.

[15]Iacopo Zabarella, *De rebus naturalibus,* (Cologne, 1590), Ch. 6, Column 252; Archangelus Mercenarius, *Dilucidationes obscuriorum locorum et quaestionum philosophiae naturalis Aristotelis eiusque interpretum,* (Leipzig, 1590), pp. 75, 83, (First edition: Venice, 1574).

[16]Cf. Benedict Pererius, S.J., *De communibus omnium rerum naturalium principiis et affectionibus,* (Rome, 1576), Book II, Ch. 3, pp. 47-48; Book V, Ch. 9-10, pp. 179, 181, 183, (First edition: Rome, 1578).

[17]Philipp Melanchthon, *Initia doctrinae physicae,* (Wittenberg, 1549), in *Melanchthoni Opera,* XIII (Halle and Braunschweig, 1856), pp. 223, 231, 244; Francis Titelmann, *Compendium physicae,* (Paris, 1545), fol. 20^{v}.

[18]Nicolaus Biesius, *De natura libri V,* (Antwerp, 1613), Book II, Ch. 2, fol. 53^{v}; Ch. 4, fol. 55^{v} (published posthumously); Julius Caesar Scaliger, *Exercitationum exotericarum liber XV,* (Paris, 1557), *Exercitatio* 61.5, fol. 96^{v}-97^{r}, *Ex.* 69, fol. 107^{r-v}, and *Ex.* 76, fol. 116^{v}.

[19]Cf. Kurd Lasswitz, *Geschichte der Atomistik vom Mittelalter bis Newton,* I (Hamburg and Leipzig, 1890; Hildesheim: Georg Olms, 1963), pp. 307-308.

into the Aristotelian system. While Aristotle believed that the elements are nothing but pairs of primary qualities,[20] Galen allowed them to be characterized as the smallest particles of a substance, and assigned to two elements (air and water) a single pair of primary qualities (cold and moist).[21] This more nearly materialistic view was echoed by many of Aristotle's sixteenth-century critics.[22]

A third, quite distinct component of the trend was a result of the religious upheavals of the century. With the advent of the Counter-Reformation, both Protestant and Catholic tended to prefer obvious piety to occasionally sophistical exegetical subtlety, and began to give more weight to Scripture in their philosophizing. In order to allow for the creation and eventual destruction of the heavens, a number of authors abandoned the Thomistic position and affirmed that the heavens and the elements are identical in substance, differing only in form. Many Church Fathers could be cited in support of this position.[23]

A fourth and final component was the rise of the philosophy, once called Neoplatonic, that is probably most accurately described as naturalistic.[24] Its salient characteristic is that Nature is personified and acts as a world soul or *Anima Mundi.* Although naturalism drew most of its adherents from beyond the pale of academia, traces of naturalism may be found in a number of works by academic philosophers of the later sixteenth century.

These four forces—classicism, Galenic Stoicism, religious turmoil, and naturalism—were perhaps not the only causes tending to change accepted opinions about the celestial matter, but there can be no doubt that such changes had begun to take place. Thus, the influential Aristo-

[20]Aristotle, *On Generation and Corruption,* Book II, Ch. 1-3.

[21]Lasswitz, *Geschichte der Atomistik,* I, *op. cit.,* pp. 233, 325.

[22]For example, Jerome Cardan, *De subtilitate rerum,* (Paris and Nürnberg, 1550), Book I, in *Opera omnia,* III (Lyon, 1663), pp. 358-359; Bernardino Telesio, *De rerum natura iuxta propria principia,* (Geneva, 1588), Book I, Ch. 1, columns 555, 559, 563.

[23]Cf. Cornelius Valerius Veteraquinas, *Physicae seu de naturae philosophia institutio,* (Antwerp, 1567), p. 26, citing St. Basil the Great. For more authorities, cf. Christopher Scheiner, *Rosa Ursina,* (Bracciano, 1630), pp. 647, 673, 679, 681, 686, 688, 693, 766.

[24]It has been so called by Samuel Purchas, *Purchas his Pilgrimage,* 2nd ed., (London, 1614), p. 14, and recently by Léon Blanchet, *Les antecèdents historiques du ≪Je pense, donc je suis≫,* (Paris, 1920), p. 69, and Robert Lenoble, *Mersenne, ou la naissance du mécanisme,* (Paris: J. Vrin, 1943), p. 5 *et passim.*

telian commentaries written by the faculty of the Jesuit College of Coimbra treated the heavens as unambiguously material, and asserted the reality of the usual astronomical spheres.[25] Andrea Cesalpino[26] and the Swiss philosopher Sebastian Verro[27] adopted similar positions (with important differences). Moreover, it appears that these writers were not alone. Works expressing similar opinions are mentioned recently by Sister Patricia Reif,[28] who, like Léon Blanchet,[29] found a massive trend towards materialism in sixteenth century natural philosophy. Are these changes evidence of a tide of reform? Not at all—in fact, the authors of these opinions felt themselves to be conservatives, holding to what they regarded as received and authoritative opinion. Aristotle was for them still prince of philosophers. It seems likely, therefore, that a deeper tide was flowing, of such magnitude that those caught in it were unaware of its motion. And indeed, the sixteenth century is so obviously a period of sweeping change that it would be strange to find philosophy unaffected by less esoteric developments.

Religion and educational policy, closely bound up as they are with men's beliefs and ideas, could not but affect academic opinions. When the former two change, the latter may be expected to do so as well. Now widespread changes in religious belief and organization, such as the Reformation produced, are likely to be accompanied by a certain amount of historical distortion and myth-making. Present concerns overwhelm genuine interest in the past, thus tending to deaden whatever sensitivity men might have had for the ideas and preoccupations of their forebears. Even the literary humanists, who were

---

[25]*Commentarii Collegii Conimbricensis Societatis Jesu in quatuor libros de coelo,* (Coimbra, 1592), Book I, Ch. 2, Question 4, Article 2, and Q. 6, Art. 3; Book I, Ch. 7, Q. 1, Art. 2.

[26]Andrea Cesalpino, *Quaestionum peripateticarum libri V,* (Florence, 1569), Book III Ch. 3-4. For his neoclassicism, see *Nouvelle Biographie Universelle,* IX (Paris, 1852-66), pp. 436-40; for naturalist tendencies, see Cesalpino, *op. cit.,* Book I, Ch. 7.

[27]Sebastian Verro, *Physicorum libri X,* (London, 1581), Book II, Ch. 32, p. 45; Book III, Ch. 4, p. 67. The latter passage shows him to be a Galenist.

[28]Sister Patricia Reif, "The Textbook Tradition in Natural Philosophy," *Journal for the History of Ideas,* 1969, *30*: 26-27.

[29]Léon Blanchet, *op. cit.,* p. 69; cf. Reif, *Natural Philosophy in Some Early Seventeenth Century Scholastic Textbooks,* (St. Louis University: Unpublished Doctoral Dissertation, 1962), pp. 83-97; 321-322.

genuinely interested in an accurate re-creation of Latin literature and culture, adopted an unreasonably polemical attitude towards the learning of the Middle Ages, especially when this attitude was exacerbated by a Protestant suspicion of the popish and perhaps idolatrous past.[30]

Sixteenth century educators, whatever their creed or personal convictions, could not have avoided facing this sort of religious bias. Since men are more easily swayed by eloquence than by subtlety of reasoning, logic gave way to rhetoric as the foundation of the academic curriculum. The Protestants were especially eager to erase the stamp of Rome's influence, and began to reform their institutions before the Catholics. But the newly-founded Jesuits were not slow to react, and by 1600 there were nearly 250 Jesuit colleges.[31] The rhetorically-based education provided by the newly-founded or reorganized schools and colleges naturally tended to produce men good at literary pursuits, well-prepared to handle classical texts, but ignorant and often scornful of pedantic logic-based philosophy.[32] Cesalpino seems to have been such a man, reviving Aristotle as a classical author and completely misunderstanding the subtleties of scholastic physics. Analogical thinking, too, was encouraged by such a curriculum. "Reading," says Bolgar, "was always a preparation for writing. Schoolboys worked notebook in hand, and they were trained to notice the detail rather than the total purport of the texts before them."[33] Literature was, for them, a model, and not a source of information.[34] Skills acquired in this way, if applied to the study of nature, could produce novel results. At worst, they found expression in the Ramists' over-fastidious organization and categorization of material, carried out under the rubric of "method."[35] At best, they resulted in the careful consideration and comparison of qualities and quantities of which Galileo's works provide an excellent example.

---

[30]Frances A. Yates, *Giordano Bruno and the Hermetic Tradition,* (London: Routledge and Kegan Paul, 1964), pp. 165-166.

[31]R.R. Bolgar, "Education and Learning," *New Cambridge Modern History,* III (Cambridge: Cambridge University Press, 1968), pp. 430-432.

[32]*Ibid.,* pp. 437-438.

[33]*Ibid.,* p. 436.

[34]*Ibid.,* p. 432.

[35]Reif, "The Textbook Tradition in Natural Philosophy," *op. cit.,* pp. 28-29.

Might not the sort of argument that led Galileo to liken the moon to the earth have led less daring thinkers to bring the heavens just a little closer?

Whether or not the materialization of the heavens can be explained in this way, one thing at least seems certain. Those introducing the change, because of their lack of real interest in logic, did not entirely understand the peripatetic philosophy as it was handed down to them. Accustomed to the newer attitudes towards learning, they read ideas into these works which were never intended by their authors. The inconsistencies which inevitably arose they took to be the flaws of an outmoded way of thinking, and felt their modifications to be simple corrections. For this reason they were unaware that they were radically re-interpreting traditional philosophy, and so innocently prepared their own ruin.

The resurgence of realism in the last few decades of the sixteenth century, mentioned above, might under other circumstances have passed as an historical curiosity. But in the context of the philosophical currents just discussed, it takes on a new significance. If the heavens are thought to be immaterial or quasi-material, there can be no question of real, solid planetary spheres. If, however, the heavens are materialized and acquire elementary qualities, such spheres become a real possibility. So, when certain astronomers objected to Fracastoro's homocentrics but refused to admit the imaginary status of their own devices, they found philosophers increasingly responsive.

These astronomers concentrated their attack upon the homocentrists' inability to account for the phenomena. The one observable consequence of homocentric spheres, they argued, would be a constant brightness of the planets. But if the planets really move with some kind of epicyclic or eccentric motion, their brightness ought to vary with a period directly related to that of their motion. The former effect is not observed, while the latter is. A variant of this argument was used by Copernicus,[36] but to little effect, as his hypotheses were usually dis-

---

[36]Nicolaus Copernicus, *De revolutionibus orbium coelestium,* (Nürnberg, 1543), Book I, Ch. 10, fol. 8^{V}.

missed as physically absurd though mathematically elegant.[37] But it was revived by Franciscus Junctinus (Giuntini) in the 1577-78 edition of his commentary on the *Sphere* of Sacrobosco.[38] Junctinus hit upon a particularly happy way of stating the argument: he did not assert the truth of any particular epicyclic or eccentric hypothesis, but used the variation in brightness to defend the general principle of eccentric motions. He thus managed, in effect, to distinguish between verification and falsification, and gave the latter a crucial role in empirical arguments. At the same time he managed to remain at least apparently within the Aristotelian fold.

The well-known Jesuit mathematician Christopher Clavius, in the third edition of his commentary on the *Sphere*, uses arguments which are quite like those of Junctinus.[39] But there is a difference: while Junctinus did not name any opponents, and appears to have been concerned mainly with the arguments of the homocentrists, Clavius was speaking directly to the physical claims of Copernicus. He had discovered that the argument can also be used against those who claim that the phenomena necessarily entail a particular theory. Conceding that the planets do in fact vary in distance in the way predicted by Copernicus's theory, he argues that one is not thereby forced to assent to that theory: "All that can be concluded from Copernicus's assumption is that it is not absolutely certain that the eccentrics and epicycles are arranged as Ptolemy thought, since a large number of phenomena can be defended by a different method."[40] He then argues that, just as the choice of hypotheses is restricted by phenomena, it is also determined by physical principles. Of two phenomenally equivalent theories, that one must be chosen which is in agreement with sound physics and, Clavius adds, with Sacred Scripture.

---

[37]Lynn Thorndike, *A History of Magic and Experimental Science,* VI (New York: Columbia University Press, 1941), Ch. 31 *passim.*

[38]Franciscus Junctinus, *Commentarius in sphaeram Ioannis de Sacro Bosco,* II (Lyon, 1578), Ch. 4, pp. 330-343; quoted in Duhem, *To Save The Appearances, op. cit.,* pp. 85-86.

[39]Christopher Clavius, *In sphaeram Ioannis de Sacro Bosco commentarius. Nunc iterum ab ipso auctore recognitus,* (Rome, 1581), pp. 436-437.

[40]*Ibid.* Translation from Duhem, *To Save The Appearances, op. cit.,* p. 94.

Duhem believed that the positions of Junctinus and Clavius differed fundamentally,[41] and it is true that their aims were somewhat different. Junctinus was apparently interested in keeping astronomy from becoming enslaved to a system favored by philosophers, while Clavius endeavored to bind astronomy and physics more closely together. But Duhem's conclusion is unwarranted. For Junctinus, perhaps unaware, insists upon the physical reality of a particular type of theory based entirely upon phenomena. Moreover, he does so despite the glaringly obvious conflict between Aristotelian celestial physics and Ptolemaic planetary hypotheses. Junctinus's intentions are relevant only to an enquiry into the causes of his having adopted certain opinions. His possible historical effect is to be assessed by examining what he actually wrote. And his opinion, as written, departs radically from the generally accepted doctrine. This simple and apparently innocuous tacit assumption, that physical truth can be decisively established *a posteriori*, destroys the very foundations of Aristotelian physics, which is supposed to rest upon prior and more certain knowledge. Clavius, in turn, was only insisting that if physics must be responsible to phenomena, then the phenomena should properly be interpreted in the light of physics. So Clavius and Junctinus were much more nearly in agreement than Duhem suspected. But Duhem's evaluation of the significance of Clavius' argument is unexceptionable. Clavius' influence must not be underestimated. His position in Rome lent authority to his words, and his commentary on the *Sphere* became a standard reference work, highly praised[42] and widely read.[43]

Clavius and Junctinus were interpreters rather than innovators. They responded creatively to ideas with which they were presented, and so helped to establish the academical respectability of arguments that had long been used by Copernicans, Paracelsians, and other extramural theorists. In this they were assisted by the various forces mentioned

---

[41]Duhem, *op. cit.*, pp. 92-96.

[42]Lynn Thorndike, *The Sphere of Sacrobosco and its Commentators,* (Chicago: University of Chicago Press, 1949), p. 42.

[43]*Bibliothèque de la Compagnie de Jésus, Prèmiere Partie: Bibliographie, par Augustin et Aloys de Backer, S.J., Nouvelle édition par Carlos Sommervogel, S.J.,* (Brussels and Paris, 1890-), lists eighteen editions between 1570 and 1611, very nearly a new edition every two years.

above, which had by that time begun to take decisive effect. First, the materialistic trend in natural philosophy, both traditional and anti-traditional, was beginning to rout the Aristotelian idealism of the schoolmen. And second, the rising interest in astronomy gave the astronomers a growing audience, and perhaps made them more bold to speak their minds. Events were leading an increasing number of them, even of otherwise bound to tradition, to attack implicitly the scholastic separation of physics from astronomy. Assisting these two tendencies were both the fully developed idea that the parallax of comets should be measured, and the perfectly timed comet of 1577 upon which such measurements could be carried out. Tycho Brahe and others did not fail to seize the opportunity, adding one more argument, perhaps the crucial one, to those urging changes in world-view.

Tycho Brahe's importance for the history of astronomy can hardly be overestimated. His world system, and his work as an observer, are familiar to all. His discoveries relating to lunar theory have recently received the attention they merit.[44] But from the point of view of the present argument, he is a key figure for quite different reasons. For he stood as if in the confluence of the two trends just mentioned. On the philosophical side, he had read, and substantially agreed with, Cardan's theory that comets are celestial,[45] and he also comments favorably upon the idea, which he attributes to Paracelsus, that the heavens are filled with fire.[46] At the very least, this shows that his investigation of the parallax of the 1577 comet was inspired by a philosophical viewpoint which was not in the Aristotelian mainstream. His astronomical theories, too, were at odds with the traditional opinions. Here, it was neither the ingenuity nor the novelty of his world-system that he regarded as his chief accomplishment. Indeed, Jones shows that both he and others were aware that similar arrangements of planets had been hypothetically considered before. This did not particularly trouble him, as he took pride in being the first to assert the physical truth of the geo-

---

[44]Victor E. Thoren, "An Early Instance of Deductive Discovery: Tycho Brahe's Lunar Theory," *Isis,* 1967, *58*: 19-36.

[45]C. Doris Hellman, *The Comet of 1577: Its Place in the History of Astronomy,* (New York: Columbia University Press, 1944; new ed., New York: AMS, 1971), pp. 92-93.

[46]Tycho Brahe, *Opera Omnia; edidit I.L.E. Dreyer,* IV (Copenhagen: Libraria Gyldendaliana, 1913-1929), pp. 382-383.

heliocentric system.[47] Tycho agreed with Copernicus' critique of the unsystematic nature of the Ptolemaic "monstrosity," and felt that the choice of hypotheses is not purely arbitrary. His system was, in his opinion, the only arrangement which could fully account for the phenomena and at the same time agree with physical and scriptural truth. This belief, far from being just another Tychonic idiosyncrasy, provided the motivation for nearly everything that Brahe did. Its crucial role is shown by a letter to Caspar Peucer, wherein he admits that he originally believed the heavens to be filled with real (and presumably material) spheres.[48] He would not have discarded this belief had his new system, in which the paths of several planets crossed that of the sun, allowed its retention. But since his system did imply the impossibility of real, distinct heavens, Tycho repudiated his former belief and enlisted the support of parallax measurements. And unlike his predecessors, he approached the tasks of observation and calculation with the hope of proving this comet to be beyond the moon, which hope caused him to insist upon unprecedented accuracy. The tacit assumptions which lie beneath his theories and measurements are very clear: 1) planetary theories reflect physical realities, and 2) the heavens must be either solid or fluid. These two points are the historical conclusions of the two major trends discussed above. Without these assumptions, Tycho's and others' investigations of cometary parallax would have been in vain.

Around the turn of the seventeenth century the effects of the changes in matter-theory and of the rise of realism in astronomy are clearly visible. The philosophers could no longer maintain a convincing distinction between astronomy and physics. Even as determined a peripatetic as Bartholomaeus Keckermann failed here. Despite his arguments explicitly intended to keep astronomy in its place,[49] he materializes the

---

[47]Christine Jones (Schofield), *The Geoheliocentric Planetary System: Its Development and Influence in the Late Sixteenth and Seventeenth Centuries,* (The University of Cambridge: Unpublished Doctoral Dissertation, 1964), pp. 34-39.

[48]Brahe, *Opera,* VII, *op. cit.,* p. 130. Jones, *op. cit.,* p. 58, has made the understandable error of taking "solid" as a translation of the Latin *"realis."*

[49]Bartholomaeus Keckermann, *Systema physicum. . . anno Christi MDCVII publicé propositum in Gymnasio Dantiscano,* Book II, Ch. 5, in Keckermann, *Operum omnium quae extant tomus primus* [*-secundus*], I (Geneva, 1614), p. 1416; *De loco disputatio,* (Danzig, 1598), Book II, Ch. 14, in *Opera,* I, p. 1794; *Systema astronomiae compendiosum,* (written before 1605), Book I, Ch. 1, in *Opera,* I, p. 1835.

heavens (opposing Scaliger) and rests his physical argument upon observable magnitudes such as the bulk and figure of the celestial bodies.[50] The position which he thus took was quite indefensible against arguments based upon the parallax of comets, and his total suppression of these arguments, despite his awareness of them,[51] can only be taken as an admission of vulnerability.

Keckermann's unsuccessful attempt to retain a logically sound division of the sciences shows, above all, the extent to which astronomy and physics had become intertwined. Although it was possible even then to separate the two, few were able to do so. Only two writers of a selection of twenty-three who published between 1595 and 1610 held physical theories which could have withstood arguments from parallax.[52] Several others distinguished physics and astronomy only as a matter of convenience or habit.[53] In these circumstances, any critical pressure brought to bear upon the usual system of natural philosophy would meet with little resistance.

At the same time, the idea that the heavens might be fluid was gaining adherents among educated men, mostly outside the universities.[54] Earlier, peripatetics could easily have mounted a convincing opposition to this position, based upon the idea that the heavens are not

[50]Keckermann, *Systema physicum, op. cit.*, Book II, Ch. 1-2, pp. 1405, 1407.

[51]Keckermann, *De loco disputatio, op. cit.*, Book II, Ch. 14, p. 1795.

[52]M. Fabianus Hippius, *Problemata physica et logica peripatetica,* (Wittenberg, 1604), Problem 5, p. 59, and Prob. 9, p. 80; Antonius Laurentius Politianus, *De numero, ordine, et motu coelorum adversus recentiores liber,* (Paris, 1606), Ch. 1, p. 9, and Ch. 20, pp. 82-86.

[53]Adriaan Metius, *Institutionum astronomicarum libri tres,* I (Franeker, 1606), *praefatio;* Georg Stampelius, *Tabulae cosmographicae,* (Frankfurt, 1609), *Tabula* VIII; Edward Wright, *The Description and Use of the Sphere,* (London, 1613), Book I, Ch. 1, p. 1; John Chamber, *Treatise Against Judicial Astrology,* (London, 1601), Ch. 20, p. 102; Claude Duret, *Discours de la verité des causes et effets, des divers cours, mouvements, flux, reflux, & saleure de la mer Ocean,* (Paris, 1600), Ch. 6, p. 32.

[54]Conrad Aslachus, *De natura caeli triplicis libelli tres,* (Siegen, Nassau, 1597), Book II, Ch. 21-23; William Gilbert, *De magnete,* (London, 1600), p. 208; Christopher Heydon, *A Defence of Judiciall Astrologie,* (Cambridge, 1603), Ch. 12 and 18, pp. 302 and 370; Nicholas Hill, *Philosophia Epicurea, Democritiana, Theophrastica,* (Paris, 1601), *passim*; Johannes Kepler, *Mysterium cosmographicum,* (Tübingen, 1596), and *De stella nova,* (Prague, 1606), *passim;* Justus Lipsius, *Physiologica Stoicorum,* (Antwerp and

really material. Now, however, owing to the tacit questioning of this idea by late sixteenth century natural philosophers, no such coherent opposition existed. Two authors maintained (though ineptly) a truly traditional position,[55] and one has been found who clearly believed in solid celestial spheres.[56] The remainder of the sample either qualify their statements to the point of meaninglessness,[57] or are entirely noncommittal.[58] These last obviously meant to remain solidly within the scholastic tradition, and they appear to have been relying upon that tradition to lend weight to their assertions. Unfortunately for them, the clear tradition no longer existed.

This circumstance was the occasion of much confusion. Authors tended to impute to their opponents the most backward opinions that could be wrung from their writings, or to explain away any weaknesses in their friends' theories. Christopher Heydon, for example, accuses John Chamber of believing in solid spheres,[59] although Chamber adopts a modest agnosticism.[60] Similarly, J.C. Lagalla, who was either a naturalist or a very late exponent of immaterial heavens,[61] was misrepre-

Paris, 1604), Book II, Ch. 11-12; Thomas Lydiat, *Praelectio astronomica de natura coeli et conditionibus elementorum,* (London, 1605), Ch. 3-4, pp. 23-32; Edward Wright, preface to Gilbert, *De magnete,* trans. P. Fleury Mottelay, (New York: Dover Publications, 1958), pp. xli-xlii.

[55]M.F. Hippius, *Problemata, op. cit.,* Problem 5, p. 59, and Prob. 9, p. 80; Laurentius Politianus, *De numero coelorum, op. cit.,* Ch. 1, p. 9, and Ch. 20, pp. 82-86.

[56]Duret, *Discours, op. cit.,* Ch. 1, p. 1.

[57]Georg Henisch, *Commentarius in Sphaeram Procli Diadochi,* (Augsburg, 1609), p. 48; Johannes Jessenius, *De anima et corpore universi,* (Prague, 1605), p. 10; Bartholomaeus Keckermann, *Systema physicum, op. cit.,* Book II, Ch. 1, p. 1406; Justus Lipsius, *Physiologia Stoicorum, op. cit.,* Book II, Ch. 13, p. 102.

[58]Peter Crueger, *Theoremata exegetica de cometis in genere,* (Danzig, 1605), in Keckermann, *Disputationes philosophicae,* (Hanover, 1606), esp. Theses 11, 12, 15-24; Rodolphus Goclenius the Elder, *Cosmographia,* (Marburg, 1603), Book I, Ch. 2, fol. A4$^{r\text{-}v}$; Georg Stampelius, *Tabulae cosmographicae, op. cit.,* Tabula VIII; XXI; John Chamber, *Treatise, op. cit.,* Ch. 20, p. 102; Adriaan Metius, *Institutionum astronomicarum, op. cit., praefatio.*

[59]Christopher Heydon, *Defence, op. cit.,* Ch. 12, p. 302.

[60]Chamber, *Treatise, op. cit.,* Ch. 20, p. 102.

[61]Julius Caesar Lagalla, *De phoenomenis in orbe lunae, novi telescopii usu a Galileo nuperrime suscitatis, physica disputatio,* (Venice, 1612), Ch. 9, pp. 37-40.

sented by John Wilkins as an ignorant believer in fluid heavens.[62] It is not often easy to determine an author's true position, or to characterize the broad divisions of opinion. What does emerge from this study, however, is that the formerly widespread belief that the heavens are of a nearly incomprehensible nature had become distinctly unpopular. This does not mean that the peripatetics' beloved quintessence is not to be found in early seventeenth century writings. It was still very much in evidence, but was usually similar in substance to the elements and was endowed with certain qualities which earlier scholastics had banned from the heavens (solidity, for example). And even if this was not always so, writers who were accustomed to thinking in these terms were likely to read them into works originally written in a different conceptual milieu (cf. Wilkins on Lagalla). The materialist trend in philosophy had mingled with the realist trend initiated by philosophically inclined astronomers such as Copernicus, Tycho, and Clavius, resulting in the synthesis of the notorious "Aristotelian" solid deferents and epicycles.

While the educated community at large was attracted by the idea of fluid heavens, most of the writings of strictly academic provenance in the period 1595-1610 either opposed[63] or ignored[64] the idea. Few of these paid any attention to measurements of the parallax of comets, an omission which was in at least one instance deliberate, as was remarked above.[65] The one author who ventured to refute the astronomers cogently argued the philosophical side of the question, but displayed a woefully inadequate knowledge of astronomy.[66] A similar ignorance,

---

[62]John Wilkins, *The Discovery of a World in the Moone,* (London, 1638), Proposition 6, p. 87, responding to Lagalla, *op. cit.*, Ch. 1, p. 15.

[63]Bartholomaeus Keckermann, *Systema astronomiae, op. cit.,* Book I, Ch. 1, pp. 1836-1837; *Systema physicum, op. cit.,* Book II, Ch. 2-3, pp. 1407-1408, 1416-1417; Laurentius Politianus, *De numero coelorum, op. cit.,* Ch. 20, p. 84; M.F. Hippius, *Problemata, op. cit.,* Problem 5, p. 59; Johannes Jessenius, *De anima et corpore universi, op. cit., pp. 10-11.*

[64]Georg Henisch, *Commentarius, op. cit.,* p. 48; Peter Crueger, *Theoremata, op. cit.,* Theses 11, 12, and 15-24; Rodolphus Goclenius the Elder, *Cosmographia, op. cit.,* Book I, Ch. 2, fol. 14^{r-v}; Georg Stampelius, *Tabulae Cosmographicae, op. cit, Tabulae* VIII, XXI; John Chamber, *Treatise, op. cit.,* Ch. 20, p. 102; Adriaan Metius, *Institutiones, op. cit.,* preface.

[65]See note 51.

[66]Laurentius Politianus, *De numero coelorum, op. cit.,* Ch. 1, p. 9, Ch. 20, pp. 83-86.

sometimes an intentional neglect, characterized the writings of the more traditionally-inclined academic authors of the day. It is therefore clear that the non-fluid celestial orbs beloved of the peripatetics would be abandoned as soon as the academics were squarely confronted with the astronomers' parallax arguments. That such a confrontation was increasingly likely is shown by the growing presence in the universities of men competent in astronomy and willing to allow it a role in physics. While such men were not unknown in the sixteenth century,[67] their number and competence at the turn of the seventeenth century was unprecedented.

Significantly, the pupils of the independent mathematicians of the sixteenth century were beginning to occupy university chairs. The professor of Latin, Greek, and Theology at Copenhagen was Conrad Aslachus, a student of Tycho Brahe.[68] He published in 1597 a work which attempted to integrate recent astronomical realism into a modified peripatetic cosmology.[69] He was soon to be joined at Copenhagen by another follower of Tycho, Christian Longomontanus. The same thing also was happening elsewhere; for example, Kepler's student and son-in-law Jacob Bartsch taught at Strassburg, and Oughtred's students are found at the English universities.

Several other authors, though members of a university, decisively rejected the scholastic doctrines concerning the heavens. David Origanus, of the University of Frankfurt on the Oder, read and substantially agreed with Gilbert's *De magnete*, and Thomas Lydiat, while at Oxford,[70] composed a fascinating cosmology in which astronomy is based upon theology and physics in a manner reminiscent of Kepler.[71] Justus Lipsius, while not an originator of any theories, provided in his *Physiologia Stoicorum* a considerable body of respectably

[67]Cf. F.R. Johnson, *Astronomical Thought in Renaissance England,* (Baltimore: The Johns Hopkins Press, 1937), p. 12; A.C. Crombie and M.A. Hoskin, "The Scientific Movement and its Influence," *New Cambridge Modern History,* IV (Cambridge: Cambridge University Press, 1970), pp. 138-139.

[68]"Kurt Aslaksen," in *Dansk Biografisk Leksikon,* Copenhagen, 1933-1944.

[69]Conrad Aslachus, *De natura caeli triplicis, op. cit.,* esp. Book II, Ch. 3-13.

[70]Thomas Lydiat, *Praelectio astronomica, op. cit., praefatio.*

[71]See W.H. Donahue, "A Hitherto Unreported Pre-Keplerian Oval Orbit," *Journal for the History of Astronomy,* 1973, *4*: 192-194.

ancient but decidely non-Aristotelian physical theory.[72] Finally, one must not forget Clavius, still teaching mathematics at the Jesuits' Collegio Romano and still publishing new editions of his *Commentary on the Sphere*. It was he who wrote the most detailed defense of astronomical realism, thus helping to make epicycles worthy of natural philosophers' attention. He went beyond that, however. In drawing attention to Tycho's comet observations and challenging philosophers to come to terms with them, he sounded a call which was in time to bring about, not (as he hoped) the modification, but the ruin of the peripatetic cosmology.

Clavius and his Jesuit colleagues, and others (as, for example, the nascent group at Copenhagen), had developed a cautiously progressive attitude which was to play a crucial role in this process of ruin. The true potential of this attitude, however, was not realized until Galileo's startling telescopic observations provided a further occasion for debate on the nature of the heaven. Academic response was swift, generally favorable,[73] and was led by Clavius and other Jesuits. Clavius inserted into the last edition of his *Commentary on the Sphere* a brief mention of the observations, with a reiteration of his challenge to the peripatetics to account for them.[74] He thus lent his considerable authority to the opponents of scholasticism. The following year, Christopher Scheiner, S.J., Professor of Mathematics at Ingolstadt, drew attention to Clavius' acceptance of the Medicean Planets in his second work on sunspots.[75] Two years later (1614) he presided over the publication of the *Disquisitiones mathematicae* of his pupil Johann Georg Locher.[76] These publications signal the debut of a group of men within the Society of Jesus who were deeply interested in reconciling natural philosophy with observations, especially of the heavens. They were initially based at the

[72]Justus Lipsius, *Physiologia Stoicorum, op. cit.*

[73]See the works contained in Galileo, *Le opere di Galileo Galilei,* III (Florence: G. Barbèra, 1890-1909), as well as those listed in the following three notes.

[74]Christopher Clavius, *In sphaeram Ioannis de Sacro Bosco commentarius,* Ch. 1, in *Opera,* III (Mainz, 1611), p. 75.

[75]Christopher Scheiner, *De maculis solaribus et stellis circa Iovem accuratior disquisitio,* in Galileo, *Opere* V, *op. cit.*, p. 69.

[76]Johann Georg Locher, *Disquisitiones mathematicae de controversiis et novitatibus astronomicis,* (Ingolstadt, 1614).

Collegio Romano and the University of Ingolstadt, but their influence quickly spread *via* the network of the Jesuit Colleges.

These Jesuit scientists were anxious to renovate rather than to build anew; hence, their natural philosophy has a distinctive tone which sets it apart from traditional philosophy and contemporary innovations alike. They do not question the legitimacy of the telescopic observations, but adopt their own explanations: while adopting a positive attitude towards celestial novelties such as new stars and "lunate Venus," they differed as little as possible from standard early seventeenth century school lore. Locher's frequent mention of epicycles[77] initially leads the reader to suspect that he somehow still believed in solid spheres, though he elsewhere denies that the planets are fixed to epicycles[78] and in a diagram has the orbs of Mars and the sun intersecting[79] (this last feature of the Tychonic system was universally supposed to rule out real or solid deferents and epicycles[80]). Clavius' recognition of the new phenomena took the form of the two insertions, mentioned above, in later editions of his *Commentary*. He did not, however, alter the text itself, which, it will be recalled, was instrumental in establishing the idea of celestial solidity. Scheiner is only slightly more explicit than Clavius, whom he cites as an authority.[81] Like his pupil Locher, he writes of orbs and eccentrics as though real,[82] but he goes on to deny the existence of solid spheres "particularly in the heaven of the Sun, and Jupiter."[83] This is wholly in accord with the theory which was at that time being proposed by Jesuits of the Collegio Romano. According to the theory, the sunspots' erratic behavior is to be explained by supposing them to be the effect of superposition or confluence of many circumsolar planets. Each of these planets can then be supposed perfectly spherical and incor-

---

[77]*Ibid., Disquisitio* 25, pp. 55-56; *Disquisitio* 44, p. 89.

[78]*Ibid., Disq.* 25, p. 55.

[79]*Ibid.,* p. 52.

[80]This point is discussed by Libert Fromondus, *Ant-Aristarchus sive orbis terrae immobilis liber unicus,* (Antwerp, 1631), Ch. 17, pp. 92-93.

[81]Scheiner, *Accuratior disquisitio, op. cit.,* p. 69.

[82]Scheiner, *Tres epistolae de maculis solaribus,* (Augsburg, 1612), Letter 3, in Galileo, *Opere,* V, *op. cit.,* p. 29.

[83]Scheiner, *Accuratior disquisitio, op. cit.,* Letter 3, p. 69: "*praesertim ad Solis, Iovisque coelum.*"

ruptible, and the heavens themselves thus remain free from decay. It seems possible that Scheiner, too, subscribed to this opinion,[84] though Galileo understands him as saying that each spot is a star.[85] None of the three stated positively what he believed to fill the heavens, despite their rejection of solid spheres. One cannot be entirely certain, therefore, that they maintained the incorruptibility of the heavens. Nonetheless, the evidence provided by Cesi's letter strongly suggests that they were trying to save the incorruptible heavens despite their newly-acquired fluidity.

There was another author, circumstantially close to Jesuit circles, who was more explicit about the contents of the celestial regions. This man, Julius Caesar Lagalla, was a physican in the papal service and Professor of Philosophy at the *Sapienza*; therefore, he was likely to have known Clavius. He wavered between belief in a fluid, incorruptible substance,[86] and Averroist immaterialism, which he conflates with Telesio's naturalism.[87] He was, however, quite certain about the solid spheres: not only are there none, but no one ever seriously believed in them.[88] For this reason, his position appears to have been much like that of the Jesuit scientists.

It can therefore be said with some confidence that, as a direct response to Galileo's telescopic discoveries, a new and distinct group has arisen, in addition to the traditional academic and the independent writers. The opinions adopted by these writers are distinct from those of the other academics in their response to celestial novelties, and from those of the independents in their adherence to traditional natural philosophy. But most importantly, this group is distinguished by the circumstantial similarity of its members, and by the apparent relatedness of their opinions. For Lagalla, Locher, and Scheiner, at crucial points in their arguments, bring in as their chief and most

[84]Letter of Federico Cesi to Galileo, 14 September, 1612, in Galileo, *Opere,* XI, *op. cit.,* p. 393. For a thorough discussion of Scheiner's work, see W. Shea, "Galileo, Scheiner, and the Interpretation of Sunspots," *Isis,* 1970, *61*: 498-519. Jean Tarde, *Borbonia sidera,* (Paris, 1621), gives a full presentation of this theory.

[85]Galileo, *Istoria e dimonstrazioni,* (Rome, 1613), Letter 3, in *Opere* V, *op. cit.,* p. 231.

[86]J.C. Lagalla, *De phoenomenis in orbe lunae, op. cit.,* Ch. 7, p. 15; Ch. 9, p. 35.

[87]*Ibid.,* Ch. 10, p. 40; Ch. 9, pp. 34-36.

[88]*Ibid.,* Ch. 7, p. 15.

dramàtic authority (because of his apparent change of opinion) their colleague Clavius, who thus assumes a curious position in the development of theories of celestial matter. It was he, as much as anyone, who bears the responsibility for the widespread insistence upon the analogy of heaven and earth and the introduction of terrestrial physical criteria into astronomy. His theory of the status of hypotheses was meant to serve as a defense of traditional astronomy against Copernicus in particular, and it seems at first as though Clavius' remarks about the 1572 nova and the telescopic observations represent a reversal of his former opinion.[89] In fact, as should be clear, he was only accepting the consequences of that same theory. Having led many of his contemporaries to adopt solid celestial spheres, he simultaneously lent impetus to the movement which was to do away with that ill-starred invention. This analysis provides one possible reason for Lagalla's failure to mention Clavius, with whom he was no doubt personally acquainted: he was largely unaware of the chain of effects which such men as Clavius and Junctinus had set in motion. Lagalla, it appears, was exclusively concerned with the philosophies of Telesio, the Averroists, and other late sixteenth century writers and the way in which they could be saved in the face of the new phenomena. Following the line of reasoning which this approach suggested, he independently arrived at the conclusion reached by Clavius, Locher, and Scheiner.

The appearance of several comets in 1618 provided a further opportunity for the Jesuit reformers to air their views. By this time, Clavius had died, but his approach to celestial events was emulated by his successors at the Collegio Romano, Gregory of St. Vincent and Orazio Grassi. Scheiner was still alive and active, but it was his pupil Johannes Baptista Cysatus who wrote on the comets. He used his mentor's work as a paradigm, and supported his argument with the most nearly continuous of all published observations of the comets. His physical theory is based upon careful observation of the head of the comet, wherein he perceived a number of irregularly shaped bright spots. Seizing upon Scheiner's theory of sunspots, he advances the hypothesis that comets are formed in exactly the same way (by the confluence or conjunction of

[89]See, for example, J.G. Locher to that effect, *Disquisitiones mathematicae, op. cit., Disquisitio* 22, p. 50.

small, individually incorruptible bodies), and are in fact sunspots which do not happen to be near the sun. The comparative rarity of comets he ascribes partly to the need of such a confluence and partly to their invisibility when distant from earth,[90] apparently forgetting that he had earlier placed at least some readily visible comets at considerable distances.[91] He was perhaps aware of this difficulty, for he adds that comets may be bodily generated and destroyed.[92] But even here he clings to the tattered remains of the incorruptible heavens, suggesting that such rare occurrences be reckoned direct acts of God.

Unlike Cysatus, Grassi does not mention Scheiner's work, and while he identifies the matter of sunspots with that of comets, he does not say whether he accepted Scheiner's explanation.[93] In calling a comet a "temporary planet,"[94] he may be hinting at a belief that comets are generated and are subject to decay, but cannot in all prudence argue this.[95] His theory is otherwise not much different from that of Cysatus.[96]

The work of Gregory of St. Vincent was unfortunately not available to the present writer for study.[97] It may have been only a single sheet of theses stated without argument and printed for the purpose of disputation. It was highly praised by the author of another work on the comets, the Fleming Erycius Puteanus,[98] who nonetheless never states that his explanation of comets agrees with Gregory's.

There was another Jesuit pupil of Clavius who briefly considers transient celestial phenomena in his *Sphaera mundi* of 1620. Joseph Blancanus very likely wrote with Scheiner's second triad of letters on sunspots

---

[90]J.B. Cysatus, *Mathemata astronomica de loco, motu, magnitudine, et causis cometae qui sub finem anni 1618 et initium anni 1619 in caelo fulsit,* (Ingolstadt, 1619), Ch. 7, pp. 75-77.

[91]*Ibid.*, Ch. 3, Proposition 7, pp. 56-57.

[92]*Ibid.*, Ch. 7, p. 77.

[93]Orazio Grassi, *Libra astronomica et philosophica,* (Perugia, 1619), p. 27.

[94]*Ibid.*, p. 20.

[95]*Idem.*

[96]Cf. Grassi, *De tribus cometis anni MDCXVIII disputatio astronomica,* (Rome, 1619), pp. 12-13.

[97]Gregory of St. Vincent, *Theses de cometis,* (Louvain, 1619). It was not available even to the compilers of the *Bibliothèque de la Compagnie de Jésus, op. cit.*

[98]Erycius Puteanus, *De cometa anni 1618,* (Louvain, 1619), *"Ad Lectorem."*

in hand, as he treats their properties in the same order.[99] Nevertheless, in emphasizing their transitory nature, he strongly suggests that, *de facto,* a certain amount of alteration takes place in the heavens. On the other hand, he considers comets to be permanent and quintessential,[100] and, by ignoring the question, agrees implicitly with the Jesuit scientists' belief in a fluid yet incorruptible aether.[101]

The most important feature of the Jesuits' explanation of sunspots and comets—though hardly the most striking—is the distinction they drew between the planets and the surrounding space. In the late sixteenth century, the standard academic doctrine was that the planets are only denser parts of the continuous and homogeneous celestial matter.[102] This was invariably the opinion of those who upheld the reality of the orbs, even in the seventeenth century, as both were features of Aristotle's celestial physics, which they were trying to preserve.[103] When Cysatus calls the constituent bodies of comets "small planets" and treats them as being quite distinct from their surroundings, he takes an apparently small but extremely important step away from the accepted view of the heavens. Despite his continuing refusal to allow alteration to take place on the celestial bodies, he made them hard, permanent globes, not radically different from the earth. He thus prepared

[99]Joseph Blancanus, *Sphaera mundi,* (Modena, 1635), Book X, Ch. 20 (misnumbered 21), p. 132. (First edition: Bologna, 1620). Cf. Christopher Scheiner, *Accuratior disquisitio, op. cit.,* Letter 3, p. 65.

[100]Blancanus, *Sphaera, op. cit.,* Book XVI, Ch. 5 (misnumbered 4), Sect. 2, pp. 160-161.

[101]Cf. *ibid., loc. cit.;* also *proemium,* p. 27.

[102]Cf. Henricus Brucaeus, *De motu primo,* (Rostock, 1604), fol. 4^{v} (first edition: Rostock, 1573); *Commentarii Collegii Conimbricensis in quatuor libros de coelo, op. cit.,* Book I, Ch. 3, Question 1, Article 4, column 71-72; Archangelus Mercenarius, *Dilucidationes, op. cit.,* p. 83; J.C. Scaliger, *Exercitationes, op. cit., Exercitatio* 61, *passim; Ex.* 76, fol. 116^{r} et alibi; Aristotle, *De coelo,* Book II, Ch. 7.

[103]D'Abra de Raconis, *Totius philosophiae, hoc est logicae moralis physicae et metaphysicae capita, atque clara compendia,* III (Paris, 1617), p. 483; Rodolphus Goclenius the Younger, *Urania cum geminis filiabus, hoc est Astronomia et Astrologia speciali,* (Frankfurt, 1615), Ch. 2, p. 12; Ch. 4, p. 27; Gilbert Jacchaeus, *Institutiones physicae,* (Leiden, 1615), Ch. 12, p. 115; and others.

the way for later theories according to which earthlike planets circle the sun, but float in an incorruptible (i.e. chemically inert) aether.[104]

In introducing this novel kind of theory the Jesuit scientists further demonstrate their peculiar position *vis-à-vis* their contemporaries in the universities. For some of the latter had been drawn by recent developments in natural philosophy and in particular by its liaison with astronomy to accept the fluidity of the heavens while retaining a belief that the stars are merely denser parts of the heavens.[105] These men were by no means intellectual leaders, though they were more willing than many of their contemporaries to take note of what was to be seen in the sky. Their writings make it clear that they were drawn with some reluctance to abandon old theories. Such reluctance is plainly visible in the work of Ambrose Rhodius, Professor of Mathematics at Wittenberg. As he was writing on an astronomical subject (one of the comets of 1618), he felt obliged to follow the opinions generally accepted by astronomers.[106] But before finally demolishing the traditional arguments, he takes a last nostalgic look at Aristotle's comet theory.[107] Franciscus Suarez, probably partly in response to the new science, looked back to the Fathers for authoritative physical theories.[108] Jean Tarde, who, while not an academic, held thoroughly scholastic views, lamely revived Scheiner's theory of sunspots, probably only to obtain royal favor, as (following Galileo's example) he names the allegedly new stars (by then rather long in the tooth) after the ruling house of France.[109] The essential difference between these men and the Jesuit scientists is conveniently shown by Grassi's own attitude. With his experimental flair[110] and his considera-

[104]Ismael Boulliau, *Astronomia Philolaica,* (Paris, 1645), Book I, Ch. 4, p. 6; Phocyllides Holwarda, *Philosophia naturalis,* (Franeker, 1651), Book III, Ch. 1, p. 123 ff.

[105]Ambrose Rhodius, *De cometa qui 1618 [per Bootem] conspectus est,* (Wittenberg, 1619), fol. B2^r; Franciscus Suarez, *Commentaria ac disputationes in primam partem D. Thomae. Tractatus secundus, de opere sex dierum,* (Coimbra, 1620), in Suarez, *Opera omnia,* III (Paris, 1856), Book II, Ch. 8, Paragraph 5-6, p. 149; Par. 16, p. 153; Jean Tarde, *Borbonia sidera, op. cit.,* Ch. 5, p. 15.

[106]Rhodius, *De cometa, op. cit.,* fol. B1^v and esp. B3^v.

[107]*Ibid.,* fols. B2^v-3^r.

[108]Suarez, *Commentaria, op. cit.,* Book I. ch. 5, pp. 29-30 *et alibi.*

[109]Tarde, *Borbonia sidera, op. cit.,* Ch. 5.

[110]Orazio Grassi, *Libra astronomica, op. cit.,* Third Weighing, Proposition 1, pp. 43-52, for example.

tion of parallax,[111] he obviously felt himself to be in the scientific vanguard, and perhaps even somewhat daring. For, taken aback by what he felt to be excessive harshness in Galileo's criticism of his work, he responds,

> Hence, that we might seem to have granted something to the observations of our friends and at the same time publicly confuted the ignorance of those for whom this instrument the telescope was of no significance, we maintained publicly that this argument ought to be presented in third place, and finally these words ought to be added by which Galileo stated that he was displeased, since we hoped that protecting from invidious calumnies this telescope . . . we might therefore deserve well of him rather than ill.[112]

This would never have been said by Rhodius. He and his cohorts would no doubt have felt more sympathy with Aristotle than with Galileo.

As was remarked in passing above, the Jesuits' comet theory was enthusiastically received by Erycius Puteanus, Professor of Latin at Louvain. There appears to be one important difference, however, between his views and the Jesuits' theories: he seems to have admitted to the heavens the property of corruptibility.[113] This must be stated with hesitation because in the section devoted to the nature of comets Puteanus, in effect, rehearses Cysatus' theory, adding only that the gathering of constituent bodies is somehow caused by light.[114] Moreover, while he had no objection to physical alterations in the heavens, he still balked at the idea that they might be similar to the earth.[115] The distinction he wished to draw between the heavens and the elements is neither sharp nor clear. In the first place, as his intended audience believed in both the incorruptibility and the separate nature of the heavens, Puteanus must have felt greater need to argue against

[111]Grassi, *De tribus cometis, op. cit.*, pp. 8-12.

[112]Grassi, *Libra astronomica, op. cit.*, First Weighing, p. 16. Translation from Stillman Drake and C.D. O'Malley, *The Controversy on the Comets of 1618,* (Baltimore: Johns Hopkins Press, 1960), pp. 80-81.

[113]Erycius Puteanus, *De cometa, op. cit.*, Book I, Ch. 5, p. 27.

[114]*Ibid.*, Book I, Ch. 6, pp. 41-42.

[115]*Ibid.*, pp. 43-45.

the former than to support the latter. And secondly, he could not have drawn a sharp distinction because he believed the heavens to possess elementary qualities[116] and to include a modicum of disorder.[117] Nevertheless, the aether is distinguished from the elements, and the basis of this distinction appears to be regularity of events: for example, Puteanus argues that were the comet made of air or fire its appearance and motion would not have proceeded in so orderly a fashion.[118]

There are two other works on the 1618 comets in which a similar theory is advanced. These are by Thomas Fienus (Fyens) and his pupil Libert Fromondus (Froimont, Froidmont), who were close friends and published their accounts together. They believed that the comets were truly generated in the heavens, and can therefore be said with confidence to have accepted the idea of alterable heavens.[119] Neither of them advances any argument for a separate celestial substance. In a curious inverted anticipation of Descartes' position, they presume the heavens to be different from the elements in the absence of compelling arguments to the contrary.[120]

It is perhaps no coincidence that Fienus, Fromondus, and Puteanus were all at the University of Louvain, as Professors of Medicine, Philosophy, and Latin, respectively. Perhaps the former two, like Puteanus, were consciously following the Jesuits' lead. Their theory, however, though significantly different from the Jesuits', was not exclusively the property of professors at this university, as it was held (as far as one can tell) by Redemptus Baranzanus of Annecy, Haute-Savoie, and also by Caspar Bartholinus, Professor of Medicine at Copenhagen. Baranzanus discusses the possible corruptibility of the heavens in a rambling series of *dubia* in which enormous quantities of hair are split, and appears to come to the conclusion that the heavens are corruptible provided that they cannot be altered by the action of the elements.[121] Bartholinus, like

[116] *Ibid.*, p. 45.

[117] *Ibid.*, p. 35.

[118] *Ibid.*, p. 33.

[119] Thomas Fienus and Libert Fromondus, *De cometa anni MDCXVIII. Dissertationes Thomae Fieni et Libert Fromondi,* (Antwerp, 1619), pp. 42-43; 132.

[120] *Ibid.*, pp. 39-40; 132.

[121] Redemptus Baranzanus, *Uranoscopia seu de coelo,* ([Geneva], 1617), Part I, Question 3, especially *Dubium* 8, p. 94.

the Louvain professors, failed to argue his position, though he associates the heavens with simplicity, transparency, tenuosity, and purity.[122] The aether, he flatly states, is corruptible to the same extent and in the same manner as are the elements.[123]

When these five authors' reasons for dividing the celestial substance from the elementary are compared, the one common feature which presents itself is that none had any particularly compelling or clear argument. For them this distinction had lost all but a kind of emotional significance. It appears that the concept of celestiality emerged from the tacit debate about the materiality of the heavens with a great pliability of meaning. For this reason it played the historical role of a transitional stage between belief in incorruptibility and the abandonment of that belief. At Ingolstadt, Scheiner, Cysatus, and others boldly advanced the idea of fluidity and wavered on corruptibility, while holding firmly to celestiality. A little later, down the Rhine in Louvain, men in all probability of like mind abandoned incorruptibility but avoided complete defection by keeping a distinct celestial substance. Later yet, men in the same tradition dropped that vestige of a rapidly vanishing cosmology; thus, Christopher Borro, S.J., a direct heir of the Ingolstadt-Collegio Romano scientists, by 1630 felt it no longer necessary or desirable to insist upon a separate celestial substance.[124] In the midst of this transition stood another former disciple of Tycho, Christian Longomontanus, who was also a colleague of Bartholinus as Professor of Mathematics at Copenhagen.

Unlike many contemporary writers, Longomontanus makes it abundantly clear what he believed the heavens to contain. Like his countryman and fellow-disciple of Tycho Brahe, Aslachus, he establishes the nature of the aether by considering the properties which observation suggests it to have. He does not, however, consider directly its similarity or dissimilarity to the elements. In itself it is imperceptible; it is known

---

[122]Caspar Bartholinus, *Praecepta physicae,* II (Strassburg, 1621), *Pars Specialis Prima,* Ch. 1, fol. A2^{v}.

[123]*Ibid.,* fol. A4^{r}.

[124]Christopher Borro, S.J., *Collecta astronomica ex doctrina P. Christophori Borri Mediolanensis ex Societate Iesu de tribus coelis aereo sydereo empyreo issu* [sic] *et studio Domini Gregorii de Castelbranco,* (Lisbon, 1631), Book V, Ch. 2, pp. 324-328.

only by its effects. He therefore calls it "a most subtle and tenuous extended substance, most similar indeed to the incorporeal and insensible."[125] This *"expansum"*, as it is the medium by which light is transmitted, penetrates all space, even to the depths of the earth.[126] Longomontanus then notes the effects of the sun and stars upon "exhalations":

> Which being so, it follows that a not inconsiderable quantity of that very subtle matter lying above the surface of the land and sea is affected by the enormous power and heat of the rays of the stars, and chiefly the sun. It is thereby converted into that air which is fitting and salubrious for animals.[127]

He identifies this air with the tenuous air said by the sixteenth century pioneers Christopher Rothmann and Jean Pena to fill the universe,[128] and then concludes with a statement which is plainly intended to place Longomontanus' theory in its correct philosophical context:

> We have hitherto been concerned with the substance of the extended matter, and were anyone to say that it is comparatively spiritual, and to be likened to Aristotle's fifth essence, he would not, in my opinion, have differed much from the truth...[129]

He is, of course, associating his aether with the "spiritus" of the Stoics and naturalists, which he considers not unlike the fifth essence as developed by Aristotle himself. Questioned on the subject, he might well have referred to the very passage from *De generatione animalium* with which Thomas Aquinas bolstered his belief in quasi-spiritual

---

[125]Christian Severinus Longomontanus, *Astronomia Danica,* I (Amsterdam, 1622), p. 43: "expansum tenuissimum et subtilissimum, quippe incorporeo et insensibili simillimum...".

[126]*Ibid.*, p. 44. A similar theory is to be found in the *Enchiridion physicae restitutae,* (Paris, 1623), Canon 199, of the French alchemist Jean d'Espagnet.

[127]Longomontanus, *Astronomia Danica, op. cit.*, p. 44.

[128]*Ibid.*, p. 45.

[129]*Idem;* cf. *op. cit.*, Vol. II, "Appendix de novis coeli phaenomenis," pp. 11-12.

heavens.[130] But there is a crucial difference between the two men's theories: Thomas regarded the heavenly bodies as mere ethereal condensations; Longomontanus gave them many properties formerly proper to the earth.[131]

This account of Longomontanus' theory of the heavens suggests that his ideas differed in important ways from those of contemporary Jesuit scientists. This is indeed so: while they agree about the incorruptible aether, Longomontanus seems ready to admit the alterability of the planets themselves. In its dual aspect, his theory provides a most instructive view of a change of meaning which a certain key term was at that time undergoing as well as the far-reaching effect of that change. The term is *alteration.*

The earlier stages of this change have already been seen in the modified element-theories of the sixteenth century. It was shown that some influential natural philosophers had repudiated Aristotle's view that each element is no more than the product of two primary qualities. These men had begun to think of the *prima materia* as a kind of material substrate rather than a logical subject, and occasionally allowed it to have its own essential qualities. K. Lasswitz sees in this development the larval stage of the mechanical or corpuscularian philosophy of the seventeenth century.[132] Both logically and historically, the outcome of this theory was the idea that the elements are not transformed but merely combine to form substances with new properties. Alteration has acquired a new meaning.

The Jesuits followed Scheiner in adopting a theory of sunspots and comets which, ironically, fits almost exactly this new concept of alteration. They had intended their theory to account for apparent alterations of the heavens in terms which, in the context of Aristotelian matter-theory, would preserve their "incorruptibility." Confronted by

---

[130]Aristotle, *On the Generation of Animals,* Book II, Ch. 3, 736 b 55-57; Thomas Aquinas, *In De coelo,* Book II, Ch. 10, text 394.

[131]Longomontanus, *Astronomia Danica* I, *op. cit.,* pp. 45-46; *ibid.,* Vol. II, "Appendix de novis coeli phaenomenis," pp. 7-8.

[132]Lasswitz, *Geschichte der Atomistik,* I, *op. cit.,* pp. 306-312; 324-331.

the new concept of alteration, as in fact they were by Galileo,[133] they could no longer maintain their idea of incorruptibility. Either the heavens are in fact corruptible, or the elements themselves must be incorruptible. Thus they appeared to the proponents of this new conception of alteration to be quibbling. From the viewpoint of traditional philosophy they cannot have appeared much better, for, having described the celestial changes in terms of one theory, they judged it incorruptible on the basis of another wholly incompatible theory. The almost inevitable result was that the Jesuit scientists, even Scheiner, abandoned the idea of incorruptibility.[134]

The Jesuit defection left other writers with comparable views in an uncertain position. In the first two decades of the century, attention had been concentrated upon the nature of the heavens as a whole, and in this context Longomontanus is seen to be in agreement with Grassi and Cysatus. His opinions take on a different appearance, however, when compared with subsequent theories. For, once the nature of alteration is brought into question, the focus of attention is shifted to the specific changes observed in the heavens, such as the generation of sunspots and comets. Here, Longomontanus' ideas are not so close to those of the early Jesuit scientists, though they are comparable to the later theories of Borro and Scheiner.[135] They are closest, however, to the ideas of d'Espagnet, who affirmed the incorruptibility of the heavens on the basis of his matter-theory,[136] and replaced the fifth essence with air[137] or fire,[138] which he regarded as a manifestation of spirit.[139] As d'Espagnet clearly belongs in what has by this date become the mass of writers

---

[133]Galileo, *Opere* IV, *op. cit.*, p. 46; III, p. 4; VI, pp. 348-350.

[134]Christopher Scheiner, *Rosa Ursina*, (Bracciano, 1630), p. 647, column 2: line 16; 673, 1: 25; 679, 2: 19; 681, 1: 9; 686, 2: 35; 688, 1: 50; 693, 1: 37; 766, 2: 15; cf. Christopher Borro, *Collecta astronomica, op. cit.*, pp. 334-335; 328.

[135]Longomontanus, *Astronomia Danica, op. cit.*, "Appendix de novis coeli phaenomenis"; Borro, *Collecta astronomica, op. cit.*, pp. 334-335; 328; 337; *et alibi;* Scheiner, *Rosa Ursina, op. cit.*, p. 662, Column 1, line 25; 631, 1, 35.

[136]Jean d'Espagnet, *Enchiridion physicae restitutae, op. cit.*, Canons 118, 236.

[137]*Ibid.*, Canons 72, 237.

[138]*Ibid.*, Canons 112-114.

[139]*Ibid.*, Canon 193.

believing in corruptible elementary heavens, Longomontanus, too, fits this category, when considered in relation to works published after 1622.

Only one other author has been found who advocated a similar theory in the mid-1620's.[140] Marcus Wendelin regarded transmutation of elements as impossible, and hence considered each element incorruptible according to the Aristotelian criteria.[141] This is the same glib paralogism used by d'Espagnet.[142] But there the similarity ends, for Wendelin insists, partly upon scriptural grounds, that the stars and planets are made of the light created on the first day, and are therefore not at all like elementary things.[143] It might be possible to show how his curious theory was a plausible application of principles whose historical antecedents are plain,[144] but this would be pointless. What is already amply demonstrated is that concepts such as "incorruptible" and "celestial" had acquired such an instability of meaning that anyone using them had to indicate his choice of definition. It is hardly surprising, then, that these concepts came to be used less and less, and were retained by the most tradition-bound peripatetics[145] or by writers who, so to speak, wished to shed some old light on a new theory.[146]

By the end of the 1620's the debate over the fluidity of the heavens was very nearly concluded. Although belief in solid spheres was not quite dead, it was, even in the universities, the opinion of a minority of authors. Of a sample of writers active 1624-29, four believed in solid

[140]Marcus Friedrich Wendelin, *Contemplationum physicarum sectiones tres,* (Cambridge, 1648). Relevant section dated 1628.

[141]*Ibid.,* Book III, Ch. 2, Thesis 1, Ecthesis 2, p. 411; also Book III, Ch. 5, Thesis 7, Ecthesis 2, p. 464.

[142]D'Espagnet, *Enchiridion physicae restitutae, op. cit.,* Canons 72, 112-114, 118, 236-237.

[143]Wendelin, *op. cit.,* Book III, Ch. 11, Thes. 1, p. 502, and Ectheses 3-4, pp. 503-504.

[144]For example, J.H. Alsted, *Scientiarum omnium encyclopaedia,* II (Leiden, 1649), Book XIII, Part 2, Ch. 1-3, pp. 116-118. First edition: *Cursus philosophici encyclopaedia,* (Herborn, Nassau, 1620).

[145]Rodriguez Arriaga, S.J., *Cursus philosophicus,* (Antwerp, 1632), section entitled *De coelo,* Book III, Ch. 42, p. 503; Marius Bettinus, S.J., *Apiaria universae philosophiae mathematicae,* II, 4th ed., (Bologna, 1645), *Apiarium* 8, pp. 26-27; 49; 72-74.

[146]Ismael Boulliau, *Astronomia Philolaica, op. cit.,* Book I, Ch. 4, p. 6, and Book I, Ch. 5, p. 10; Phocyllides Holwarda, *Philosophia naturalis, op. cit.,* Book III, Ch. 1, p. 123; Book III, Ch. 4, 5, 7, *passim.*

spheres[147] and one suggested them as a strong possibility.[148] All but one, Faber, had a university post. The remaining eight believed in fluid, and for the most part in elementary, heavens.[149] The first group, already dwindling, might be further depleted by eliminating Burgersdicius, whose true opinion is presented in a less textbookish work published five years after his *Sphaera*.[150] Of the half-dozen authors between 1630 and 1650 who favored solid spheres, three were mindlessly repeating doctrines of such men as the Coimbra commentators and Clavius;[151] others mingle their peripateticism with nominalism[152] or naturalism.[153]

[147]Franco Petrus Burgersdicius, *Sphaera Ioannis de Sacro Bosco,* (Leiden, 1626), *passim;* Johannes Rudolphus Faber, *Cursus physicus,* (Geneva, 1626), *Pars Specialis,* Book II, Ch. 2, Question 1, pp. 167-168; George Hakewill, *An Apologie of the Power and Providence of God in the Government of the World,* (Oxford, 1627), Book II, Ch. 2, Sect. 2, p. 80; Book III, Ch. 9, Sect. 2, pp. 244-5; Adam Tanner, *Universa theologia scholastica, speculativa, practica,* I (Ingolstadt, 1626), Disputation VI, Question 3, *Dubium* 2, Paragraph 66, Column 1702.

[148]Raphael Aversa, *Philosophia metaphysicam physicamque completens quaestionibus contexta,* 2 vols. (Rome, 1625-27), quoted in Thorndike, *History of Magic, op. cit.*, VII, p. 394.

[149]Andrea Argoli, *Astronomicorum libri tres,* in *Ephemerides exactissimae caelestium motuum ad longitudinem almae urbis, et T. Brahe hypotheses,* I (Padua, 1648), Book I, Ch. 3, pp. 6-7, (first ed. of *Astronomicorum* in *Novae caelestium motuum ephemerides,* Rome, 1629); Caspar Bartholinus, *Enchiridion physicum,* (Strassburg, 1625), Book III, Ch. 2, pp. 139-140; cf. *Praecepta physicae,* II, *op. cit., Pars Specialis Prima,* Ch. 1, fol. A4[r]; Descartes, *Le monde,* (Paris, 1664), in *Oeuvres,* ed. C. Adam and P. Tannery, XI (Paris: J. Vrin, 1967), pp. 50-54; Libert Fromondus, *Meteorologicorum libri sex,* (Antwerp, 1627), Book I, Ch. 2, Article 1, p. 3; Pierre Gassendi, *Exercitationes paradoxicae adversus aristotelicos,* (Grenoble, 1624), preface, in Gassendi, *Opera Omnia,* III (Lyon, 1658), p. 102; [Francis Godwin], *The Man in the Moone,* (London, 1638), p. 54; Johannes Nicolaus Smogulecki, *Sol illustratus ac propugnatus,* (Freiburg/Breisgau, 1627), pp. 52-83; M.F. Wendelin, *Contemplationum physicarum, op. cit.*, Book III, Ch. 2, Thesis 1, Ectheses 5-6, pp. 413-414.

[150]Burgersdicius, *Collegium physicum, op. cit.*, pp. 111-112.

[151]Thomas Carleton (alias Compton), *Philosophia universa,* (Antwerp, 1649), *De coelo,* Book I, Ch. 3-4, pp. 399-400; Ioannes Elephantutius, *Universi orbis structura,* (Bologna, 1637), pp. 66, 72; Placidio de Titis, *Physiomathematica seu coelestis philosophia,* (Milan, [1650]), Book I, Ch. 1-2, pp. 3, 9.

[152]Joannes Poncius, *Philosophiae ad mentem Scoti cursus,* (Lyon, 1672), Disputation 22, Question 4, no. 31, p. 617; Q. 5, no. 55, p. 621, (first edition: Rome, 1643).

[153]Alexander Ross, *Commentum de terrae motu circulari,* (London, 1634), Book I, Sect. 1, Ch. 2, p. 9.

These positions bear within themselves the seeds of their own destruction. Only a return to a belief in immaterial or quasi-material heavens could offer a serious alternative to the fluid-heaven theory. This alternative was in fact adopted by one team of authors,[154] but by this time their audience was evanescent.

Other concepts, such as "incorruptibility" and "celestial" proved more durable. They had acquired a kind of ambiguity whereby their meanings, while surviving a change of context, themselves suffered change. Of these terms it can at least be said that by 1630 they had been thoroughly discussed, so that those who used them (and they, too, had become a minority) were either playing upon their ambiguity, or using them thoughtlessly. What this means is that those who were seriously interested in the nature of the heavens were agreed that they are not quintessential, and that they are either filled with some very subtle airy fluid or are empty. Once the initial departure from a consistent Aristotelian position had been made, this theory of a subtle but elementary fluid is the closest stable alternative.[155] Henceforth, serious debate was to take place almost entirely within this context, and centers on such questions as the causes of planetary motion, and the region and motion of the fixed stars.

---

[154]Giraldus and Arnoldus Botius, O.F.M., *Philosophia naturalis reformata,* (Dublin, 1641), Book II, Sect. 2, Ch. 1, p. 262.

[155]In partial confirmation of this interpretation, attention is drawn to Fromondus' change of opinion: by 1627, he had abandoned celestial matter, and was quoting Copernicus on the nature of gravity. See Libert Fromondus, *Meteorologicorum, op. cit.,* Book I, Ch. 2, Article 6, pp. 120-121.

COMMENTARY: DUHEM AND DONAHUE

John L. Heilbron

University of California, Berkeley

Dr. Donahue's learned paper is chiefly an inventory of opinions about the nature of the heavens advanced in the years 1600 ± 30. He deserves warm thanks for his collecting, which obliged him to look into some sixty writers, many of them highly obscure and anything but amusing to read. He divides these writers into samples active at different periods, and reports their answers to such questions as whether celestial matter is solid or liquid, perfect or corruptible, distinct from or similar to the four sublunary elements. The technique would be more effective were the samples defined, the criteria of choice specified, the context of the opinions supplied, the inventory more systematic (and perhaps even tabulated), and, above all, the analytical categories cleaner.[1]

Rather than offer opinions about the merits and flaws of the inventory, which is open to inspection, I shall concern myself with the historiographical tradition in which, in my opinion, Dr. Donahue misplaces it. Donahue regards his paper as a rectification and extension of Pierre Duhem's familiar thesis,[2] which finds the dynamics in the development of early modern astronomy in the ancient distinction between the roles of physicist and mathematician. This distinction, which allowed the mathematical astronomer to ascribe to the heavens motions impossible according to the accepted principles of physics, derived from demon-

[1]E.g., "materiality" as applied to the heavens sometimes means body as opposed to spirit (pp. 255-256), sometimes celestial matter as opposed to sublunary elements (pp. 246-247); what "quasi-material" (p. 251) signifies I do not know.

[2]*To Save the Phenomena. An Essay on the Idea of Physical Theory from Plato to Galileo,* tr. E. Doland and C. Maschler (Chicago: Univ. Press, 1969). First French edition, 1908.

strated failure to construct an exact physical astronomy. No doubt many, perhaps most astronomers regarded their freedom as a *pis aller*; for Duhem, however, it was the true liberation of science, which (in his opinion) should aim only at exact and economical description.[3] I shall return to this point.

Duhem claims that, except in the school of Averroes, astronomy jealously and successfully maintained its independence until the sixteenth century, when it found itself more and more restricted by pushy physicists, over-inquisitive theologians, homocentric revivalists, equant-hating Copernicans, aristotelianized Jesuits, and Keplerian bumblers who mixed up mathematics and philosophy and made a cosmic mystery of the straightforward business of astronomy. Under these pressures, according to Duhem, astronomers—and particularly those who followed the Jesuit Clavius—came to accept the reality of the crystalline celestial spheres. This new realism caused astronomers both Copernican and traditional to lose sight of the nature of science, and prepared the way for the confrontation between Galileo and the Church.

Dr. Donahue accepts Duhem's periodization, locates a transition to realist astronomy in the last third of the sixteenth century, and construes the opinions in his inventory as *responses* to the new state of affairs. Hence his belief that he is bringing Duhem's account forward in time. But he does nothing of the sort: an extension of Duhem would study the relation between *exact* astronomy and physics, for example in the use of Kepler's laws;[4] whereas Donahue surveys the opinions of natural philosophers (some of whom appear to be low-grade pedagogues or even cranks) on the nature of celestial matter. The background against which he should set these opinions is not the *new* suppositious constraining of the mathematical astronomer but the *old*

---

[3]Duhem, *Aim and Structure of Physical Theory,* tr. P.P. Wiener (Princeton: Univ. Press, 1954), pp. 19-23. First French edition, 1906; English tr. reprinted by Atheneum, 1962.

[4]As, e.g., J.L. Russell, "Kepler's Laws of Planetary Motion, 1609-1666," *British Journal for the History of Science,* 1964, *2*: 1-24.

speculative peripatetic natural philosophy. There he would have found forerunners for every opinion he has discovered.[5]

However Donahue may relate to Duhem, Duhem's categories fail to do justice to early modern astronomy. In particular, the methodological alternatives he offers—naive realism or crude instrumentalism—cannot capture the logical, much less the psychological function of the central Copernican doctrine, that astronomy must restrict itself to circles which rotate uniformly about their own centers. It was this doctrine, as we know, that caused Copernicus to attack Ptolemy's elegant equant and which, as Professor Wilson persuasively argues,[6] may have brought him to the heliostatic system. Should we trace his insistence on this point, as some do, to a naive realism, to a knowledge vouchsafed to him as a believing hermeticist or neoplatonist or pythagorean?[7] Or should we take it to be a normalizing principle, or even a definition, of astronomy, the observance of which renders the science not truer but merely more consistent?

The texts suggest that, on the logical order, the demand for proper circular motions was more than a definition and less than a literal transcription of the workings of the world. It was rather an epistemological principle, specifying the nature of knowledge in mathematical astronomy and what is more, how that knowledge may be acquired. Following Ptolemy (*Almagest,* II.3), Copernicus takes reduction to proper circular motions as an "axiom...set up and assumed" (*De Rev.,* IV.2[8]), as a "principle of the art" (*ibid.,* V.2), as the "first principle of regularity of movement" (*ibid.,* Dedication). Relaxing or violating the principle, as

[5]A luxuriant crop of such opinions was harvested by Ch. Scheiner, *Rosa ursina sive sol* (Bracciano, 1630), pp. 626-784, which includes this delicious bit from Bellarmine: "Voluit [deus] enim ut Caelum ipsum esset Palatium solis, in quo libere perambularet & operaretur." One can of course obtain any amount of speculative peripatetic natural philosophy from Duhem's *Système du monde* and Thorndike's *History of Magic and Experimental Science.*

[6]Supra, pp. 28-39.

[7]For recent bibliography and criticism see E. Rosen, "Was Copernicus a Hermetist?" in *Historical and Philosophical Perspectives of Science,* ed. R.H. Steuwer (Minneapolis: Univ. Minnesota Press, 1970), pp. 163-71.

[8]Quotations from *De Revolutionibus* are from the translation of C.G. Wallis in vol. 16 of *Great Books of the Western World.*

by introducing the equant, might save the phenomena, but at the price of denaturing and dehumanizing astronomy, of making it "amazing" (IV.2) and "foreign" (Dedication), "neither sufficiently absolute nor sufficiently pleasing to the mind."[9] Copernicus takes it for granted that only by ascribing circular motions to the heavens can we understand their periodicity: "it is only the circle [he says] which can bring back what is past and over with" (I.4). And to complete our understanding, the circles must rotate properly; for if they moved irregularly we should have to posit an "inconstancy of the motor virtue." And why not? Because the concept makes the "mind shudder" (*ibid.*).

No doubt a realist cosmology underlies this insistence that only proper circular motions confer intelligibility. It is natural for spherical bodies like planets to turn in a (proper) circle (I.4); it is because mathematicians have ignored this principle that they have failed to find "a more certain ground for the motions of the *machina mundi*" (Dedication). But although the circles are representations, they are by no means blueprints of the celestial spheres. As Copernicus well knew, the astronomer could not in principle identify with certainty the true character of any given celestial motion: as he says, "things which take place through an epicycle can take place through an eccentric circle," and no unexceptionable way existed for choosing between them (IV.1; cf. III.12, IV.3, and *Almagest,* III.3). The mathematician must invoke proper circular motions for the art; and he must use only such motions—even though he cannot know which obtain—in order that his necessarily incomplete representation contain nothing contrary to the heavens or unintelligible to the mind.

Another case may make clear my thought. As one knows, most physicists towards the end of the last century believed the goal of their science to be the reduction of all physical phenomena to the laws of mechan-

[9]Copernicus, "Commentariolus," in *Three Copernican Treatises,* tr. E. Rosen (New York: Dover, 1959), p. 57. Cf. G. Fracastoro, *Homocentrica* ([Venice], 1538), p. lv: "si vero Eccentricis melius quidem demonstrare videbantur, sed inique & quodammodo impie de divinis illis corporibus sentiebant, situsque illis ac figuras debant, quae minime Coelum deceant."

ics.[10] Their practice was far from uniform, some being content with differential equations like Lagrange's and their mathematical solutions, others requiring in addition "full-bodied conception and pictorial imagery and mechanical models of what was going on."[11] J.J. Thomson, Maxwell's successor one removed as professor of physics at Cambridge, may be taken as representative of English physicists. He worked both with Lagrange's formulation and with special full-bodied models, particularly of the electromagnetic field, radiation, and the structure of the atom. Yet he emphasized that none of the extravagant models he devised could be said to be true of nature; for, as he showed, if one such model existed then, in principle, any number could with diligence be found.[12] In a word, Thomson's models, like Copernicus' circles, were not unique, and therefore not exclusively true. Nonetheless Thomson's model-making, again like Copernicus', reflected a realist bias: the world in fact must be so constructed that its physical phenomena can all in principle be described by the laws and concepts of classical physics.[13]

A third parallel between Thomson and Copernicus is that, while recognizing the unavoidable limitations of his method, Thomson insisted that physics must search for and rest with mechanical reductions. "When we have done this we have got a complete explanation of any

[10]"L'esprit de Descartes plane sur la Physique moderne, que dis-je? il en est le flambeau: plus nous pénétrons dans la naissance des phénomènes naturels, plus se développe et se précise l'audacieuse conception cartésienne relative au mécanisme de l'Univers: 'Il n'y a dans le monde que de la matière et du mouvement.' " Thus A. Cornu, in the opening address to the Congrès international de physique, held in Paris in 1900; see the Congrès' *Rapports* (4 vols. Paris: Gauthier-Villars, 1901), *4*, p. 7. Cf. the keynote speech of E.L. Nichols, "The Fundamental Concepts of Physical Science," in *Congress of Arts and Science. Universal Exposition, St. Louis, 1904,* ed. H.J. Rogers (8 vols. Boston & New York: Houghton-Mifflin, 1905-07), *4*, p. 19: "It is the object of such speculation [i.e., physical theory] to place all phenomena upon a mechanical basis."

[11]O. Lodge, "George Francis Fitzgerald, 1851-1901," in Royal Society of London, *Yearbook* (1902), p. 254.

[12]Thomson, *Applications of Dynamics to Physics and Chemistry* (London, 1888), pp. 12-15. Cf., J.C. Maxwell, *A Treatise on Electricity and Magnetism* (Oxford: Clarendon, 1892), *2*, p. 470.

[13]As in Thomson's inference of the existence of an ether in *On the Light Shown by Recent Investigations of Electricity on the Relation between Matter and Ether* (Manchester, 1907), reprinted in Smithsonian Institution, *Annual Report* (1908), 233-44.

phenomenon."[14] And conversely: a theory which stops short of full mechanical reduction—compare Copernicus on reduction to proper circles—is at best a stop gap and more likely a screen of ignorance. Consequently Thomson and others, like Lord Rayleigh, who shared his epistemological views, could never bring themselves to accept Bohr's quantum theory, which Thomson regarded as a cowardly excuse for shirking the task of the physicist, for quitting before satisfying the mind.[15]

There is a fourth point of similarity. For Copernicus, reduction to proper circles recommended itself on heuristic as well as on epistemological grounds. It was while implementing this program, he said, that he had "occasion for working out the mobility of the earth and some other ways by which regularity and the principles of the art might be preserved" (V.2). Similarly for Thomson reduction to mechanics, and particularly to mechanical models, was justified and even demanded as a guide to discovery.[16] "[It] suggests extensions, suggests advances in a way that purely analytical investigations seldom do...; there is no doubt that a mechanism we can distinctly picture the working of is enormously easier to reason about than one of whose structure we know nothing."[17] And like Copernicus Thomson could justify his methods by a splendid discovery of his own, namely the electron, to which he came earlier than his rivals by following up the consequences of a mechanical picture of the cathode rays.[18]

Max Planck also worked out a position similar to Copernicus'.[19] Planck was one of the few physicists active in the generation before 1900 who doubted the possibility of full reduction to mechanics. He took this

---

[14]Thomson, *Applications,* p. 15. Cf. *Dictionary of Scientific Biography,* s.v. "Thomson, J.J."

[15]Thomson, *The Atomic Theory* (Oxford: Clarendon, 1914), p. 27; Lord Rayleigh, *The Life of John William Strutt, 3rd baron Rayleigh* (Madison & London: Univ. Wisconsin Press, 1968), p. 356.

[16]Thomson, *Notes on Recent Researches in Electricity and Magnetism* (Oxford: Clarendon, 1893), p. vi.

[17]G.F. Fitzgerald, *Scientific Writings,* ed. J. Larmor (Dublin: Hodges & Figgis; London: Longmans Green, 1902), p. 311.

[18]Thomson, "Cathode Rays," *Philosophical Magazine*, 1897, *44*: 293.

[19]Cf. E. Brachvogel, *Nikolaus Copernicus (1473-1543) und Aristarch von Samos (ca. 310-230 v. Chr.)* (Braunsberg: Ermländ. Zeitungs-und Verlagsdruckerei, 1935), pp. 34-45.

position chiefly because he believed in the unrestricted validity of the second law of thermodynamics, which the mechanical theory interpreted as valid only statistically. He later accepted that theory, which helped him to the discovery of the quantum; but he kept his conviction that the goal of physics was the discovery of absolute and invariable laws.[20] Now if the real world obeys such laws, its processes must be causal; and hence (according to Planck) the physicist must assume the reign of strict causality in designing his abstract world pictures. Planck recognized perfectly that his own chief discovery rendered the old mechanistic conception of causality untenable; that no proof that strict causality controlled the real world could be given; and that many younger physicists believed natural processes to be ultimately indeterministic.[21] Nonetheless he cleaved to causality as a minimal reflection of an objective world, independent of and, at bottom, inaccessible to the human mind;[22] as an epistemological principle ("[Indeterminists] are compelled to set a limit to their impulse for knowledge"[23]); and as a guide to discovery. "The law of causality cannot be demonstrated any more than it can be logically refuted: it is neither correct nor incorrect; it is a heuristic principle;. . .it is the most valuable pointer we possess in order to find a path through the confusion of events, and in order to know in what direction scientific investigation must proceed. . ."[24]

Planck began to make his philosophical views public in 1908, in an apparently gratuitous attack on Ernst Mach, whose chief sin (according to Planck) was the doctrine that physics aimed solely at economy of thought, at saving the phenomena in the easiest and most pleasant manner, at subjectivism and even at solipsism, and not at the absolute

[20]Planck, "Scientific Autobiography," in *Scientific Autobiography and Other Papers,* tr. F. Gaynor (New York: Greenwood Press, 1968), pp. 32, 35, 39, 46.

[21]Planck, "The Philosophy of Physics," tr. W.H. Johnstone, in Planck, *The New Science* (New York: Meridian, 1959), pp. 246-49, 261, 282-83. The theme is exhaustively treated by H. Vogel, *Zum philosophischen Wirken Max Plancks* (Berlin: Akademie-Verlag, 1961), pp. 175-202.

[22]Planck, "Philosophy of Physics," pp. 249-50; "The Concept of Causality," *Proc. Phys. Soc.,* 1932, *44*: 529-39; "Where is Science Going?" tr. J. Murphy, in Planck, *The New Science,* pp. 41-42.

[23]*Ibid.,* p. 289. Compare Thomson on Bohr and Copernicus on Ptolemy.

[24]*Ibid.,* p. 290.

laws of the real world.[25] Now among those Mach identified as kindred spirits was none other than Pierre Duhem who, like Mach, turned to history for ammunition against the importation of metaphysical preconceptions into science and who, even more than Mach, deplored the program of mechanical reduction. *To Save the Phenomena* strives to show that the method by which the bellwether of the mathematical sciences, planetary astronomy, reached its early-modern perfection was precisely the method of economy urged by Mach and practiced by Duhem. The little book is a masterful polemic in the same cause which provoked Duhem's amusing caricature of the "broad and shallow" English mind,[26] and his endless inventory of the paralogisms of James Clerk Maxwell.[27]

To Save the Phenomena first appeared in the *Annales de philosophie chrétienne* for 1908. By then, as Duhem recognized, the English conquest of the last bastion of right (and clear) thinking, French physics, was far advanced. During the nineties the seductive methods and piecemeal science of Maxwell and Thomson had entered the Sorbonne, whence they spread abroad recommended for their fertility, for their appeal to intuition and common sense, for their "soupçon of truth."[28] Duhem did not expect to reverse the trend, but to provide evidence to which his countrymen could appeal when they regained their senses.[29]

To Save the Phenomena treats Copernicus and Galileo as generally retrograde for attempting to impose realist preconceptions on astronomy, and as momentarily forward looking for beginning the unification of sublunary and celestial physics.[30] (They thereby showed that they had glimpsed the principle of economy of thought.) In Duhem's other and major historical writings they fare rather better, winning high

[25]Planck, "Die Einheit des physikalischen Weltbildes," *Phys. Zs.*, 1909, *10*: 62-75.

[26]Duhem, *Aim and Structure,* pp. 69-99. Cf. Kurt Hübner, "Duhems historische Wissenschafts-theorie und ihre gegenwärtige Weiterentwicklung," *Philosophia Naturalis*, 1971, *13*: 81-97, p. 86.

[27]Duhem, *Les théories électriques de J. Clerk Maxwell* (Paris: Hermann, 1902).

[28]Duhem, *Aim and Structure,* pp. 87-91, 101-02, 319; A. Rey, *La théorie physique chez les physiciens contemporains* (Paris: Alcan, 1907), p. 48; H. Bouasse, "Le rôle des principes dans les sciences," *Rev. gén. sci.*, 1898, *9*: 561.

[29]Duhem, "Notice sur les titres et les travaux scientifiques," in Société des sciences physiques et naturelles de Bordeaux, *Mémoires*, 1917, *1*: 1: 157.

[30]*To Save the Phenomena,* pp. 63, 109, 114-17.

marks for the scientific acumen that caused them to jettison the Renaissance and to take up physics precisely where the Parisian scholastics had left it in the fourteenth century.[31] To this odd position also Duhem came not as an historian, but as a propagandist: a devout Catholic as well as a leading physicist, he wished to show that the science prized by the anti-clerical Third Republic had been nurtured by the Church.[32] But this is a matter that neither time nor Donahue's interesting paper gives me warrant to pursue.

[31]E.g., *Études sur Léonard da Vinci,* 3e série: *Les précurseurs parisiens de Galilée* (Paris: Hermann, 1913). Cf. *To Save the Phenomena,* p. 60: "[the] doctrine of physical method [of the Parisian scholastics] far surpassed in truth and profundity all that was going to be said on this subject until the middle of the nineteenth century."

[32]See, e.g., the remarkable letter from Duhem to Père Bulliot, 21 May 1911, in Hélène Pierre-Duhem, *Un savant français, Pierre Duhem* (Paris: Plon, 1936), pp. 165-67; and H.W. Paul, "The Crucifix and the Crucible: Catholic Scientists in the Third Republic," *Catholic Historical Review*, 1972, *58*: 195-219.

IX

THREE RESPONSES TO THE COPERNICAN THEORY: JOHANNES PRAETORIUS, TYCHO BRAHE, AND MICHAEL MAESTLIN *

Robert S. Westman

University of California, Los Angeles

1. *The Wittenberg Legacy*

Attitudes toward scientific change and innovation can provide an exceptionally revealing index of the resiliency of a generation, one dimension along which its theoretical and methodological inclinations can be discerned. The period between 1537 and 1550 witnessed the birth of a group of scholars none of whom were old enough to remember the publication of *De revolutionibus orbium coelestium* nor to have known any of the main protagonists during the early period when the theory was first received. Besides the three figures to be treated in this

*It is with pleasure that I gratefully acknowledge the helpful criticisms of this paper by Amos Funkenstein (UCLA), Peter Machamer (Ohio State) and Elizabeth Eisenstein (University of Michigan). A memorable lunch-time discussion with Charles Kennel (UCLA) helped to crystallize my understanding of the importance of scientific generations. Father Martin F. McCarthy (Vatican Observatory) gave generous encouragement and assistance for my research in Rome and Joshua Lipton (UCLA) provided some preliminary translations of a number of texts analyzed in this study. All errors of fact or judgment are, of course, entirely my own.

paper, I would include in this generational group: Christopher Clavius (1537-1612), William Gilbert (1540-1603), Thomas Digges (ca. 1546-1595), Giordano Bruno (1548-1600), Christopher Rothmann (1550-1597) and Francois Viète (1540-1597), although the list might conceivably be lengthened. What unites this small group of astronomers, mathematicians and philosophers, beyond the fact of age, is neither a common nationality nor a common university nor a common speciality (viz. astronomy) but rather a shared intellectual legacy. By contrast with the first generation of astronomers who countenanced the new set of Copernican cosmological and astronomical claims, men born in the period from roughly 1497 to 1525 and who responded to the new theory from the late 1540's to about 1570, the initial phase of reception was not marked by intense debates and polemics such as would impel Bishop John Wilkins to warn opponents of the Copernican theory in the much later post-Condemnation year of 1638:

> 'Tis an excellent Rule to be observed in all Disputes, That Men should give *soft Words* and *hard Arguments*; that they would not so much strive to *vex*, as to *convince* the Enemy.[1]

For the first generation which received the theory of Copernicus the urgency to form a commitment to an entirely new "research program" or "paradigm," if you prefer, did not exist. And it did not exist simply because the planetary models and parameters of *De revolutionibus* were not perceived as a new *Weltanschauung*—i.e., a new set of claims about the order of the planets and the nature of motion—but merely as a useful set of auxilliary mathematical hypotheses and tables to be exploited by the practitioners of geostatic astronomy. At the University of Wittenberg, the leading German educational institution of this period, a major center of the Reformation and the first port of entry for the Copernican theory, all students who took courses in the natural sciences would be familiar with the name of Copernicus and the barest rudiments of his principal ideas. At the same campus, a Master's candidate would not only encounter Copernican planetary models and values in his textbooks but he would be openly encouraged to read *De revolutionibus* itself. Through the energetic work of Philipp Melanchthon (1497-1560) and his disciples Erasmus Reinhold (1511-1553) and Caspar Peucer (1525-

[1] *A Discourse Concerning a New Planet* (London, 1684), "To the Reader."

1604), men whose *working* lives began before the publication of Copernicus' work, a moderate and pragmatic methodological consensus was bequeathed to the generation of men whose productive intellectual years commenced in the 1570's.[2]

What the "Wittenberg Interpretation" (as I have called it) omitted by its silence, rather than by overt censorship, however, was any reference to the fact that here, for the first time, one encountered a *system* of planetary motions in which the periods of the planets increased proportionately with the magnitude of their radii from the sun and where no distances could be altered without affecting all other ones. This "symmetria" of the new theory, its relating of planetary distances to a common unit of measurement, was deeply consonant with the widely-pervasive Graeco-Renaissance values of balance and harmony, an architectonic criterion found so extensively in almost all areas of Renaissance cultural activity from art and architecture to music, medicine, theology and magical theory.[3] Here, indeed, one had a principle of *necessity* by which the planets were ordered in place of the quasi-arbitrary principle of the contiguity of the spheres whereby, for a given planet, the absolute distances might still be expanded and contracted (while holding constant the ratio of maximum and minimum distances) without necessarily affecting the magnitudes of *all* other spheres. The only member of the first generation to recognize—indeed, to exult in—the systemic feature of Copernicus' theory was George Joachim Rheticus (1514-1574). It was Rheticus, in his *Narratio Prima,* who had initially convinced the aging Copernicus to publish his life's work; yet the immediate impact of Rheticus' *Report* on the Wittenberg technicians was not to reproduce the conversion experience that he himself had undergone.

[2]These ideas are discussed more fully in Robert S. Westman, "The Melanchthon Circle, Rheticus and the Wittenberg Interpretation of the Copernican Theory," *Isis*, 1975, *66*: 165-193; and in a shorter version, "The Wittenberg Interpretation of the Copernican Theory," in O. Gingerich (ed.), *The Nature of Scientific Discovery* (Washington, D.C.: Smithsonian, 1975), pp. 393-429.

[3]Cf. Leo Spitzer, *Classical and Christian Ideas of World Harmony,* (ed.) A.G. Hatcher, Baltimore: Johns Hopkins Press, 1963. I have analyzed this concept in Kepler's thought ("Kepler's Theory of Hypothesis and the 'Realist Dilemma'," *Studies in History and Philosophy of Science,* 1972, *3*: 233-264) and with respect to Renaissance architecture, magic and astronomy in "Method and Metaphysics: The Yates Thesis Reconsidered," *Hermeticism and Science,* Los Angeles: William Andrews Clark Memorial Library, 1975.

For while Rheticus' adoption of the new theory was greatly influenced by his close, personal relationship with its author, his Wittenberg colleagues could not appreciate the deeper emotional issues involved. It was not until the next generation that Rheticus' cosmological vision was re-discovered.

In this paper, I shall confine myself to three astronomers—Johannes Praetorius, Tycho Brahe and Michael Maestlin—whose *diverse* responses reveal to us some of the conservative and innovative tendencies in the second major phase of the reception of the Copernican theory.

2. *The Generation of the 1570's*

By the 1570's there begins to emerge a new appreciation of the work of Copernicus.[4] This development is by no means completely uniform in its chronological unfolding yet a distinct pattern already is clearly present. It amounts, in brief, to a revaluation of Book I of *De revolutionibus* accompanied by a further articulation of the planetary models and a growing critical awareness of how the previous generation had underplayed and ignored the theory's cosmological import. As Rheticus lay dying in 1574, his vision of the Copernican *system* was, for the first

[4]My argument here departs considerably from Pierre Duhem's conceptualization in his *To Save the Phenomena* (trans. E. Dolland and C. Machler, [Chicago: University of Chicago Press, 1969]). Duhem's philosophical interest is really in the status of physical theories and, therefore, he uses as his sole principle of organization the fictionalist-realist dichotomy. This, in turn, leads him to establish the period of the reform of the calendar (1582) as the crucial turning point from a fictionalist to a realist view of hypotheses in astronomy. While Duhem's work presents us with a valuable first approximation, it overlooks the fact that there were other features of the Copernican theory which attracted the interest of subsequent thinkers, such as the reduction of the annual element to the earth's orbit and Copernicus' attempt to replace the equant by uniform, circular motions. As I argue in this paper, concern with these other characteristics of the theory define a second phase in its reception, a phase which Duhem's orientation caused him to overlook. In his commentary on the present paper, Professor Machamer seems to have had difficulty, at first, in putting aside his own Duhemian spectacles (pp. 346-347). For a recent example of the Duhemian viewpoint, supplemented by useful new manuscript material, see Owen Gingerich, "From Copernicus to Kepler: Heliocentrism as Model and as Reality," *Proceedings of the American Philosophical Society*, 1973, *117*: 513-522.

time, beginning to be understood. And with the revitalized awareness of the Copernican arguments about planetary order and distances, the size of the universe and the dynamics of falling bodies, the discussions begin to impinge upon disciplines outside of astronomy. This trend is particularly evident with Digges, Gilbert and Bruno, none of whom were formally associated with universities, although it is present to some degree in other major figures of this generation. Accompanying the growing consciousness of Copernicus' serious intent to propound a realist astronomy, there develops a growing suspicion that the author of *De revolutionibus* was not the author of *Ad lectorem*. This feeling was frequently reinforced by a strong oral tradition, whose lines of transmission are not yet fully worked out, a tradition which asserted that Andreas Osiander was the true source of the spurious preface.[5] And finally, the 1570's were marked by two well-known celestial events, the appearance of a supernova in 1572 and a brilliant comet in 1577. Appearing at a time when the network of astronomers was just commencing to grasp the broader meaning of the Copernican hypothesis, the observations and interpretations of these two great occurrences were swept into a new framework, a framework in which the *choice* between the Ptolemaic and Copernican theories took on competitive significance for the first time in the eyes of contemporaries. While full-scale cosmological debate does not erupt until the 1590's, by which time the Tychonic system was a serious contender, the lines of battle can already be discerned in the misty intellectual fields of the 1570's.

3. Johannes Praetorius' Articulation of the "Wittenberg Interpretation"

No matter how radically a scientific generation may reject the work of its predecessors, it will always preserve a large measure of the earlier achievement, although sometimes more than it is willing to admit. This is especially true of the "technological" side of astronomy. If progress in

[5]For an initial discussion see Ernst Zinner, *Enstehung und Ausbreitung der Coppernicanischen Lehre* (Sitzungsberichte der physikalisch-medizinischen Sozietät zu Erlangen, 74), (Erlangen, 1943), pp. 451-454; A. Bruce Wrightsman "Andreas Osiander and Lutheran Contributions to the Copernican Revolution," Unpublished Doctoral Dissertation: University of Wisconsin, 1970.

astronomy is evident anywhere, it must be in the gradual improvement in techniques of observation and the mathematics of mensuration. Certain problems such as the measurement of parallax and the solution of visual anomalies persist across the historical divides that mark the great theoretical upheavals in spite of the obvious fact that new theories will make new predictions and new technologies will dredge up new evidence for theoretical speculation.[6]

Johannes Praetorius (1537-1616) represents the most conservative of the three figures under consideration in this paper. His interests were not only heavily "technological" but he also bears most forcefully the banners of the Wittenberg Interpretation. Moreover, he was considered by Tycho Brahe to have been one of the leading "mathematici" of his time and his views are of interest here because they show how the earlier interpretation of the Copernican theory persisted and was further elaborated within the universities in the last quarter of the sixteenth century.[7]

Praetorius was born in Joachimsthal in Northwest Bohemia in 1537. Virtually nothing is known of his early life except that he studied at the University of Wittenberg, perhaps as early as 1555.[8] He was certainly there in 1560 since he obtained a copy of *De revolutionibus* at about that time from Paul Eber (1511-1589), a former classmate of Rheticus and professor of theology at Wittenberg.[9] There can be little doubt that he would have studied astronomy under Caspar Peucer, Melanchthon's son-in-law and a prime architect of the Wittenberg Interpretation, as well as Sebastian Theodoric of Winsheim, the professor of lower mathe-

[6]For a view which stresses the theory-dependence of technologies, see Thomas S. Kuhn, *The Structure of Scientific Revolutions,* 2nd edition (Chicago: University of Chicago Press, 1970), pp. 35-42.

[7]On Praetorius, see Tycho Brahe's letter to Thaddeus Hagecius, 1 Nov. 1589, *Tychonis Brahe Dani Opera Omnia,* ed. I.L.E. Dreyer, VII (Copenhagen, 1924), pp. 206-207. Hereafter cited as "TBOO."

[8]On Praetorius, cf. Heinrich Kunstmann, *Die Nürnberger Universität Altdorf und Böhmen,* (Köln, Graz: Böhlau, 1963), p. 147 ff. There is no evidence of when Praetorius entered the university but, since he would have been 18 in 1555, that is probably a reasonable early date.

[9]Zinner, *op. cit.,* p. 454.

matics.[10] In 1562, at the age of 25, he went to Nürnberg where he spent about seven years constructing globes, astrolabes and other astronomical instruments. An early eighteenth-century biography of Praetorius lists eight instruments in the Nürnberg Library with dates ranging from 1562 to 1566.[11] Praetorius left Nürnberg in 1569 and travelled to Prague and Vienna where he met the Hungarian humanist, Andreas Dudithius (1533-1589). From there he went with Dudithius to Cracow where he came in contact with Rheticus[12] and his reputation spread quickly to the court of the Emperor Maximillian II who sought him out as Court Mathematician.[13] In March, 1571, Sebastian Theodoricus of Winsheim, by then lecturing in higher mathematics at Wittenberg, was asked by the Academic Senate to transfer his full energies to the teaching of medicine—an indication that the "scientific role" in this period was still broadly defined.[14] Caspar Peucer and the Senate both addressed letters to Praetorius in Cracow in which they implored him "lovingly" (*amanter*) to return once again to his old school.[15] Praetorius then succeeded Theodoricus as lecturer in mathematics. A fragment of a lecture given by Praetorius at Wittenberg is still preserved today in the Vienna Nationalbibliothek, entitled "Certain Little Annotations on the Hypotheses of Peucer Given at the Academy of Wittenberg in 1572."[16] The notes are rather elementary and were undoubtedly based upon

[10]Sebastian Theodoricus' dates are unknown but we do know that he obtained his M.A. in 1544 and was admitted to the Faculty of Arts in 1545. By 1560, he was lecturing also in "higher mathematics." In 1571, he transferred to the Faculty of Medicine (Cf. Walter Friedensburg, *Geschichte der Universität Wittenberg,* (Halle: Max Niemeyer, 1917), p. 280.

[11]Sigismundus Iacobus Apinus, *Vitae Professorum Philosophiae Qui A Condita Academia Altorfina ad hunc usque diem claruerunt...* (Nürnberg and Altdorf: Tauber, 1728), p. 20.

[12]Zinner, *op. cit.,* p. 424.

[13]Kunstmann, *op. cit.,* p. 148.

[14]Joseph Ben-David says virtually nothing about the "scientific role" in sixteenth-century universities (*The Scientist's Role in Society: A Comparative Study,* [Englewood Cliffs, N.J.: Prentice Hall, 1971], p. 55 ff.).

[15]Apinus, *op. cit.,* pp. 22-23. The letters are reprinted in this work.

[16]Codex 10641, fols. 1^r - 11^r.

Peucer's *Hypotheses astronomicae* (1571)[17] but they show clearly Praetorius' early acceptance of Peucer's astronomical framework. While still at Wittenberg, he observed the supernova of 1572 which he believed to be a "Meteor" in the aetherial region.[18] Finally, in 1576, he accepted a position as "Mathematicus" at the Nürnberg University of Altdorf where he taught until his death in 1616. During this period he published very little yet authored over twenty treatises, many of which are still extant,[19] together with several personally-annotated books which include copies of *De revolutionibus* and Rheticus' *Narratio Prima.*[20] From about 1580 onward, Praetorius wrote a series of lectures on the planetary hypotheses in which his use and interpretation of the Copernican theory is clearly revealed. It is to these works that we shall now turn.

Praetorius' viewpoint appears quite consistently in a variety of manuscript treatises. Several representative remarks, however, are to be found in his *Hypotheses Astronomicae* which he began to compose in March, 1592.[21] After a considerable period of time, he informs us, not a few astronomers noticed that Ptolemy's calculations deviated from the appearances as set forth in the planetary tables. As a result, new tables and emendations were proposed by Geber, Al-Battānī, Arzachel, Prophatius Judaeus, Thābit ibn-Qurra, Alphonsus and, finally, Nicholas Copernicus, all of whom followed in the footsteps of Ptolemy, *princeps Astronomorum.*[22] Praetorius continues:

> Of all those who followed Ptolemy, Nicholas Copernicus alone deservedly stands forth. For he wished his astronomy to be

[17] A copy of this work is listed by Petrus Saxonius, a pupil of Praetorius who succeeded him at Altdorf, in a catalogue of his works and mss. This catalogue is reprinted in Zinner (*op. cit.*, pp. 428-448). I am unaware that the work in question was published in 1570, the date given by Saxonius (Item No. 41, p. 430). Unfortunately, the book appears to be missing from the Schweinfurt Stadtarchiv where I searched for it.

[18] *TBOO,* p. 157.

[19] Most of Praetorius' MSS. are located today in the Erlangen Universitätsbibliothek and the Schweinfurt Stadtarchiv.

[20] These works are both located in Schweinfurt. Apinus's list of Praetorius' MSS. is incomplete (*op. cit.*, pp. 18-19).

[21] Erlangen MS. 816.

[22] *Ibid.*, fol 2^r. See G.J. Toomer, "Prophatius Judaeus and the Toledan Tables," *Isis,* 1973, *64*: 351-355.

eternal and that all observations, both past and future, might be in agreement [with his theory], while not a few others who adhered to his opinions reflected on changes [in his theory] and, each for his own reasons wished to insert further assumptions.... Now, just as everyone approves the calculations of Copernicus (which are available to all through Erasmus Reinhold under the title *Prutenic Tables*), so everyone clearly abhors his hypotheses on account of the multiple motion of the earth. Here, in the present work, we shall not discuss at length the freedom of astronomers to form hypotheses, for elsewhere this matter is discussed more fully.[23] Nevertheless, having omitted all ambiguities, we wish to say that inasmuch as we can properly approach astronomical hypotheses, and in those very hypotheses which are to be discussed here, we follow Ptolemy, in part, and Copernicus, in part. That is, if one retains the suppositions of Ptolemy, one achieves the same goal that Copernicus attained with his new constructions. And, in this way, we shall demonstrate the foundations and origin of the new calculations transmitted to us since Copernicus and more accurately discussed in the *Prutenic Tables*.[24]

The moderate attitude reflected in this passage, so characteristic of the Wittenberg school, was not merely the result of an a priori commitment to an abstract methodological position. As a working astronomer, Praetorius believes that the new theory possesses important theoretical and pragmatic advantages. His main task, therefore, is to study the Copernican models in great detail and to strive to adapt them to the Ptolemaic framework.

Although this goal is constantly reiterated throughout Praetorius' works, it is most clearly worked out in two treatises: *Theoricae Planetarum per homocentrepicyclos respondentes placitis Copernici* (dated: 25 March 1592)[25] and *Compendiosa enarratio Hypothesium Nic.*

---

[23]Cf. Praetorius' later lecture entitled *Theoriae Planetarum Anno 1605,* (Schweinfurt MS. H 73), fol 3^{r}.

[24]Erlangen MS. 816, fol 2^{v} - 3^{v}. This theme is echoed in practically all of Praetorius' lectures. Cf. e.g. *Theoriae Planetarum Anno 1591,* Erlangen MS. 814, fols. 43^{r}, 45^{r}.

[25]Erlangen MS. 814, fols. 70^{r} - 86^{v}.

Copernici, Earundem insuper alia dispositio super Ptolemaica principia (dated: 18 January 1594–29 September 1594).[26] And we are not surprised to learn that what most impresses Praetorius is the principal axiom underlying the Copernican hypotheses: to wit, the rejection of the equant and its replacement by models which employ only eccentrics, epicycles and deferents or various combinations of these devices. Erasmus Reinhold had called this the "Axioma Astronomicum"[27] and, as we shall see later, Tycho Brahe will share this abhorrence for the equant as well. Praetorius considers eliminating the word "Eccentric" altogether because of its affinity with the equant but then decides to retain it "because by the introduction of new words obscure fictions are brought into the tables."[28] But if the word is retained, the model itself must be extirpated.

> Just as in a whirlpool we say that the same anomalous motion can be explained by diverse reasons, so with Ptolemy. For he proposed an eccentric and referred its regular motion not to its own center but to another point of equality. Copernicus, however, succeeded in replacing this hypothesis, which was absurd, by means of an eccentric with an epicycle so that he might have [a device with] equal motion about its own center.[29]

The first of the two treatises mentioned above opens with a discussion of a bi-epicyclic solar model and proceeds to a similar device for the moon.[30] A few leaves later, Praetorius introduces a tri-epicyclic model for the three superior planets, Mars, Jupiter and Saturn, and for Venus. Mercury is then loaded with four epicycles.[31] In Figure 1 we see a diagram entitled: "The Theories of the Three Superior Planets According to Copernicus. The Same is Explained by a Concentric with three

[26] *Ibid.*, 87^{r} ff.

[27] On this Axiom, cf. Owen Gingerich, "The Role of Erasmus Reinhold and the Prutenic Tables in the Dissemination of the Copernican Theory," *Studia Copernicana* VI (*Colloquia Copernicana II*), (Warsaw: Ossolineum, 1973), p. 58 and Plate IIa; and Westman, *op. cit.*, "Wittenberg Interpretation."

[28] Erlangen MS. 814, fol. 52^{v}.

[29] *Ibid.*

[30] *Ibid.*, fol. 70^{r} ff.

[31] *Ibid.*, 73^{r} - 75^{v}.

THEORIA TRIVM SVPERIORVM, SECVNDVM
Copernicum. Eademq explicata per concentricum
cum tribus epicyclis, retentis principijs Ptolemaicis;
terra sc: in centro Zodiaci, et immobili, collocata.

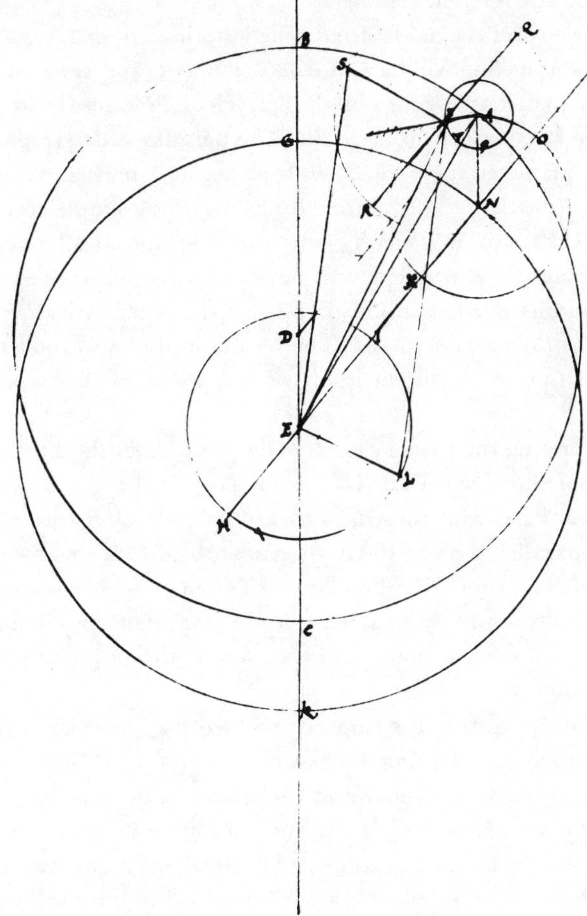

Figure 1. Johannes Praetorius' equant-less model for the superior planets "according to Copernicus" but "while retaining the principles of Ptolemy." From *Theoricae Planetarum per homocentrepicyclos respondentes placitis Copernici* (1592), courtesy of the Erlangen Universitätsbibliothek MS. 814, fol. 79v.

epicycles, while retaining Ptolemaic principles; The Earth is located at rest in the center of the Zodiac."[32] It is not my purpose to examine this model in detail nor with respect to its predictive capabilities but simply to point out a few salient features which relate to the commutability of helio- and geostatic reference frames.

If we first interpret the model from a heliostatic perspective, as does Praetorius, then the Sun will be at *E,* the Earth at *L* and the apparent position of the planet at *F,* riding on the epicyclet *FTP* whose center *A* is defined by the large eccentric circle *BC*. The parallax is determined by the sun-earth-planet triangle *EFL*. While Copernicus himself had been content to describe the eccentric circle *BC* by use of the simple eccentric with center *D,* Praetorius is clearly determined to uproot all eccentric points and replace them with epicyclic devices. Hence, he adds epicycle *OAFX* with radius *AN* equal and parallel to the eccentricity *DE* and deferent radius *EN* parallel to *DA*. The resultant motions produce the same effect as the simple eccentric, as we can see from the parallelogram *DANE*.

Now, interpreting the model geostatically, *E* becomes the Earth and the direction of the mean sun is 180° from *L* as seen from *E*. At this point, however, Praetorius constructs a third epicycle *QSR* with radius *SF* equal and parallel to the radius of the sun's orbit *EL*. If we now complete the parallelogram *ESFL,* then the earth's motion will be projected, as it were, onto the newly-constructed epicycle *QSR*. The apparent position of the planet from *E* now will be at *S* and the parallax will be defined by triangle *SEF*.

Praetorius' deep study of the Copernican theory appears throughout his university discourses. During the first nine months of 1594, he wrote a series of lectures on the Copernican hypotheses. Chapter 2 concerns the order and distances of the planets and, in Figure 2, we notice that the heliocentric scheme is no longer treated merely as a qualitative aid to the mind but as a series of nested spheres with varying dimensions. Thus, the maximum, minimum and mean relative distances of the planets are given. With an absolute value for the earth-sun distance,

[32] *Ibid.*, 79^{v}.

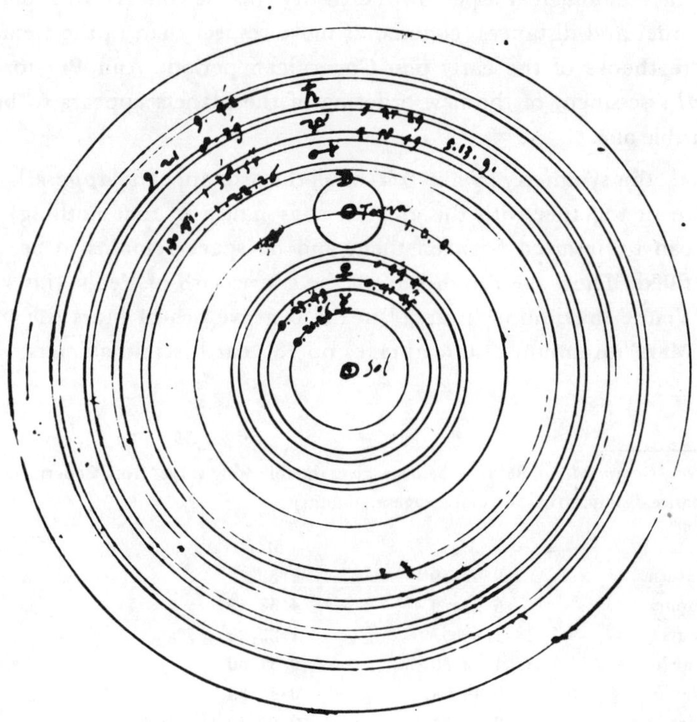

Figure 2. The relative values of the planetary distances according to Copernicus. From Johannes Praetorius, *Compendiosa enarratio Hypothesium Nic. Copernici, Earundem insuper alia dispositio super Ptolemaica principia* (1594). Courtesy of the Erlangen Universitätsbibliothek MS. 814, fol. 93^{r}.

one could then obtain the actual thicknesses of the orbs.[33] At the top of the diagram Praetorius has written a paraphrase of a famous passage in *De revolutionibus* I:10: "Autem omnium residet Sol tanquam in centrum mundi quiescens et omnia illustrans."[34] Already, we can observe that the cosmological aspect of the theory, i.e. its concern with planetary order and distances, commands more respect than in the elementary textbooks of the early post-Copernican period. And Praetorius' *initial* assessment of the new ordering of the planets appears to be a favorable one:

> . . . this symmetry [*simmetria*] of all the orbs appears [*apparet*] to fit together with the greatest consonance so that nothing can be inserted between them and no space remains to be filled. Thus, the distance from the convex orb of Venus [i.e. Venus' maximum distance] to the concave orb of Mars [i.e. Mars' minimum distance] takes up 730 earth semidiameters,

[33] *Ibid., Compendiosa,* 93^r ff. Praetorius gives the following values from Copernicus for the relative distances (in the usual sexagesimal units):

	Max.	*Min.*
Saturn	9°42′ 39″	8°39′
Jupiter	5 27 29	4 58 49
Mars	1 38 57	1 22 26
Earth	1 2 30	0 57 30
Venus	0 44 4	0 41 35
Mercury	0 23 43	0 21 26

From these ratios, he derives the following values for the absolute distances (94^v):

	Max.	*Min.*	(Expressed in Earth radii) *Mean*
Saturn	11077	9877	10477
Jupiter	6233	5687	5960
Mars	1883	1569	1726
Earth	1179	1105	1142
Venus	839	791	815
Mercury	415	408	430

[34] *Ibid.,* 93^v.

> in which space the great orb contains the moon and earth and moving epicycles.[35]

There follows, however, a most extraordinary consequence of Praetorius' general assumption regarding the commutability of the Copernican and Ptolemaic coordinates. In a section entitled, "Reductio harum hypothesium ad principia Ptolemaica," he proposes a new scheme *more astronomico* in which the earth is at rest, the sun revolves about the earth and the sun's orb bears two epicycles.[36] In Figure 3 we see that the author has clearly experienced some indecision. The outer epicycle is labelled "Mercurij Veneris" and carries the symbol for Venus, while the inner epicycle reads "Veneris Mercurij" with the planetary symbol for Mercury! Perhaps unconsciously, Praetorius has introduced the old order of the inferior planets into his figure and then, realizing his error, replaces it with the new Copernican ordering. But the great "X" slashed through the scheme well states Praetorius' sentiments on this proto-Tychonic cosmology. The main difficulty, in fact, was the same one that Tycho had stumbled over a decade earlier, namely the delicate issue of establishing the parallax of Mars and hence the distance of its sphere from the Sun. The problem is strikingly expressed by Praetorius in the following passage:

> It is certain that if we were to transfer the Sun with the surrounding orbs of Venus and Mercury to the place of the Earth [on the Copernican theory] and, in like manner, the Earth, with the moon following at its feet, to the place of the Sun then . . . it would be necessary to add epicycles of the size of the great orb [to the already unequal distances] with the result that there would occur a great confusion of the orbs (especially with Mars). . . . Thus, according to Copernicus, Mars' maximum distance from the Earth ought to be 3044 [Earth radii] and its minimum 427 [Earth radii]; yet, this simply cannot be

[35] *Ibid.*, 94^{v}. See also Note 33 above.

[36] *Ibid.*, 94^{r}. It is possible, although not certain, that this bit of speculation was inspired by Praetorius' reading of Tycho Brahe's *De mundi aetherei recentioribus phaenomenis*, a copy of which had been sent to him on 18 September 1588 and today located in the Universitätsbibliothek Breslau (Cf. Wilhelm Norlind, *Tycho Brahe: en levnadsteckning*, [Lund, 1970], p. 124). I should like to thank Victor E. Thoren for this reference.

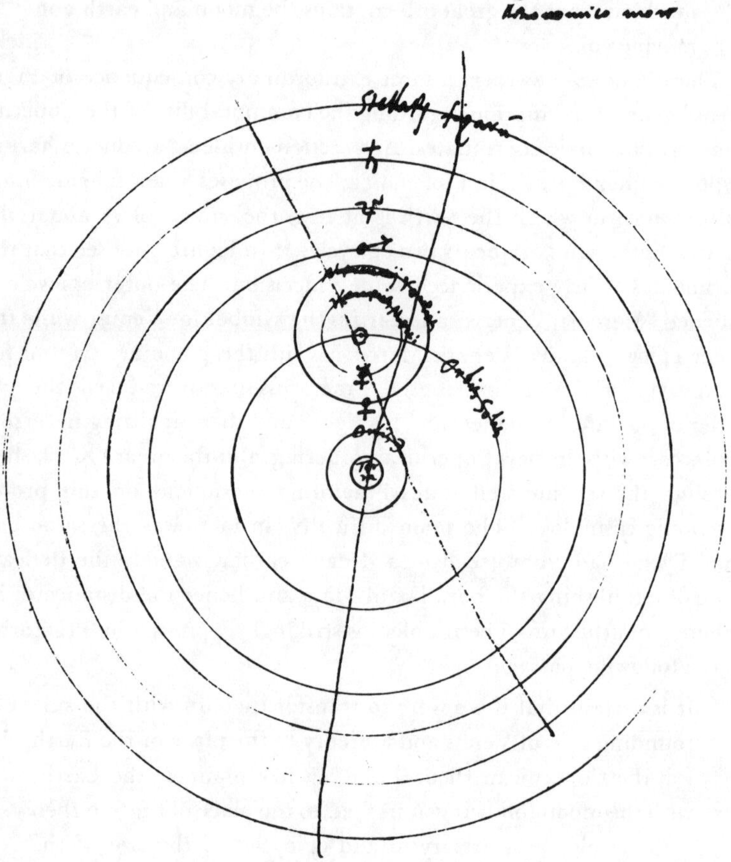

Figure 3. Praetorius' proto-Tychonic system. Notice that the orb of Mars does not intersect the solar orb, although Praetorius' values for the maximum and minimum distances of Mars indicated that this would occur. From *Compendiosa enarratio Hypothesium Nic. Copernici, Earundem insuper alia dispositio super Ptolemaica principia* (1594). Courtesy of the Erlangen Universitätsbibliothek MS. 814, fol. 94[r].

> allowed because it would then occupy not only the Sun's orb but also the great part of Venus. . . .[37]

The new Copernican arguments for the ordering of the planets were beginning to cause unexpected difficulties.

Praetorius' solution, which returns to the old Ptolemaic arrangement, reveals to us some interesting conceptual moves. In the first place, his geocentric scheme roughly doubles the distance of Saturn from the earth (Figure 4). He justifies this step by appealing to Copernicus.

> Copernicus has shown that the sphere of the fixed stars has an immense distance not only with respect to the Earth's semi-diameter but also to the great orb; hence, the sphere of Saturn is an almost immense distance from the region of the fixed stars.

And now, if Saturn's orb can be so greatly expanded, then:

> Nothing prohibits us, having changed the boundaries of revolution [*mutatis revolutionum terminis*], from making Mars' orb greater [*altiorem*] so that it will not invade the territory of the Sun. . . . Thus, we may argue as follows: The ratio of Mars' minimum to its maximum distance from Earth is 427 to 3044. Wherefore, if we now postulate that its minimum distance is contiguous to the Sun's sphere—i.e. 1180 semi-diameters, its maximum distance 8405 and its mean distance about 4792—then these distances, i.e. the orb of Mars with respect to the other eccentrics and epicycles, will produce the same revolutions with regard to one another as with Copernicus, and therefore, the same motion will be elicited from both.[38]

The great Copernican cosmological ordering, which had inspired Rheticus and Kepler to such lofty metaphysical dithyrambs, resonating with their deeply-rooted Neoplatonic assumptions, fails to drive Praetorius into the Copernican camp. Much as he admires the systematicity of the new theory, he will retreat back into the old fortress throwing up a wall of *ad hoc* defenses in his wake.

[37] *Ibid.*, 98^{v}.

[38] 97^{r}; 99^{v}.

Figure 4. Praetorius' final cosmological scheme: geocentric. From *Compendiosa enarratio Hypothesium Nic. Copernici, Earundem insuper alia dispositio super Ptolemaica principia* (1594). Courtesy of the Erlangen Universitätsbibliothek MS. 814, fol. 100^{v}.

Where the young Kepler had discovered in Copernicus' necessary ordering of the planets a sign of God's sacred intentions revealed to him through the proportionalities of the five Platonic solids, Praetorius can barely stifle a yawn. In April, 1598, he writes to Herwart von Hohenberg after reading Kepler's *Mysterium Cosmographicum*.

> I started to read this book with great expectations but I must truly confess that. . . I became more and more languid until I was frustrated of all hope. And if one seeks to know the reason for this, I can reply in no other way than to say that I understood only the smallest part of these matters and I think that this [work] departs somewhat from the definition of Astronomy or, rather, that it pertains to Physics, which surely cannot treat Astronomy in such matters. . . . It is necessary (I think) for the astronomer to apply his teachings in the following way: such that the phenomena perceived with the eyes and sense agree with one's hypotheses as if such changes of motion were guided by certain causes. But that speculation [*speculatio*] of the regular solids, what, I beg, does it offer to Astronomy? It can (he says) be useful for marking the limits or defining the order or magnitude of the celestial orbs, yet clearly the distances of the orbs are derived from another source, i.e., *a posteriori,* from the observations. And, having defined these [distances] and shown that they agree with the regular solids, what does it matter?[39]

How indeed could Kepler's discovery seem anything but foolish to a man who believed that,

> . . . the astronomer is free to devise or imagine circles, epicycles and similar devices although they might not exist in nature. . . . The astronomer who endeavors to discuss the truth of the positions of these or those bodies acts as a Physicist and not as an Astronomer—and, in my opinion, he arrives at nothing with certainty.[40]

---

[39]Johannes Kepler, Gesammelte Werke, XIII (Munich: C.H. Beck, 1945) pp. 205-206. Praetorius' own copy of the *Narratio Prima* shows that he was more interested in the treatment of precession rather than in any of Rheticus' philosophical claims (Schweinfurt, No. 6501).

[40]*Planetarum Theoriae Inchoatae* (August 1605), Schweinfurt MS. H 73, fol. 2^v.

And these sentiments came from an astronomer who knew that Copernicus had not written the anonymous preface to *De revolutionibus.* Indeed, on one of his two copies of *De revolutionibus,* Praetorius or the previous owner, Paul Eber, has written the words, "Andr. Osiandri, (ut aiunt)."[41] On the Yale University copy (1543) there appears a note in Praetorius' hand which reads: "Rheticus affirmed that this preface was added by Andreas Osiander. But it was rejected by Copernicus. The title also was changed by the same person against the author's will, for it ought to be *De revolutionibus mundi.* Osiander, however, added *orbium coelestium.*"[42] And the owner of the 1543 edition of *De revolutionibus* at Trinity College, Cambridge has written under *Ad lectorem* as follows: "Praetorius the Nürnberg Mathematician thought that this letter was from Andreas Osiander. He condemned him because he revealed the secrets of astronomy." We can only conclude, however, that had Osiander not written the preface, Praetorius still would have supplied it in practise.

With Praetorius, therefore, we find a deepened articulation of the Wittenberg interpretation of the Copernican hypothesis. Where Caspar

---

[41]Schweinfurt No. 6796, fol. 1^{v}. This information clearly originated with Rheticus and came either through Eber, who was a good friend of Rheticus, or directly to Praetorius when he met Rheticus in Cracow between 1569 and 1571.

[42]In a letter to Herwart von Hohenberg in 1609, the wording and content of which directly parallels the annotation cited from the Yale University copy, Praetorius reports that Osiander had changed the title of Copernicus' work from *De revolutionibus orbium mundi* to *orbium coelestium*; and Copernicus was evidently unhappy about the change. Praetorius explains that Osiander was responsible for changes both in his addition of the *praefatiuncula* and in the title change: "Ait prefationem priorem in Copernicum, incerti esse autoris. Eam autem prefixit Andreas Osiander, quondam conciniator Norimbergensis. Nam ipso curante Copernici opus primum Norimbergae impressum est. Et primae aliquot paginae ad Copernicum missae sunt. Sed paulo post Copernicus diem suum obijt antequam totum opus videre potuit. Serio autem Rheticus affirmabat, Copernico plane displicuisse illam Osiandri praefationem, imo non mediocriter irritum fuisse. Et verisimile est nam aliam eius mentem fuisse, ac quod se fert praefatio illa, ex dedicatoria Epistola et ipsius libro apparet. Titulus etiam citra mentem autoris ab eodem immutatus fuit debuit enim esse: De revolutionibus orbium mundi et fecit Osiander: orbium coelestium: etc." (Zinner, *op. cit.*, p. 454). Hans Blumenberg believes that the title change, like the addition of the preface, was designed to underplay the realist claims of the new theory (*Die Kopernikanische Wende,* [Frankfurt am Main: Edition Suhrkamp, 1965], pp. 94-95).

Peucer, Praetorius' teacher, had merely hinted at some possible ways in which the new theory might be accommodated to the Ptolemaic viewpoint, Praetorius makes this the cornerstone of his own investigations and lectures. Students of Praetorius from about 1580 onward, so far as we know, would be exposed to extensive details of the Copernican models, the theory of precession and a sense of controversy about the ordering of the planets. They would learn to use the *Prutenic Tables* but they would also receive the old admonition that the astronomer should stay clear of the physicist's domain.

4. Tycho Brahe and the Wittenberg Interpretation

Just as Polish students in the fifteenth century often cut a route to the Italian universities of the south to further their studies in law, medicine or astronomy, so it was that Danish students of the sixteenth century, who had frequented Paris two hundred years earlier, gradually began to pursue their educations at the German universities.[43] The locus of astronomical activity had shifted. In March, 1562, a young Danish nobleman, named Tycho Brahe, and his tutor, Anders Sorensen Vedel, rode into Leipzig where the former was supposed to take up the study of law.[44] Fortunately for the course of subsequent astronomical progress, there exists no evidence of the fulfillment of this intention. Instead, Brahe immediately acquainted himself with the professor of mathematics at Leipzig, Johannes Homelius (1518-1562), who had come to Leipzig well recommended from the University of Wittenberg where he had once been a pupil of George Joachim Rheticus and a classmate of Caspar Peucer.[45] The close intellectual bonds which united sixteenth-century academics were sometimes complemented, if not reinforced, by marital connections much as was conducted the contemporary diplomacy of Europe. Caspar Peucer, for example, had married a daughter of Philipp Melanchthon and Homelius became the son-in-law of Melanchthon's close friend, Joachim Camerarius (1500-1574).

[43]See J.L.E. Dreyer, *Tycho Brahe: A Picture of Scientific Life and Work in the Sixteenth Century* (New York: Dover, 1963), p. 15. First published in 1890.

[44]*Ibid.*, p. 16.

[45]Karl Heinz Burmeister, *Georg Joachim Rhetikus, 1514-1574,* I (Wiesbaden: Guido Pressler, 1967), p. 72.

Although Homelius died three months after Tycho's arrival, the 15-year-old lad learned from him, or his pupil Bartholomeus Scultetus, the method of dividing a cross-staff by transversals.[46] And it is not unlikely that Tycho first heard of Copernicus through Homelius, once a student of Reinhold and Rheticus, since the latter, in his unpublished lectures on Proclus' *Hypotyposis astronomicarum positionum,* makes frequent use of Copernicus' planetary values.[47] In any case, this was only the beginning of Tycho's contacts with the ideas of Copernicus and the astronomers of the Wittenberg Circle.

From April to September 1566, he was in Wittenberg where he made the acquaintance of Peucer. Amidst extensive travels during the next ten years, until he settled on the island of Hveen, Tycho revisited Wittenberg twice for short periods in 1568 and 1575; Cassel, in 1575, where he met and befriended Wilhelm IV, Landgrave of Hesse-Cassel (1532-1592) and, in the same year, he passed through Saalfeld where the son of Erasmus Reinhold showed him the manuscripts of his father.[48] Tycho's contacts with Wittenberg, however, were only the prelude to his deeper understanding of the Copernican theory. That fuller appreciation dates from the period of his middle twenties when he observed the New Star of 1572 and concomitantly acquired a first-hand knowledge of the new theory by the study of *De revolutionibus* itself. (*See Appendix*).

About 400 years ago, in September 1574, Tycho commenced a series of astronomical lectures at the University of Copenhagen upon the urging of the Danish king and a number of friends. Although Tycho himself had never completed a formal university education, the types of problems which he chose to examine and his attitude toward the Copernican theory, clearly reflect his extensive contacts with Peucer and the University of Wittenberg.

---

[46]Dreyer, *op. cit.,* p. 330.

[47]*Dictata in Hypotheses Procli. . . Anno 1557, 8 Februarj,* Schweinfurt MS. H 178. DR Yale (1543), referred to in footnote 42, also contains some annotations by Homelius. On fol. 63^{V}, for example, Praetorius has written the words "Homelij censura" above a note written in a hand different than his own. In general, Homelius' entries reveal a strong interest in the Copernican precession model as well as in the mechanism of libration (67^{V}), the latter concern perhaps reflecting the influence of Reinhold.

[48]Dreyer, *op. cit.,* pp. 23, 83; *TBOO,* III, p. 213.

> In our time, however, Nicholas Copernicus, who has justly been called a second Ptolemy, from his own observations found out something was missing in Ptolemy. He judged that the hypotheses established by Ptolemy admitted something unsuitable and offensive to mathematical axioms; nor did he find the Alphonsine calculations in agreement with the heavenly motions. He therefore arranged his own hypotheses in another manner, by the admirable subtlety of his erudition, and thus restored the science of the celestial motions and considered the course of the heavenly bodies more accurately than anyone else before him. For although he holds certain [theses] contrary to physical principles, for example, that the Sun rests at the center of Universe, that the Earth, the elements associated with it, and the Moon, move around the Sun with a threefold motion and that the eighth sphere remains unmoved, he does not, for all that, admit anything absurd as far as mathematical axioms are concerned. If we inspect the Ptolemaic hypotheses in this regard, however, we notice many such absurdities. For it is absurd that they should dispose the motions of the heavenly bodies on their epicycles and eccentrics in an irregular manner with respect to the centers of these very circles and that, by means of an irregularity, they should save unfittingly the regular motions of the heavenly bodies. Everything, therefore, which we today consider to be evident and well-known concerning the revolutions of the stars has been established and taught by these two masters, Ptolemy and Copernicus.[49]

This passage could well have been written by Praetorius.[50] The goals of the two astronomers are identical: the *restoration* of Ptolemaic astronomy by the conversion of equant-less Copernican models into the old reference frame. The planetary motions are to be analyzed, Tycho

---

[49]*TBOO,* I, p. 149. Quoted and translated in Kristian Peder Moesgaard, "Copernican Influence on Tycho Brahe," *Studia Copernicana* V (*Colloquia Copernicana* I), (Warsaw: Ossolineum, 1972), p. 32. I have modified Moesgaard's translation in several places. In the following section, my indebtedness to Moesgaard's excellent study will be self-evident from the citations.

[50]See above, pp. 292-293.

writes, "according to Copernicus' opinion and numbers but referring everything to a stable Earth, rather than as he [Copernicus] supposed, i.e., by three motions [of the earth]. . . ."[51] In his lecture series, Tycho reports, he had planned to demonstrate how the coordinate transformation which he had shown for the sun and moon could also be applied to the remaining planets but for the fact that, "I was inclined to make a certain trip to Germany. . . ."[52] And furthermore, he tells us, his exposition surpasses, ". . . the hypotheses set down, in vain, in a certain book, published recently by Peucer and Dasypodius. For these men apply the calculations of Copernicus unsuitably to the Ptolemaic and Alphonsine hypotheses."[53] No doubt, Tycho was bothered by Peucer's continued adherence to the equant in his *Hypotheses astronomicae,* in spite of the latter's aim to show that the Copernican hypotheses could be "accommodated" to the geostatic universe.[54] Like Praetorius, therefore, Tycho seems to have been motivated not only by the desire to follow in the footsteps of his Wittenberg masters, but to supersede them. At any rate, some such feelings may have existed in Tycho when he set out from Copenhagen in the Winter of 1575 to inspect the extant papers of Peucer's old teacher, Erasmus Reinhold.

From the published edition of Brahe's manuscripts, Christine Jones Schofield and Kristian P. Moesgaard, respectively, have been able to demonstrate that Tycho's commitment to an equantless astronomy is evident at least as early as 1574; and in a manuscript of 1585, *De Marte,* Tycho can be seen inverting the Copernican model for Mars.[55] In

[51] *TBOO,* I, p. 172. Partially quoted in Moesgaard, *op. cit.,* p. 32. I wish to take exception to Moesgaard's contention that *after* Tycho, ". . . astronomers were from now on free to make use of Copernican astronomy without troubling about his cosmology. . ." (p. 33). This was a tradition that clearly antedated and nurtured Tycho himself.

[52] *TBOO,* I., p. 173.

[53] *Ibid.* Tycho's reference is to Peucer's *Hypotheses astronomicae seu theoriae planetarum ex Ptolemaei et aliorum veterum doctrina ad observationes Nicolai Copernici et canones ab eo conditos accomodate opera et studio Casparis Peuceris,* Wittenberg, 1571, Cf. footnote 17 above and Westman, *op. cit.,* "Wittenberg Interpretation" for further discussion of this work.

[54] *Ibid.*

[55] Christine Jones Schofield, "The Geoheliocentric Planetary System: Its Development and Influence in the Late 16th and 17th Centuries," Unpublished Doctoral Dissertation: Cambridge University (Newnham College), 1964, p. 35 ff.; Moesgaard, *op. cit.,* p. 32 ff.

Inuerſio Hypotheſis Copernianæ in motu ♂, vt centrum vniuerſi TERRA quieſcens iuxta veterum ſententiam occupet, et nihilominus motus æquales circulorum propria ſua reſpiciant centra, repudiata Ptolemaica discohærentia.

Hæc ratio inuertendi Copernianam Hypotheſin ſufficit tribus ſuperioribus planetis ♄ ♃ ♂.

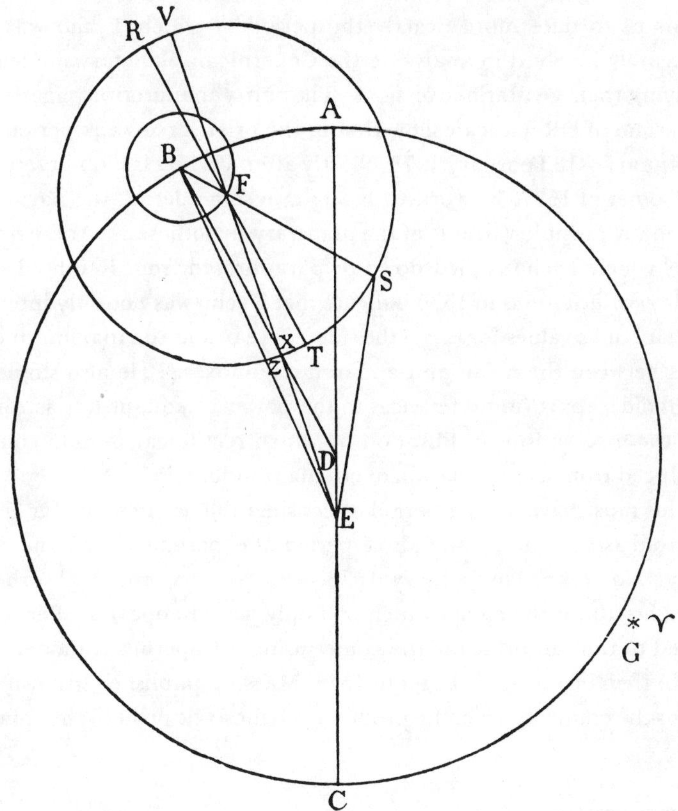

Figure 5. Tycho Brahe's geocentric inversion of the Copernican model for Mars. From *De Marte* (1585) in *Tycho Brahe Opera Omnia*, Vol. V, p. 284.

Figure 5, we observe that Tycho's inversion uses an eccentric-cum-epicycle device where Praetorius (Figure 1) had chosen to use the equivalent bi-epicyclic construction. The planet at *F* rides on the epicyclet *BF* which represents Copernicus' replacement of the equant. The circle described by radius *FS* is the annual element, the projection of the earth's motion, which, in fact, is more clearly shown in Praetorius' diagram.

With the recent identification of a second copy of *De revolutionibus* which belonged to Tycho (*Appendix*), we have new evidence which now allows us to date more exactly the period in which Tycho was most intensively involved in analyzing the Copernican planetary models and studying their geostatic inversions. The thirty manuscript pages bound at the end of DR Vatican show that in the brief three-week period from 27 January—13 February 1578, shortly after Tycho's last observations of the Comet of 1577, he worked through, in some detail, the parameters and many possible variants of the planetary hypotheses.[56] And extensive notes which Tycho copied down or paraphrased from Reinhold's copy of *De revolutionibus* in 1575 indicate that Tycho was not only interested in Reinhold's values for (e.g.) the sidereal year and the maximum equations between the mean and apparent equinoxes.[57] He also singles out Reinhold's approving references to the power of equant-less astronomy and the mechanisms of libration by which rectilinear motion could be produced from a combination of circular motions.[58]

This must have been a period of considerable excitement for the 31-year-old astronomer; it was also a period of experimentation and receptivity. No doubt, in January 1578, he was very much absorbed in thoughts about the comet which had only just disappeared. Perhaps he hoped to find an orbit for the comet using a Copernican model which might then be inverted. Later in 1578, Maestlin published his own analysis of the comet in which he proposed a heliostatic orbit in the sphere of

[56] *De Revolutionibus* Vatican 1543, fols. 197^{v}- 210^{v}. Hereafter all references to copies of *De revolutionibus* will be abbreviated "DR."

[57] *Ibid.*, Book III, Ch. 7, fol 73^{r}; 14, 81^{r}.

[58] *Ibid.*, 15, 84^{v} - 85^{v}; 20, 91^{v} - 92^{v} and especially 4, 67^{v} and the same page in DR Prague 1566.

Venus while allowing the earth to move.[59] It is possible, as I have suggested elsewhere, that Tycho's own geoheliostatic analysis of the comet may owe some inspiration to Maestlin's solution, for the two are quite similar.[60] But if Tycho's interest in the planetary models in early 1578 was primarily motivated by the search for a cometary orbit, there is no direct evidence of it amongst the new manuscript material.

One of Tycho's aims, at any rate, was certainly to provide himself with a convenient paraphrase and consolidation of Copernicus' theories which were scattered inconveniently throughout the lengthy *De revolutionibus.* At the same time, such a summary would enable him to visualize more easily the geostatic inversions. In Figure 6, we see, as an example, his construction for the three superior planets, drawn on 13 February 1578. Unlike the diagram for Mars in 1585 (Figure 5), the center of the epicyclet is located on the line of apsides rather than on the eccentric and the sun's orbit is sketched in by a dotted line. The caption below reads: "The Semi-diameter OK in this place, for the Three Superior Planets, is equal to the *Orbis Magnus*"; and above: "The Theories of the Three Superior Planets Accommodated to the Immobile Earth." Added below this, self-consciously, Tycho has written: "The Idea for this New Hypothesis came to me on the 13th of February in the Year 1578."[61] Over the next three days, Tycho continued to develop further heliostatic devices with geostatic translations. On 14 February he shows a bi-epicyclic model for Venus with the epicyclet on the eccentric and adjacent to this, a tri-epicyclic device for Mercury.[62] In all of this activity, Tycho's main purpose seems to have been the improvement of the "unsuitable" applications of Copernican theories to the "Ptolemaic and Alphonsine hypotheses," such as those set forth by Peucer. One result was a further heightening of his appreciation for Copernican geometry and particularly the Reinhold-Rheticus variant of

[59]See Robert S. Westman, "The Comet and the Cosmos: Kepler, Maestlin and the Copernican Hypothesis," *Studia Copernicana V* (*Colloquia Copernicana I*), (Warsaw: Ossolineum, 1972), pp. 7-30, reprinted in J. Dobrzycki (ed.), *The Reception of Copernicus' Heliocentric Theory,* (Dordrecht, Holland: Reidel, 1973), pp. 7-30.

[60]*Ibid.*, p. 25.

[61]DR Vatican, fol. 208^{r}.

[62]*Ibid.*, 208^{v} - 209^{r}.

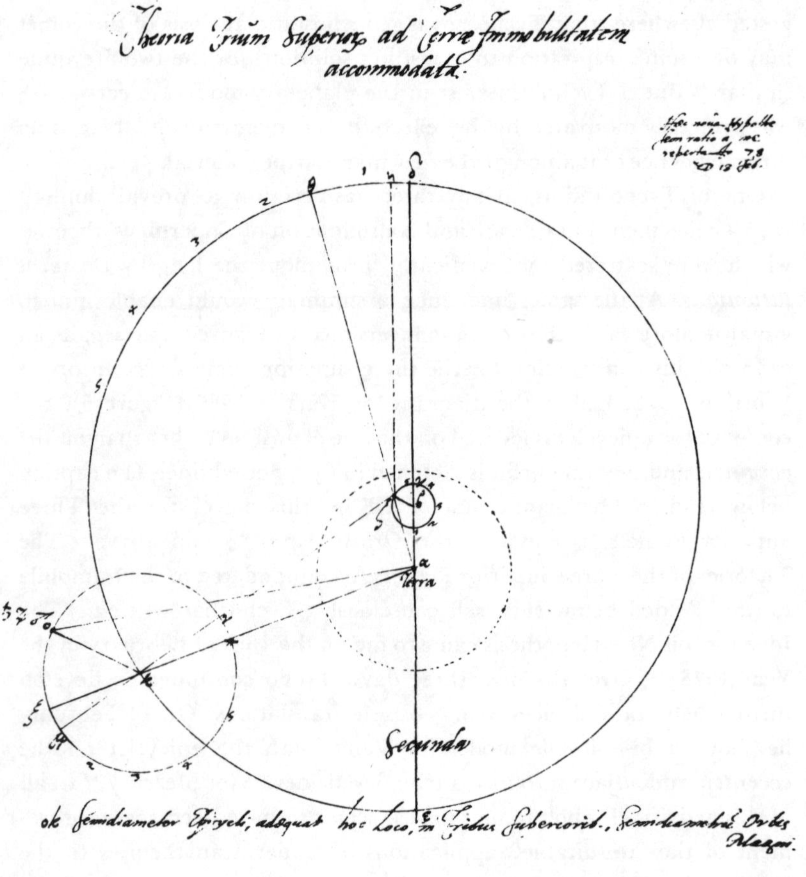

Figure 6. Tycho Brahe's model for the superior planets "accommodated to the immobility of the Earth" (13 February 1578). From Tycho Brahe's working notes in *De revolutionibus* (1543). Courtesy of the Biblioteca Apostolica Vaticana, fol. 208[r].

the Wittenberg Interpretation which praised Copernicus' "liberation from the equant."[63]

5. Tycho and Copernican Cosmology

It should be recognized that Copernicus' preference for an astronomy of circles revolving about their own centers possessed more than a kinematic status. One of the primary aims of *De revolutionibus* Book I is to show that natural motion, which is circular, represents a principle inherent in spheres and the parts of spheres.[64] From an historical viewpoint, Copernicus' reasoning reveals a deep immersion in Aristotelian categories and modes of argumentation, yet neatly re-directed for his own uses. Chapter 1 defends the sphericity of the world from a combination of final causes (the sphere is an "integral whole and needs no joints"; its volume is "especially suitable" for conserving and comprehending things); formal and material causes (all the planets have this same form; drops of water are delimited by the sphere as well).[65] The second and third chapters argue that the Earth is also spherical although the irregularities of its surface tend somewhat to modify its universal roundness."[66] Chapter 4 now introduces the principal axiom of astronomy: "The Motion of the Celestial Bodies is Regular, Circular and Everlasting—Or Else Compounded of Circular Motions," but this axiom of motion is then immediately "justified" by the nature of the sphere.

> . . . the motion of a sphere is to turn in a circle; by this very act expressing its form, in the most simple body, where beginning and end be discovered or distinguished from one another, while it moves through the same parts in itself.[67]

---

[63]The phrase is used by Rheticus in his *Narratio Prima* (trans. Edward Rosen in *Three Copernican Treatises* (3rd edition, New York: Octagon Books, 1971), p. 135.

[64]Cf. Alexandre Koyré, *La révolution astronomique* (Paris: Hermann, 1961), pp. 62-63.

[65]Nicholas Copernicus, *On the Revolutions of the Heavenly Spheres,* trans. Charles Glenn Wallis in *Great Books of the Western World,* XVI (Chicago: Encyclopaedia Britan-nica, 1952), I, 1, p. 511. Hereafter cited as *GBWW*.

[66]*Ibid.*, 2, p. 511.

[67]*Ibid.*, 4, p. 513.

Copernicus then explains further that since a simple celestial body cannot be moved non-uniformly by a single sphere ("the mind shudders" at an inconstant motor virtue), it follows that the inequalities of motion are produced either because they *appear* to occur or because the spheres bearing the planets have different axes of rotation.

Now, in all of this discussion, there is an evident desire by Copernicus to show that his concepts are not arbitrary. There is a *reason* why the planetary motions are composites of circles. And while his reasons may not strike some modern philosophers as *good* reasons, it is clear that Copernicus believed, as we might say, that he had achieved some initial probability or conceptual "entrenchment" for his basic claims.[68] After all, every innovator who bears the slightest hope that his ideas might be recognized and received, must be prepared to justify the intelligibility of his concepts in addition to the empirical adequacy of his hypotheses. Where Aristarchus left only an interesting idea, so far as we know, Copernicus bequeathed a highly-articulated theory of the universe—articulated more strongly, we should say, from a kinematic viewpoint but not without some definite pretensions toward dynamic explanations. Yet it is obvious that reasons which may seem sufficient to the innovator will not necessarily raise a congruent sense in others.

We have seen, thus far, that the physical "grounding" of the "Astronomical Axiom" in the form and physical nature of the sphere[69] could hardly fail to win the assent of many sixteenth-century astronomers eager to reject the equant. Beginning with Chapter 5, however, Coper-

[68]Cf. Gerd Buchdahl, "History of Science and Criteria of Choice," in Roger Stuewer (ed.), *Historical and Philosophical Perspectives of Science* (Minnesota Studies in the Philosophy of Science, 5), (Minneapolis, Minn.: University of Minnesota Press. 1970), p. 207.

[69]In his interesting commentary on this paper, Professor Machamer raises the issue of whether the equant, as discussed by Copernicus, ought to be interpreted as a mathematical or as a physical model. I confess that the text of *De Rev.* I.4 *per se* is not completely clear on this matter. In a very important paper, which bears directly on this problem and which appeared after the present study, Professor Noel Swerdlow claims that Copernicus unquestionably regarded the spheres as material and that equant motion about an axis other than the diameter of a sphere would constitute violation of a "mechanical principle" ("The Derivation and First Draft of Copernicus' Planetary Theory: A Translation of the *Commentariolus* with Commentary," *Proceedings of the American Philosophical Society,* 1973, *117*: 424 f.; 434 ff.). Swerdlow's interpretation seems to proceed chiefly from the

nicus raises a further, and more dangerous implication: "Now that it has been shown that the Earth too has the form of a globe, I think we must see whether or not a movement follows upon its form and what the place of the Earth is in the universe."[70] This chapter merely suggests, on the basis of an argument from relative motion, that it is *plausible* to entertain the possibility of the earth's motion, a possibility which is not inconsistent with the appearances. By Chapter 8, however, having shown earlier that the assumption of a sufficiently immense, but finite, universe allows for the possibility of a circular motion of the earth consistent with observations, Copernicus now draws clear implications from his earlier arguments.

> Why therefore should we hesitate any longer to grant to it [the Earth] the movement which accords naturally with its form, rather than put the whole world in a commotion—the world whose limits we do not and cannot know? And why not admit that the appearance of daily revolution belongs to the heavens but the 'reality' belongs to the Earth? And things are as when Aeneas said in Virgil: "We sail out of the harbor, and the land and the cities move away."[71]

In systematically-censored copies of *De revolutionibus* which I have examined, this passage is always one that is carefuly expunged.[72]

---

reasonable assumption that Copernicus' frame of reference was formed by Regiomontanus' *Epitome of the Almagest* and Peurbach's *New Theories of the Planets* where an elaborate apparatus of spheres, axes, rotations and inclinations is employed to account for the various motions. According to Swerdlow, Copernicus probably "took for granted" the reader's familiarity with such models. Now it is clear that *later* commentators on Copernicus, such as Rheticus, Tycho, Maestlin, Praetorius and others, did read *De rev.* as a new system of solid spheres—which lends support to Swerdlow's contention. However, it does not explain why Copernicus, when he laid down the basic principles of his new astronomy in Book I of *De rev.,* did not more explicitly enunciate this claim. Whatever Copernicus' actual intentions might have been, the expression of the "Astronomical Axiom" in *De rev.* I.4 can be read either as a purely mathematical statement or as a physical statement (in conjunction with circumstantial evidence).

[70] *GBWW, op. cit.,* I, p. 514.

[71] *Ibid.,* 8, p. 519.

[72] Such "corrected" copies may be found in *at least* the following libraries: Stanford University (1566); Rome Observatory (1566); Biblioteca Nazionale di Roma (1566) (Shelf No.: 201. 39. i. 18); Hale Collection, Mt. Wilson Observatory Library (1566): The State

Clearly, it was one matter for the planet to possess circular revolutions with respect to the center of its epicycle because a sphere must rotate about its own diameter, yet quite another thing to draw the same inference with regard to the rotation of the Earth. Copernicus, however, does precisely that and he bolsters his contention by arguing that rectilinear motion is merely an accident, a motion unnatural to a sphere but attainable temporarily by its parts.[73] Gravity, then, is a tendency of matter to unite in the form of a globe.[74]

Tycho's annotations on Chapters 5-9 of Book I are by no means unfavorable toward these Copernican arguments, at least in the sense that he nowhere criticizes them.[75] This is quite remarkable in view of the fact that his 1574 lectures recoil at Copernicus' "physical absurdities" and there exist a multitude of well-known statements in his writings and correspondence from the late 1580's which are unequivocally Aristotelian.[76] The Tycho of the middle 1570's, however, shows a less polemical, defensive visage than the same man a decade later. The latter has a new system to defend not only against Ptolemaean opponents but also against others who claim to have discovered the geoheliocentric theory. The former entertains a more receptive and flexible stance. As an example of this openness, we may cite his gloss in DR Vatican on Copernicus' definition of gravity:

> By this reason, therefore, the Elements do not desire the Center of the universe but rather the center of their globe, just as like seeks like; and the only reason why the stars and the Earth are globes is because they are moved by their form.[77]

Library of Victoria, Melbourne, Australia (1566); Warsaw University Library (1566) (Shelf No. 28.20.2.1065); Trinity College Library, Cambridge University (1543); Universitätsbibliothek Wien (1543); Sjögren Library, Royal Swedish Academy of Engineering Sciences, Stockholm (1543); Poznań, Bibl. Poznanskiego Towarzystwa Przyjacioł Nauk (1566); University of Oklahoma (1566); Biblioteca Nazionale di Palermo (1543; 1566); Biblioteca Riccardiana di Firenze (1566); University of Rochester (1566); Biblioteca Publica Municipal, Porto, Portugal (1543); Biblioteca Nacional de Lisboa (1566).

[73] *GBWW, op. cit.,* I, 8, p. 520.

[74] *Ibid.,* 9, p. 521.

[75] DR Prague is more heavily annotated.

[76] Cf. Moesgaard, *op. cit.,* p. 48 ff.

[77] DR Vatican, I, 9, fol 7^r.

And in DR Prague, he comments:

> If all motions of Terrestrial objects are composed of straight motions away from the center and circular motions, then this argument—that a simple body has only a simple motion—which others hold [to be true], will be invalid. I anticipate this argument. He [Copernicus] replies. . .[78]

While space does not permit us to offer further examples, we may conclude generally that Tycho offers a fair hearing to the physical arguments of Copernicus. His annotations show him exploring and paraphrasing the inner logic of Copernicus' theses without formulating any rejoinders. There seems little question, however, that he was never quite persuaded by these arguments.

If there exist any doubts regarding Tycho's appraisal of Copernican dynamics, there can be few concerning his evaluation of the famous claim that the planets are ordered so uniquely that any change in either the periods or distances would upset the whole universe. Like Rheticus and Praetorius, Tycho is struck by the *symmetria* of the theory. Rheticus himself had written: ". . .there is something divine in the circumstance that a sure understanding of celestial phenomena must depend on the regular and uniform motions of the terrestrial globe alone."[79] Tycho has underlined and glossed this passage as follows: "The reason for the revival and establishment of the Earth's motion." And in the margin next to the sentence where Copernicus speaks of the symmetry of the world, Tycho adds:

> The testimonies of the planets, in particular, agree precisely with the Earth's motion and thereupon the hypotheses, assumed by Copernicus are strengthened [*confirmatur*].[80]

On perhaps the most famous page in Copernicus' work, the page where he displays the heliocentric scheme, Tycho shows that he fully understands and admires the empirical consequences of the interconnexity of the orbits.[81] Consider now Figure 7. Here we see that the mean sidereal

---

[78]DR Prague, I, 8, fol, fol 6^{V}.

[79]*Narratio Prima* 1566, fol. 202^{V}. Tycho cross-references this passage on fol iij^{r} of both DR Prague and DR Vatican. Cf. *Appendix*, p. 344.

[80]DR Prague, fol. 10^{r}; cf. 9^{r}.

[81]DR Vatican, fol 9^{V}.

NICOLAI COPERNICI

net, in quo terram cum orbe lunari tanquam epicyclo contineri diximus. Quinto loco Venus nono menſe reducitur. Sextum denique locum Mercurius tenet, octuaginta dierum ſpacio circũ currens. In medio uero omnium reſidet Sol. Quis enim in hoc

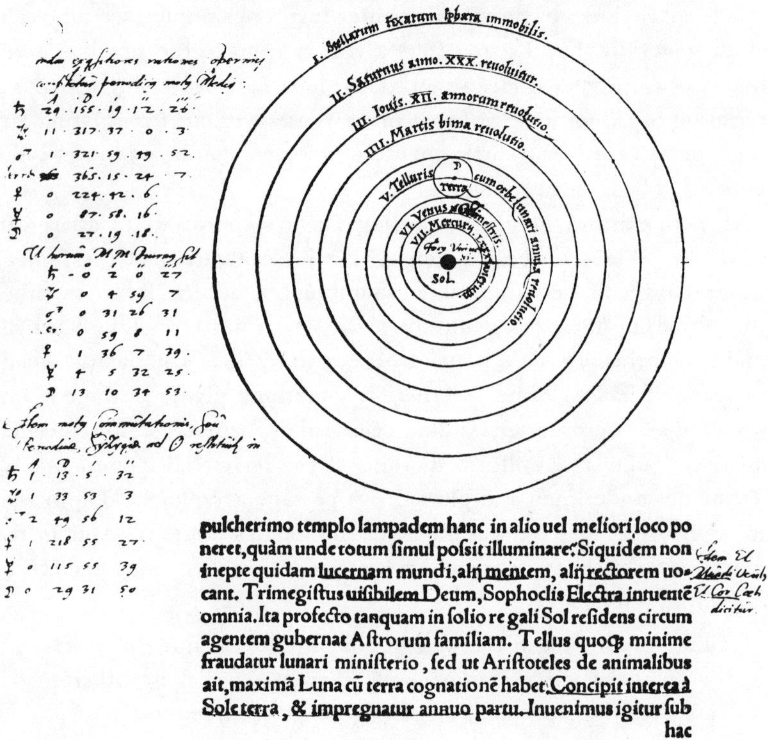

pulcherimo templo lampadem hanc in alio uel meliori loco poneret, quàm unde totum ſimul poſsit illuminare? Siquidem non inepte quidam lucernam mundi, alij mentem, alij rectorem uocant. Trimegiſtus uiſibilem Deum, Sophoclis Electra intuentẽ omnia. Ita profecto tanquam in ſolio regali Sol reſidens circum agentem gubernat Aſtrorum familiam. Tellus quoque minime fraudatur lunari miniſterio, ſed ut Ariſtoteles de animalibus ait, maximã Luna cũ terra cognationẽ habet. Concipit interea à Sole terra, & impregnatur annuo partu. Inuenimus igitur ſub hac

Figure 7. Tycho Brahe's notes in one of his copies of *De revolutionibus* (1543) showing the "exquisitiores rationes" according to which the planetary orbs are arranged by their mean sidereal periods in the Copernican system. Courtesy of the Biblioteca Apostolica Vaticana, fol. 9^{v}.

periods of the planets are displayed "according to the more exquisite reasons of Copernicus." These periods, it will be noticed, increase continuously from the sun to the furthermost planet, Saturn. In the column beneath the first we see that the mean daily motions decrease in a continuous pattern from the sun. And in the third column we have the "motion in commutation" which gives values for the synodic periods from which the sidereal periods can be derived. It is interesting to compare these numbers with the copy of *De revolutionibus* owned by the Scottish astronomer Duncan Liddel (See Figure 8), who was evidently impressed by the same feature of the Copernican theory and transcribed this data from Tycho's volume.[82]

Throughout this analysis, Tycho recognizes the predominant role of the sun, *Focus universi* (Hearth of the Universe), in the various planetary theories. Once the sun's component has been removed, then the periods fall into a "more exquisite" order. Tycho celebrates the sun's preeminence with a eulogy that he adds in the margins of DR Prague next to Copernicus' own Hermetic dithyramb.

> Phoebus sits in the middle; His Golden Crown Shines, King and Emperor of the World, as it were, governing Temenos with his sceptre,
> In whom all celestial virtues reside, as Jamblichus and others do testify.
> And Proclus says of the Sun's aspect,
> That all celestial virtues unite and terminate in One,
> Which then are disseminated into this World by the Sun's fiery celestial breath.[83]

Yet there would be a price to pay for the new-found harmony. If the Sun were to be granted this imperial role as governor of the planets and the harmonious continuity of the periods were to be preserved, then the

[82]See *Appendix*, pp. 343-344. In Tycho's copy the numbers are written in the left-hand margin with the heliocentric diagram; in Liddel's copy, as can be seen, the numbers are written on the inserted folio. With two very slight exceptions, the wording and numbers are identical. By 1589 or 1590, Liddel was lecturing on Tycho's system at the University of Rostock.

[83]DR Prague, I, 10, fol 7^r.

NICOLAI COPERNICI
net, in quo terram cum orbe lunari tanquam epicyclo contineri
diximus. Quinto loco Venus nono menſe reducitur. Sextum
deniq; locum Mercurius tenet, octuaginta dierum ſpacio circũ
currens. In medio uero omnium reſidet Sol. Quis enim in hoc

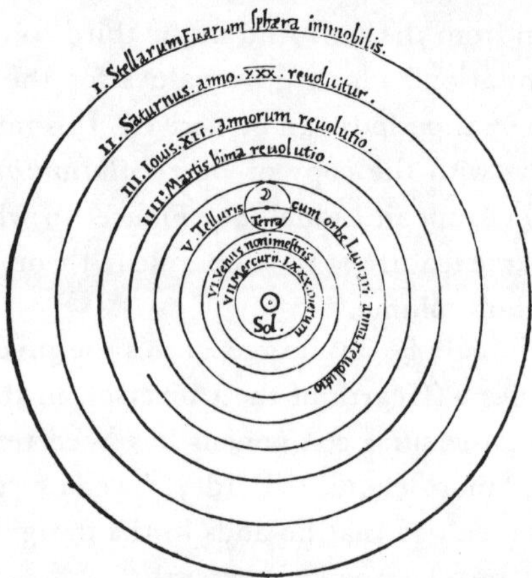

I. Stellarum Fixarum ſphæra immobilis.
II. Saturnus anno XXX. reuoluitur.
III. Iouis XII. annorum reuolutio.
IIII. Martis bima reuolutio.
V. Telluris cum orbe Lunari annua reuolutio
VI. Venus nonimestris.
VII. Mercurii LXXX. dierum
Sol.

pulcherrimo templo lampadem hanc in alio uel meliori loco po
neret, quàm unde totum ſimul poſsit illuminare? Siquidem non
inepte quidam lucernam mundi, alij mentem, alij rectorem uo•
cant. Trimegiſtus uiſibilem Deum, Sophoclis Electra intuentẽ
omnia. Ita profecto tanquam in ſolio regali Sol reſidens circum
agentem gubernat Aſtrorum familiam. Tellus quoque minime
fraudatur lunari miniſterio, ſed ut Ariſtoteles de animalibus ait,
maximam Luna cum terra cognationẽ habet. Cõcipit interea à
Sole terra, & impregnatur annno partu. Inuenimus igitur ſub
hac

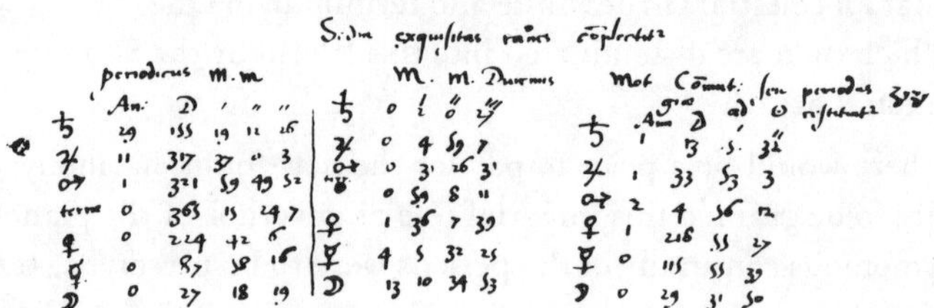

Figure 8. Notes on *De revolutionibus*, fol. 9^{v}, and interfoliated page from the Scottish astronomer and physician, Duncan Liddel (1551-1613). The annotations are virtually identical with those on Tycho Brahe's copy of *De revolutionibus* (see Figure 7) from which it was probably copied. Courtesy of the Aberdeen University Library.

earth must be allowed to move. This was a step that Tycho hesitated to take. But out of his dilemma there was born a new cosmological system.

6. *The Genesis of the Tychonic System*

J.L.E. Dreyer has written: "The idea of the Tychonic system was so obvious a corollary to the Copernican system that it almost of necessity must have occurred independently to several people. . . ."[84] Actually, a principal component of the theory had long been available from Martianus Capella's brief statement in *On the Marriage of Philosophy and Mercury* ("The Stars of Mercury and Venus. . .do not go around the earth at all, but around the sun in freer motion") as well as in Simplicius' Commentary on *De caelo.*[85] And in his cosmological discussion of the order of Venus and Mercury, Copernicus refers to the Capellan order as follows:

> Wherefore I judge that what Martianus Capella—who wrote the *Encyclopedia*—and some other Latins took to be the case is by no means to be despised. For they hold that Venus and Mercury circle around the sun as a center; and they hold that for this reason Venus and Mercury do not have any further elongation from the sun than the convexity of their orbital circles; for they do not make a circle around the earth as do others, but have perigee and apogee interchangeable [in the sphere of the fixed stars]. . . . Thus the orbital circle of Mercury will be enclosed within the orbital circle of Venus—which would have to be more than twice as large—and will find adequate room for itself within that amplitude.[86]

We can see at once that, by comparison with the original statement of

[84]J.L.E. Dreyer, *A History of Astronomy From Thales to Kepler,* (2nd edition, New York: Dover, 1953), p. 367. First published in 1905.

[85]Capella, VIII. 857, 859 and Simplicius, 519.9-11, translated in M.R. Cohen and I. Drabkin, *A Source Book in Greek Science* (Cambridge, Mass.: Harvard, 1966), p. 107. Tycho knew Simplicius' Commentary (cf. DR Vatican, fol. iiij^{V} and Figure 11 above). For a recent discussion of the inner planets in pre-Ptolemaic astronomy, see Walter Saltzer, "Zum Problem der inneren Planeten in der vorptolemäischen Theories," *Sudhoffs Archiv,* 1970, *54*: 141-172.

[86]*GBWW, op. cit.,* I, 10, pp. 523-24.

Capella, Copernicus' interpretation articulates the theory in more detail and at the same time thereby enhances the probability of developing it beyond the purely schematic level. Three decades later, this is precisely what happened.

The first published diagram of the "Capellan" theory appeared 30 years after *De revolutionibus* in an elementary textbook of astronomy entitled: *Primarum de coelo et terra institutionum quotidianarumque Mundi revolutionum Libri Tres* (Venice, 1573). The work was dedicated to Prince Stephen Batorý of Transylvania and intended especially for young gymnasium-level students, such as Batorý's grandson. Its author, a professor at the University of Cologne, named Valentin Naibod, describes himself as "Physicus et Astronomus."[87] Little is known about him except that he was born in the early sixteenth century, wrote at least two astrological works and was murdered in Venice sometime after 1573.[88] Near the end of Book I, there appears an interesting chapter with the title, "Of Various Opinions Concerning the Order of the Celestial Orbs."[89] It cites familiar classical authorities, such as Plato, Cicero, Pliny and Ptolemy, on the question of whether Venus and Mercury are above or below the sun.[90] Then follows a diagram (See Figure 9) captioned, "The System of the Principal Parts of the Universe according to the Opinion of Martianus Capella." Typical of such schematic representations of cosmological systems, designed as illustrations to aid the mind rather than as accurate descriptions, the ratios of the orbital radii are distorted. The superior planets lie equal distances from one another and the epicycle of Venus is constructed equal to the lunar orb. Yet the main idea is unmistakable: Mercury is located closer to the Sun and together the two inferior planets are borne around it whilst the Sun encircles the Earth. Furthermore, there can be no doubt that Naibod was working from Copernicus' text because he follows the Capellan schema with another to demonstrate the Copernican system

[87]Book I, Ch. 1, fol 1^r.

[88]Cf. *Allegemeine Deutsche Biographie,* pp. 242-243.

[89]Naibod, *op. cit.,* fols. 39^v - 42^r.

[90]*Ibid.,* 40^r.

PRIMVS. 41

Syſtema maximarum vniuerſitatis partium ex ſententia Martiani Capellæ.

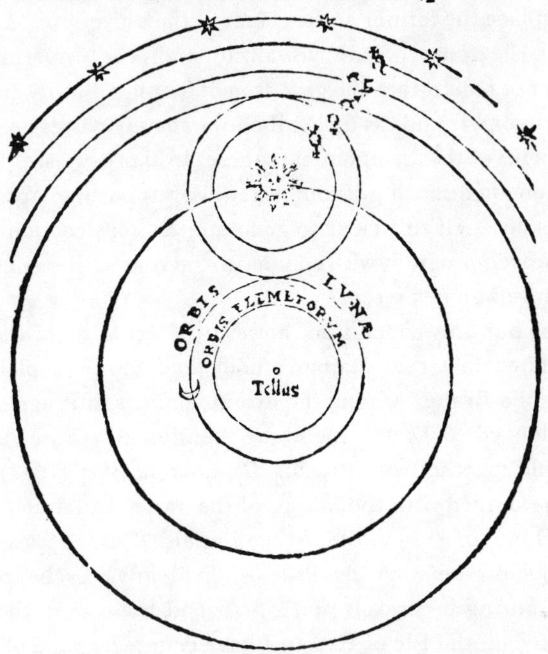

Hinc ſumpta occaſione ſummus uir Copernicus, Saturnum quoque & Iouem Martemque atque adeo Lunam unà cum incluſis ſuo circulo elementis, ad eundem Solem tanquàm uniuerſitatis centrum comparat, atque ſic extremis mundi mænijs, quibus ſtellæ ratæ inhærent, firmatis immotisq; , circum Solem ſimiliter immo-

F *tum*

Figure 9. The Capellan Universe. From Tycho Brahe's copy of Valentinus Naibodus' *Primarum de coelo et terra institutionum quotidianarumque Mundi revolutionum Libri Tres,* (Venice, 1573), fol. 41r. Courtesy of the Statní Knihovna Československé socialistické republiky—Universitní Knihovny, Prague.

(Figure 10). Here we notice quite plainly that Naibod has merely flipped over the lunar orb with the earth so that it now occupies the place formerly taken up by the Sun and the inferior planets. And the latter now replace the former at the center of the universe.

What was the result of this visualization of the Copernican text? History does not record that the grandson of Stephen Batorý entered the jousts of astronomers and, as for Naibod, he merely passes on to discuss further elements of the astronomical sphere. In short, we see that a hint or speculation mentioned without extensive supporting arguments is usually never by itself sufficient to generate the construction of a new scientific research program with competitive status. Other conditions of receptivity must also be present.

If this were our only conclusion, however, it would be of less interest than the related fact that Naibod's book had found a place in the library of Tycho Brahe. Among the extant volumes in Prague which at one time belonged to Tycho, the above mentioned treatise can be discovered bound together with Proclus' *De Sphaera liber* (1547) in a vellum binding stamped with the initials of the owner and date of acquisition: "T.B.O. 1576."[91] In addition, the initials "T.B." appear twice at the beginning of the work by Proclus. Evidently, Tycho purchased many books during his travels of 1575-76 and these were then bound when he settled on the Isle of Hveen. I have counted a total of 33 books which are bound and stamped with the same designation "T.B.O. 1576," which suggests that Tycho had a successful book-buying spree.[92]

Now it would be entirely too facile to conclude from this evidence that the Tychonic system owed its sole origins to a schematic diagram in an introductory astronomical text. Nonetheless, it would not be too far-fetched to suggest that the idea had been implanted in his mind at some level, as early as 1576; and that when he studied Copernicus' summary of Capella, Naibod's diagram would assist in heightening the impact of

---

[91]T.B.O. = Tycho Brahe Ottonidis (son of Otto). Cf. F.J. Studnička, *Prager Tychoniana* (Prague: Kön. Böhm. Gesellschaft der Wissenschaften, 1901), p. 29. The volume is No. 28 in Flora Kleinschnitzová's *Ex Bibliotheca Tychoniana Collegii Soc. Jesu Prague ad S. Clementem,* (Uppsala: Almquist and Wiksells Boktryckeri, 1933), p. 19.

[92]In Kleinschnitzová, these are Nos.: 15, 19, 21, 22, 26, 27, 29, 34, 35, 39, 40, 43. Probably many of the books were purchased at Frankfurt am Main (Cf. Dreyer, *Tycho Brahe, op. cit.*, p. 81).

L I B E R

*tum atque tanquàm in medio mundo conſiſten-
tem lampadem,præter Venerem & Mercurium,
etiam Saturnũ, Iouẽ & Martem, unà cum glo-
bo telluris & huic incumbentibus elementis
Lunaque conuertens,tam paruo numero orbium
omnes omnium ætatum cœli apparentias, maxi-
ma cum laude & admiratione doctornm ità ſal-
uat,ut ante ipſum nullus,quamuis multi ſphęra-
rum decadem,alij etiam hendecadẽ aſſumerent.*

Syſtema vniuerſitatis de ſententia ſum-
mi viri Nicolai Copernici Torinenſis.

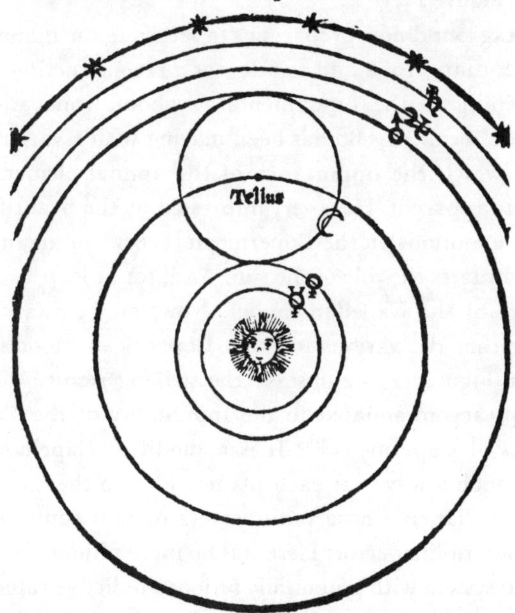

Figure 10. The Copernican System. From Tycho Brahe's copy of Valentinus Naibodus' *Primarum de coelo et terra institutionum quotidianarumque Mundi revolutionum Libri Tres,* (Venice, 1573), fol. 41^{V}. Notice that the diameter of the moon's orb is constructed by the artist equal to the orb of Venus in Figure 9.

the text. Confidence in our interpretation is increased by Tycho's glosses on the Capella passage. In DR Prague, he writes:

> If it be postulated that these two epicycles of the inferior planets have their center in the sun (as, in fact, appears agreeable to others), then, in other respects, the same things will follow [as on the Copernican theory]. For it will be necessary to grant that such a calculation corresponds to the observations and appearances and that the certitude of these hypotheses thereby are strengthened.[93]

And, in DR Vatican, he has inserted a small diagram in the left margin which shows the orbits of Venus and Mercury encircling the moving sun.[94] (See Figure 11.)

This background now prepares us to return to the manuscript notes of January-February, 1578 and to attempt a reconstruction of the circumstances leading to the fundamental Tychonic innovation. For three weeks, as we recall, Tycho has been playing with a variety of planetary models in which the importance of the annual element has become increasingly apparent. He is very impressed by the beautiful ordering of the periodic motions on the Copernican theory, an arrangement which again accentuates the role of the sun. And, for at least two years, he has been aware of the Capellan model. Now, on February 17, 1578, he brings together the various inverted Copernican models into a single scheme. In Figure 12, we observe the striking results: "The Sphere of Revolutions accommodated to the Immobility of the Earth from the Hypotheses of Copernicus."[95] It is a modified Capellan system, augmented in such a way that each planet rides on the annual epicycle of the superior planets whose radii vectors remain constantly parallel to the earth-sun radius vector. Here was no mere philosophical conjecture, but a world system with potentially serious predictive value.

Tycho Brahe, however, was a cautious man. Isolated in his observatory, with his dog, assistants and occasional visitors—although always in

[93]DR Prague, I, 10, fol 8^{V}.

[94]Above the diagram, Tycho writes: "Accordingly, Albumasar considered a heaven with Venus, Mercury and the Sun alone. And he tried to save the appearances of Venus and Mercury by the magnitude of their epicycles." (DR Vatican, I, 10, fol 8^{V}, lines 15-19).

[95]*Ibid.*, fol 210^{V}.

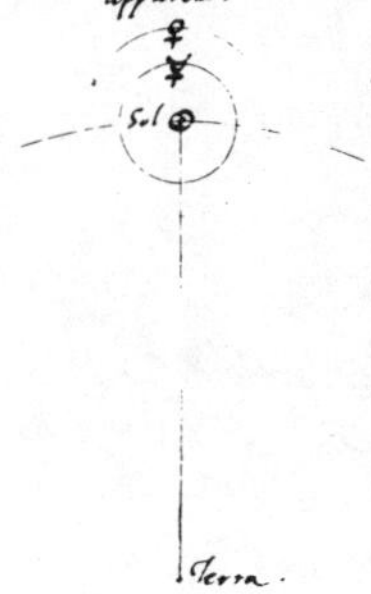

NICOLAI COPERNICI

ingens ille Veneris epicyclus occuparet, ſi circa terrā quietam uolueretur. Illa quoq; Ptolemæi argumentatio, quòd oportuerit medium ferri Solem, inter omnifariam digrediētes ab ipſo, & nō digredientes, quàm ſit imperſuaſibilis ex eo patet, quòd Luna omnifariam & ipſa digrediēs prodit eius falſitatem. Quā uero cauſam allegabunt ij, qui ſub Sole Venerem, deinde Mercurium ponunt, uel alio ordine ſeparant, quod non itidem ſeparatos faciunt circuitus, & à Sole diuerſos, ut cæteri errantium, ſi modo uelocitatis tarditatisq; ratio non fallit ordinem? Oportebit igitur, uel terram non eſſe centrum, ad quod ordo ſyderum orbiumq; referatur: aut certe rationem ordinis nō eſſe, nec apparere cur magis Saturno quàm Ioui ſeu alij cuiuis ſuperior debeatur locus. Quapropter minime contemnendum arbitror, quòd Martianus Capella, qui Encyclopædiam ſcripſit, & quidam alij Latinorum percalluerunt. Exiſtimāt enim, quòd Venus & Mercurius circumcurrāt Solem in medio exiſtentem, & eam ob cauſam ab illo non ulterius digredi putant, quàm ſuorum conuexitas orbium patiatur, quoniam terram nō ambiunt ut cæteri, ſed abſidas conuerſas habent. Quid ergo aliud uolunt ſignificare, quàm circa Solem eſſe centrum illorū orbiū? Ita profectò Mercurialis orbis intra Venereum, quem duplo & amplius maiorem eſſe conuenit, claudetur, obtinebitq; locum in ipſa amplitudine ſibi ſufficientem. Hinc ſumpta occaſione ſi quis Saturnum quoq;, Iouem & Martem ad illud ipſum centrū conferat, dummodo magnitudinem illorum orbium tantam intelligat, quæ cum illis etiam immanentem contineat, ambiatq; terram, non errabit, quod Canonica illorum motuum ratio declarat. Cōſtat enim propinquiores eſſe terræ ſemper circa ueſpertinum exortum, hoc eſt, quando Soli opponuntur, mediante inter illos & Solem terra: remotiſſimos autem à terra in occaſu ueſpertino, quando circa Solem occultantur, dum uidelicet inter eos atq; terram Solem habemus. Quæ ſatis indicant, centrum illorū ad Solem magis pertinere, & idē eſſe ad quod etiā Venus & Mercurius ſuas obuolutiones conferunt. At uero omnibus his uni medio innixis, neceſſe eſt id quod inter conuexum orbem Veneris & concauum Martis relinquitur ſpacium, orbem quoq; ſiue

Figure 11. Tycho Brahe's commentary and diagrammatic interpretation of Copernicus' summary of Martianus Capella's arrangement of the orbs of Venus and Mercury with respect to the Sun. *De revolutionibus,* fol. 8[v]. Courtesy of the Biblioteca Apostolica Vaticana.

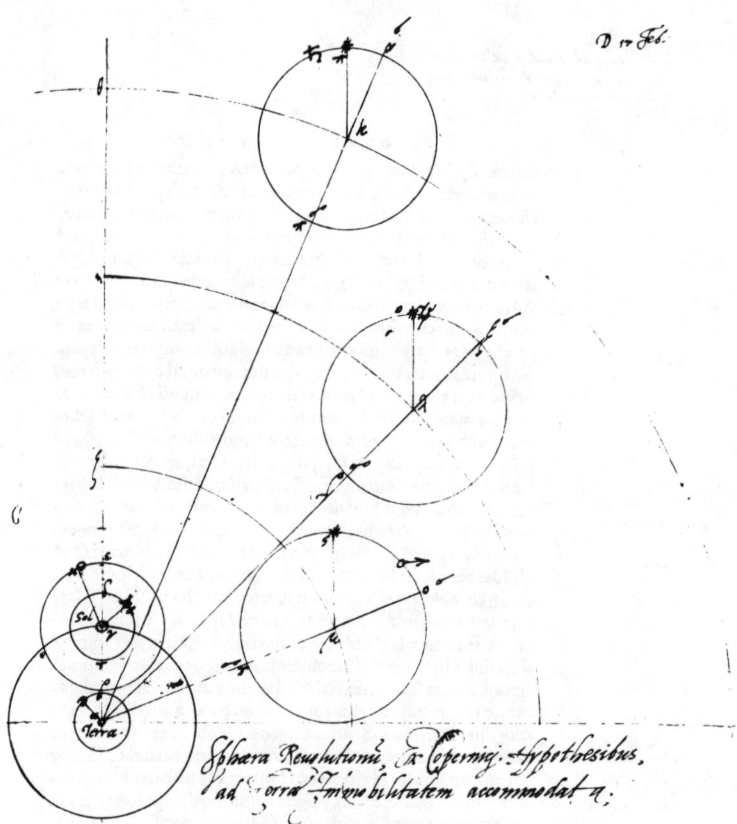

Figure 12. Tycho Brahe's modified Capellan System, February 17, 1578. While Venus and Mercury revolve about the Sun as center, the three superior planets, Mars, Jupiter and Saturn, have the Earth as the center of the revolutions of their deferents. The radii of the prominent epicycles of the three outer planets are all constructed equal to the radius of the Sun's orb. Notice that the orb of Mars does not intersect the Sun's orb. In his final version (ca. 1583), Tycho will center the revolutions of all the planets around the Sun with Mars' orbit intersecting the Sun's. This was a step that Praetorius could not take (see Fig. 4). *De revolutionibus,* working notes, fol. 210^{V}. Courtesy of the Biblioteca Apostolica Vaticana.

correspondence with the European astronomical community—Tycho had time to be sceptical about whether his theory actually corresponded to the observations.

> I was still steeped in the opinion, approved and long-accepted by almost all, that the heavens were composed of certain solid orbs which carried round the planets, and . . . I could not bring myself to allow this ridiculous penetration of the orbs; thus it happened that for some time this, my own discovery, was suspect to me.[96]

This "ridiculous penetration of the orbs" had been brought about by Tycho's evaluation of a large Martian parallax which, in turn, had caused him to make the radius of Mars' orbit slightly smaller than the diameter of the sun's orbit in order to account for the smaller distances of Mars at opposition.[97] From 1578 onward, however, his work with comets had slowly convinced him that the crystalline spheres could not tolerate such bodies passing through them and this recognition had led to his famous rejection of the spheres themselves. The system which he finally published reflects this subsequent period of criticism and re-examination. It places the earth at the exact center of the celestial sphere, the sun's orbit concentric to the earth (with Mars' orbit intersecting the sun's) and the planetary orbits concentric to the sun.[98] The caption over this final version stresses not only its superiority over the Ptolemaic and Copernican theories but also that ". . . all [these hypotheses] correspond most fittingly to the celestial appearances."[99] Tycho's modern, critical attitude toward his own innovation eventually permitted it to compete seriously with the theory out of which it had arisen.

7. *Michael Maestlin: A "Conservative" Copernican*

Of the three figures who are discussed in this paper, Maestlin's

---

[96]Tycho Brahe to Caspar Peucer, 13 Sept. 1588, TBOO, VII, p. 130. Quoted and translated in Jones, *op. cit.*, p. 58.

[97]Cf. Dreyer, History, *op. cit.*, p. 363.

[98]*TBOO*, IV, p. 158.

[99]*Ibid.*

personal ties with Wittenberg were the least direct. Born in 1550, in Göppingen (Swabia), he matriculated at the University of Tübingen in 1568.[100] In the same year, Phillip Apianus (1531-1589), son of the famous Peter Apianus (1495-1552), was dismissed on grounds of heresy from his position at the Catholic University of Ingolstadt and, in 1569, took up a teaching post at Tübingen. Nothing is known of Phillip's attitude toward the Copernican theory but it seems clear that his orientation toward astronomy, like his father's, was fairly "technological"—the construction of celestial globes, maps, sundials and other instruments. The books which Apianus used in his lectures—Proclus' *Sphaera,* Peurbach's *Theoricae Novae Planetarum,* Sebastian Theodoric of Winsheim's *Sphaera* and Peter Apianus' *Cosmographia*—were hardly different than the texts which Maestlin would have read at Wittenberg.[101]

But Maestlin was no ordinary student. On 5 September 1571, he completed for publication an edition of Reinhold's *Prutenic Tables* almost exactly one month after he had received his M.A.[102] One year earlier, on 6 July 1570, while still a student, he had aquired a copy of the 1543 edition of *De revolutionibus,* as he wrote, "from the widow of D.D. Victorinus Strigelius, sister of D. Theodore Snepfius."[103] At the age of 22, he wrote a brief treatise on the supernova which appeared in 1572 and showed that already he had studied the cosmological portions of *De revolutionibus* quite carefully.[104] By the following year, he had assumed

[100]The best recent biography of Maestlin, on which some of the following details are based, is in Richard A. Jarrell, "The Life and Scientific Work of the Tübingen Astronomer, Michael Maestlin, 1550-1631," Unpublished Doctoral Dissertation: University of Toronto, 1971, pp. 10-44.

[101]See Westman, "Wittenberg Interpretation," *op. cit.*

[102]Maestlin's own published copy is located today in the Schweinfurt Stadtarchiv with a number of personal notes written by both him and Praetorius.

[103]Schaffhausen Stadtbibliothek, last folio. Although Strigelius had authored a brief astronomical treatise (*Epitome doctrinae de primo motu aliquot demonstrationibus illustrata*: Leipzig: Vogelin, 1564) both he and Snepfius were theologians at Tübingen. Maestlin's note reflects the close ties of intermarriage among the Tübingen faculty.

[104]Cf. Westman, "Comet and Cosmos," *op. cit.,* p. 9. In a forthcoming article, I show that Maestlin had been persuaded to accept the main tenets of the Copernican theory as early as 1570-72 on the basis of the new argument from planetary order (Cf. "Michael

the position of mathematical tutor at the Stift in Tübingen. Over the next few years, he remained in Tübingen and its environs,[105] but in 1576 became deacon of a church in Backnang, a post which he held until 1580, when he became Ordinary Professor of Mathematics at the University of Heidelberg. He did not return to the University of Tübingen until 1584. In short, Maestlin's formative intellectual period and initial exposure to Copernican ideas occurred approximately during the same time span as Tycho's, viz. 1570-1580. And Maestlin's extensive commentaries on *De revolutionibus* will also provide us, as Tycho's, with valuable insights into his response to the Copernican theory.

Perhaps the most striking general feature of the notes is their consistently approving tone. Unlike Tycho's annotations, where there hardly ever occurs a direct comment on the sentiments of the text, Maestlin never seems to have had any doubts. Both Brahe and Praetorius, for example, had noted the identity of Osiander, but Maestlin possessed strong suspicions even before he learned in 1589 of the true authorship of *Ad Lectorem.*[106]

> This preface was fastened on by someone, whoever its author may be, (for indeed, the words and thinness of style reveal them not to be those of Copernicus)...and no one will approve of these [hypotheses] at first glance from concern for novelty, in spite of his opinion, but rather one first reads, rereads and then one judges. Hereafter, indeed, [if one accepts this letter] it will be the case that a proof cannot be contradicted. For the author of this letter, whoever he may be, while he wishes to entice the reader, neither boldly casts aside these hypotheses nor approves them but rather he imprudently

Mästlin's Adoption of the Copernican Theory," *Studia Copernicana, Colloquia Copernicana* IV).

[105]Jarrell has successfully exploded the oft-repeated opinion that Maestlin travelled to Italy between 1571 and 1576 and there taught Galileo the Copernican theory (*op. cit.*, pp. 20-24).

[106]On fol ij^r, Maestlin wrote that he had learned of Osiander's identity from a letter in a book which he had purchased from the widow of Phillip Apianus (d. 1589). In 1613, Maestlin wrote to Kepler that Apianus' widow "as a result of certain occult insinuations (undeserved and against me) sold every copy which had survived, and allowed me to inspect whatever books which remained." (G.W., XVII, p. 58).

> squanders away something which might better have been kept silent. For the disciplines [of astronomy] are not made stronger by shattering their foundations. Wherefore, unless this [hypothesis of Copernicus] is defended better than by this person, he labors in vain, because there is much weakness in his meaning and reasoning. And therefore, I cannot approve of all the simple and confused parts of this letter, much less defend it.[107]

In this passage, Maestlin accepts, by implication, a realist interpretation of Copernicus' views and criticizes the author of the *praefatiuncula* for trying to defend *De revolutionibus* by "shattering the foundations" of astronomy. This is a promising start and raises at once our expectations that Maestlin will provide new physical arguments in place of Osiander's "weak" and "confused" defense. Such anticipations, however are not fulfilled.

Maestlin was a slavish admirer of the physical arguments set forth in Book I. Ardent admirers, however, do not always make the best advocates for a new scientific hypothesis. Two examples will suffice. In Chapter 8, Copernicus uses a famous line from Virgil's *Aeneid* to argue that inhabitants of the earth, like passengers on a ship moving on a very tranquil sea, will see the motions of the stars and planets as "an image of their own."[108] And furthermore, he claims, the earth will carry around with it all objects floating in the air, ". . . whether because the neighboring air, which is mixed with earthly and watery matter, obeys the same nature as the Earth or because the movement of the air is an acquired one. . . ."[109] On these passages, Maestlin comments: "He [Copernicus] resolves the objections which Ptolemy raises in the Almagest, Book I, Chapter 7." In fact, Copernicus has merely articulated arguments which Ptolemy rejects and tries to establish them, as we have seen earlier, in a theory of gravity not uncomonly held during the Renaissance. Here again, Maestlin shows himself to be a faithful disciple.[110] In his important preface to Rheticus' *Narratio Prima*, pub-

---

[107]DR Schaffhausen, fol 1^v.

[108]fol 6^r; *GBWW*, p. 519.

[109]*Ibid.*

[110]For further examples, cf. DR Schaffhausen, I, 5, fols 3^{r-v}.

lished together with Kepler's *Mysterium Cosmographicum* in 1596, he clearly paraphrases Copernicus' dynamical arguments:

> What is the Earth, what is the ambient Air with respect to the entire vastness of the world? They are a point, or little points, and if anyone should say that they are smaller, he would be correct. This being the case, do you not reckon that the Philosopher will say that the argument from a small particle or this little point to the entire World is feebly constructed? We cannot be certain, therefore, that this is the center of the most Spacious World simply from the evidence of these things which approach this little point or flee from it. Indeed, these heavy and light objects approach their proper place assigned by Nature according to what Copernicus eruditely calls an "affinity," in Book I, chapter 9. It is also quite credible that it [this affinity] resides in the Sun, moon and other shining bodies as well because its efficacy lies in its roundness.[111]

Maestlin's position is perhaps best summarized by a parenthetical aside which he makes in the course of a very lengthy annotation in the Preface: ". . . as for Astronomy, Copernicus in fact wrote this entire book not as a natural philosopher [physicus] but as an astronomer [astronomus]."[112] Now we know that he cannot have meant this statement in the sense of Osiander's preface. What he intends to say here is that Copernicus' system describes reality, but it is a reality which will only be grasped by the "Mathematici" and "Astronomi." The truth of the new system springs from its mathematical simplicity and the excellent correspondence of its hypotheses with observations. These criteria, as we shall see, were sufficient to cause Maestlin to adopt the Copernican theory and to defend it on this basis alone. Referring to the argument from harmony in one of his annotations, Maestlin writes:

> Certainly this is the great argument, *viz.* that all the phenomena as well as the order and magnitudes of the orbs are bound together [*conspirare*] in the motion of the earth.[113]

[111]G.W. I, p. 83.

[112]DR Schaffhausen, iiij^{r}.

[113]*Ibid.* Contrasting the necessary ordering of the planets on the Copernican hypothesis with their arbitrary ordering in the "usual hypotheses," Maestlin writes in his introduction

And further amplifying:

> This book concerns the true order of the world and because the annual sidereal motions and the phenomena are saved most exactly it thus proves that this hypothesis is more correct than the other.[114]

This leads Maestlin to his decision between the two major world systems:

> . . . moved by this argument, I approve of the opinion and hypotheses of Copernicus, which same thing some would do, as I see it, if they did not fear the displeasure of other persons who know the hypothesis of the earth's immobility to have been confirmed for a long time, as can be seen in the Commentary which Oswaldus Schreckenfuchsius wrote on the *Sphere.*[115]

Here we encounter a remarkably interesting episode in the choice between two competing theories. The same feature of the Copernican system which had led Praetorius to juggle the size of Mars' sphere in order to defend the old geostatic order, and which had spurred Tycho to find a compromise between the two world theories, now proves adequate to induce a transformation of scientific loyalty in Maestlin. Is this an example of the dramatic "gestalt switch" of which Kuhn writes in

---

to the *Narratio Prima:* "Etenim Copernici hypotheses omnium Orbium et Sphaerarum ordinem magnitudinem sic numerant, disponunt, connectunt et metiuntur, ut nihil quicquam in eis mutari aut transponi, sine totius Universi confusione, possit; quin etiam omnis dubitatio de situ et serie procul exclusa manet. Écontra in hypothesibus usitatis, numerus sphaerarum incertus est. Alij enim novem, alij 10, alij 11 sphaeras numerant, nec adhuc conuenit numerus. Ordo ibidem est dubius: definita distantia, praeter ☉ et ☽, nulla dari, nedum demonstrari potest: De Venere, Mercurio et Sole lis nondum composita est, nec componetur unquam." (G.W. I, p. 83).

[114] *Ibid.*

[115] *Ibid.* In his *Commentaria in Sphaeram Ioannis de Sacrobusto* (Basel, 1569), p. 36, Schreckenfuchs had written: "All kinds of debate can be stirred up on the subject of the earth's motion. We shall find such discussions in the book of Nicholas Copernicus, a man of incomparable genius. I would have every right to call him the world's miracle were I not fearful of thereby offending certain men who, however correctly, hold excessively to judgments handed down by the ancient philosophers." Quoted and translated in Pierre Duhem, *To Save the Phenomena, op. cit.,* p. 90. There is absolutely no evidence here or in Maestlin's writings of this period that there was any serious opposition to his views.

The Structure of Scientific Revolutions?[116] Is Maestlin living in a different scientific world or does his research respond, in Kuhn's words, "as though that were the case"?[117]

Ironically, Maestlin's "research program" more closely resembles Praetorius' orientation than it does the work of his own pupil, Kepler. This is not to deny the crucial influence exerted by Maestlin on his famous student nor to deny his important work in Copernican mathematical astronomy,[118] but merely to characterize the differences in what each considered the proper areas of scientific inquiry—differences which are obscured by labelling both as "Copernican." For Maestlin, *in spite of his clear support of Copernican dynamical arguments and his apparent belief in the existence of solid spheres continues to treat astronomical problems in his own work as though they were distinct from the new physical issues raised by adherence to the new theory of the universe.* Much as a typical Ptolemaic astronomer of this period, who first recites the catechisms of Aristotelian physics and then proceeds to the "real business" of saving the phenomena, Maestlin showed considerable competence and innovative talents as a mathematical astronomer but little happiness for the strange "physical" investigations of his student. Thus, in 1597, Maestlin writes to Kepler concerning a diagram in the *Mysterium Cosmographicum* where Kepler further develops his theory of a solar force radiating out to the planets, but where Maestlin catches an error in labeling the equant point. After rectifying and explaining the difficulty, Maestlin adds a note of kindly disapproval.

> I do not reject this speculation about the motive spirit and virtue. Yet I fear lest it be too obscure if extended further. . . . I am really afraid that it may bring the loss and certainly the ruin of all Astronomy along with it. I consider that such specu-

[116]Kuhn, *op. cit.*, p. 111 ff.

[117]*Ibid.*, p. 117.

[118]Cf. Westman, "Comet and Cosmos," *op. cit.* and Maestlin's *De Dimensionibus orbium et Sphaerarum Coelestium, Iuxta Tabulas Prutenicas, ex sententia Nicolai Copernici* (1596). The latter represents the type of consolidation of Copernicus' models which we have seen already in Tycho's MSS. notes in DR Vatican. For a recent, excellent analysis of Maestlin's understanding of Copernican astronomy see Anthony Grafton, "Michael Maestlin's Account of Copernican Planetary Theory," *Proceedings of the American Philosophical Society*, 1973, *117*: 534-550.

> lation ought to be carried on sparingly and with total moderation. To say, truly, what I feel: I do not reject this idea, but really my assent is reluctant for, as far as I can see, many contrary things oppose it.[119]

Maestlin does not object to Kepler's new and astounding "speculations" but they simply do not feel right to him. It is sufficient if the astronomer can discover hypotheses from which no false conclusions might be deduced.

> The truth is consonant with the truth and from the truth there follows nothing but the truth. And, if in the process, something false and impossible follows from the received opinion or hypotheses, then the defect must be concealed in the hypothesis. If therefore, the hypothesis of the earth's immobility were true, then the true would also follow from it. But in the [customary] Astronomy, there follow many discrepancies and absurd things—as much in the arrangement of the orbs as in the motions of the planetary orbs. Here there exists a defect in this hypothesis.[120]

Employing the same form of argument, Maestlin defended as true his interpretation of the Comet of 1577, on the grounds that there were no contradictory observations and that the geocentric theory could not save the phenomena without great observational discrepancies.[121] Furthermore, and in the forward to his edition of Rheticus' *Narratio Prima* (1596), published jointly and appropriately with that great pivotal work in the history of astronomy, Johannes Kepler's *Mysterium Cosmographicum,* Maestlin publicly announced his adherence to the Copernican theory in the form of a defense of Kepler's polyhedral hypothesis. The orbital interconnexity of the Copernican system which had privately and so powerfully impressed Maestlin over two decades earlier is now justified by attributing the order found in the world to the intentions of a geometrizing, architectonic Renaissance God.

---

[119]Maestlin to Kepler, 9 March 1597, G.W. XIII, p. 111.

[120]DR Schaffhausen, iij^{V}.

[121]Cf. Westman, "Comet and Cosmos," *op. cit.,* 23 f. It is significant that Maestlin never attempted to translate his Copernican solution into a geostatic reference frame.

> ...our Kepler furnishes us with a most ingenious discovery from Geometry whereby he establishes the certain, determinate number and order of the celestial orbs and spheres and—what is most important—the determinate proportionality of the magnitudes and motions [of the spheres] to one another; and...he shows that, in the creation of the World, the Most Perfect Creator God constructed, extended, disposed, adorned and ordered the moving celestial spheres according to the proportions of the five, regular, geometrical solids—known after another fashion by all the most famous geometers. And he confirms his opinion neither by subtle logical disputations nor by trifling and dubious old wives' tales, much less by strange conjectures violently forced for his purpose, but rather by genuine and most proper reasons derived as much from the Nature of things [*ex rerum Natura*] as from Geometry—and which cannot be contradicted. Of these [reasons] the best is that most eloquent and sweetest harmony and the very consonant accord of astronomical calculation derived from observations with the distances of the five regular solids.[122]

As Maestlin himself recognized, Kepler's attempt to justify the theory of Copernicus was entirely within the tradition of Rheticus. "What would Rheticus have done," Maestlin comments in the margin of his published edition of the *Narratio Prima,* "if he had noticed that divine geometry of the five regular solids about which Master Kepler discourses"?[123] We can imagine that Rheticus would have been quite pleased—as pleased as was Maestlin of his student's discovery. Yet, in a sense, Maestlin never moved much beyond Rheticus in comprehending the need for a new physics. If Kepler owed his excellent training in mathematical astronomy to Maestlin, his efforts to establish the science of astronomy on different methodological foundations were little appreciated by his Tübingen professor.

[122]Johannes Kepler, *Gesammelte Werke,* I (Munich: C.H. Beck, 1938), p. 82.

[123]G.W. I, p. 116.

8. *Summary and Conclusions*

In a famous passage from his *Scientific Autobiography,* Max Planck wrote:

> A new scientific truth does not triumph by convincing its opponents and making them see the light, but rather because its opponents eventually die, and a new generation grows up that is familiar with it.[124]

Written at the very end of his long life, Planck spoke not as an historian generalizing about the reception of past scientific theories, but as a great scientist who had experienced in his youth the real dismay of non-recognition for his own theories.

> It is one of the most painful experiences of my entire scientific life that I have but seldom—in fact, I might say, never—succeeded in gaining universal recognition for a new result, the truth of which I could demonstrate by a conclusive albeit only theoretical proof. . . . It was simply impossible to be heard against the authority of Ostwald, Helm and Mach.[125]

Our study suggests a more moderate interpretation of the role of generations in the reception of the Copernican theory. The generation of astronomers contemporary with the publication of the new theory did not reject everything outright but rather focused upon certain features which fitted their own needs: especially, the articulation of equant-less models and the borrowing of new parameters. Rheticus, who played no small part in publicizing these same theoretical and calculational advantages was overwhelmed by another property of the theory—its demonstration of a unique, systematic ordering of the planets—and here, his very enthusiasm no doubt hindered further immediate recognition of the theory's cosmological core. In short, as a publicist for Copernicus' *cosmology,* Rheticus was initially an utter failure.[126]

By the time that Rheticus and those who ignored his views had died, there had grown up a new generation which now inherited Reinhold's

[124] tr. Frank Gaynor, (New York: Philosophical Library, 1949), pp. 33-34.

[125] *Ibid.*, p. 30.

[126] This interpretation is set forth in my "The Melanchthon Circle, Rheticus and the Wittenberg Interpretation of the Copernican Theory," *op. cit.*, pp. 189-190.

consolidation and recalculation of Copernican data as well as the partially-fulfilled Wittenberg program to convert the Copernican models into earth-centered devices. Less hindered by past preconceptions and curious to examine what had been previously omitted, this second generation—most of whom had been born in the decade of the 1540's—looked at the cosmological arguments of *De revolutionibus* with a fresh gaze. This new inspection, however, did not produce Rheticus-like conversions nor did it lead to the "triumph" of the new theory.

By facing issues which had not been countenanced by their predecessors, the "Generation of the 1570's" began to re-focus the attention of the astronomical community onto new problems. When seriously considered, the Copernican contentions for a re-ordering of the planets and the great role of the sun in the planetary models could not easily be dismissed. Out of these considerations came a new system—the geo-heliocentric. While this system offered sturdy opposition to both the Ptolemaic and Copernican theories, nonetheless, it incorporated the Copernican ordering of the inferior planets. And, with Tycho's rejection of the solid spheres, an important step was taken toward the unification of astronomy and physics on a new basis. By the late 1580's, we can begin to speak with some *historical* accuracy of choices between different research programs.

Until Kepler, there were no strong defenses of the Copernican theory on dynamical grounds. Maestlin, for all his outstanding abilities as a mathematical astronomer and supporter of the theory of Copernicus, failed to understand or was disinclined to develop further the physical implications of the hypothesis which he espoused. In a sense, he still reflected the split between cosmology and astronomy which had been the hallmark of the Wittenberg Interpretation. If the conditions were "ripe" for a Kepler, it still required the genius of Kepler himself to recognize and to respond imaginatively to the logical insufficiency of Copernicus' arguments. For Kepler, there was a "realist dilemma"; for Maestlin, there was none. Yet, if Newton had stood on the shoulders of giants, Kepler stood on the shoulders of a generation of very capable mortals.

Appendix

A Note on Sources: The Prague and Vatican Copies of De revolutionibus

Until very recently, little has been known of Tycho's early attitude toward the Copernican theory in the 1570's save for his work on the New Star, the extant lecture at the University of Copenhagen in 1574-75, a few scattered letters and his treatise on the Comet of 1577.[1] But in the last two years, some very fundamental material has come to light which now allows us to sharpen the focus of our historical lenses on the early period of Tycho's career and to illuminate areas that were previously shrouded in darkness. In brief, two copies of *De revolutionibus* have been identified—one in the Jesuit College Library in Prague's Clementinum (today part of the Czechoslovak State Library) by Dr. Zdenek Horský; the other in the Manuscript collection of the Biblioteca Apostolica Vaticana by Professor Owen Gingerich—both of which bear extensive marginal notes in the same hand, although lacking the owner's autograph.[2] And since the presentation of this paper, I have discovered still another member of this family of interesting copies, located at the library of the Université de Liège in Belgium.[3]

[1]These works have been analyzed most recently by Christine Jones Schofield, *op. cit.*, pp. 34-94 and Kristian P. Moesgaard, *op. cit.*, pp. 31ff.

[2]The Prague copy with Tycho's notes is one of three examples of the 1566 edition in the Clementinum; there is also one copy of the 1543 edition. Unless otherwise specified, *DR Prague* will refer to the copy with Tycho's notes (Shelf No. 14B 16 - Tres M 11). In addition to the copy analyzed in this paper, the Vatican lists two copies each of the 1543 and 1566 editions in its catalogue of printed books. It had been known for some time by Dr. H.M. Nobis (Deutsches Museum, Munich) that there existed a copy of *De revolutionibus* in the Vatican MSS. collection (it is also listed by P.O. Kristeller, *Iter Italicum*, II [London: Warburg, 1967], p. 419) but it was Dr. J. Dobrzycki (Warsaw) who first alerted Professor Gingerich to its existence. The identification of this volume was described by the latter in a paper entitled, "The Astronomy and Cosmology of Copernicus," read before a session of the Extraordinary General Assembly, International Astronomical Union in Warsaw on 4 September, 1973 (published in George Contopoulous [ed.], *Highlights of Astronomy*, III [Dordrecht: Reidel, 1974], pp. 67-85). Professor Gingerich has also described his findings in "Copernicus and Tycho," *Scientific American*, 1973, *229*: 86-101.

[3]I first announced this discovery publicly in a commentary on a paper read by Owen Gingerich ("New Light on the Reception of the Copernican Doctrine") at the History of

Horský, who published a valuable facsimile of the Prague copy in 1971 together with a short commentary, argues convincingly that the book belonged at one time to Tycho Brahe. And Gingerich, on the basis of a comparison of the handwriting and annotations between the Prague facsimile and the Vatican original, has conjectured that the former was probably a derivative from the latter.[4] Beyond the question of the identity of the author of the annotations and the relationship between the two copies, however, there remains a difficulty endemic to the use of marginal notations in historical research, *viz.* the question of dating. Because of the importance of this new evidence to our analysis, a separate discussion is now merited.

Let us first consider Horský's arguments in support of the identification of Tycho as the source of the annotations in DR Prague.[5] This proof rests upon several foundations. First, there is the entry of the Prague Jesuit College librarian who, in 1642, wrote on the frontispiece, "Ex Biblioteca et Recognitione Tichoniana" which provides a reliable authority both for the original ownership of the book as well as for the authorship of the handwritten commentaries. Secondly, a comparison of the commentator's hand with letters and manuscripts known to have been authored by Tycho shows a strong resemblance. Thirdly, from other writings of Tycho, it is certain that he had access to the second edition of *De revolutionibus* (1566) since he refers in one of his works to the edition of Rheticus' *Narratio Prima* that was jointly published with the Basel impression.[6] Fourthly, his references to astronomical writings of the 1560's and 1570's, such as those by Petrus Nonius, Hieronymus Cardanus, Peter Ramus, Johannes Stadius and John Dee, are men who appear frequently in his correspondence and writings. And finally, the only corrections made by the annotator in Copernicus' star catalogues

Science Meeting in Norwalk, Connecticut on October 27, 1974. Professor Gingerich and I intend to publish a joint study comparing the three Tycho Brahe copies as well as demonstrating their connections with several other annotated copies of *De rev.* (forthcoming in *Centaurus*).

[4] *Ibid.,* Gingerich.

[5] Zdenek Horský, "Copernicus' Writings on the Revolutions of the Celestial Spheres with Marginal Notes By Tycho Brahe," (Prague: Pragopress, 1971), pp. 12-13. This article is also translated into Czech, German, French and Russian and accompanies the facsimile edition.

[6] *TBOO,* II, pp. 31, 443.

are for two stars in Cassiopeia, the very constellation in which Tycho had observed the supernova of 1572.

Beyond Horský's claims, we can argue as follows: In the first place, there were few astronomers in this period who could have written notes of such high quality. Numerous diagrams and arguments, for example, go beyond a mere paraphrasing of the text. Indeed, of the approximately 300 examples of the first and second editions of *De revolutionibus* about which I have information, it can be inferred safely that the scientific productivity of an astronomer is usually proportionate to the quantity and quality of the annotations which he leaves on his copy. The outstanding exception is Kepler. His copy, now located in Leipzig, is largely barren of annotations. Nevertheless, his public support of Copernican ideas far outshone the efforts of any of the heavy sixteenth-century annotators. The writers mentioned above may be eliminated, of course, because they are already referred to by the annotator in the third person. With these removed from consideration, the already-small population of possible candidates becomes even smaller. Only four remaining figures would seem to qualify: Michael Maestlin, Christopher Clavius, Nicolai Reymers Ursus and Christopher Rothmann. On the basis of a handwriting comparison, which I have undertaken, between works and letters of the above mentioned writers and DR Prague,[7] there can be no doubt that Horský's identification is correct. And, furthermore, since I can confirm from personal observation that the handwriting on DR Prague and DR Vatican is identical, Gingerich's claim can also be substantiated with confidence.

---

[7]The works compared include: Maestlin's notes on his copy of *De revolutionibus* (See Footnote No. 103); Christopher Rothmann's *Astronomia: in qua hypotheses Ptolemaicae ex hypothesibus Copernici corriguntur et supplentur*. . . (ca. 1586) Murhardsche Bibliothek der Stadt Kassel und Landesbibliothek, Astronom MS. 4°11; various letters of Christopher Clavius in the Archives of the Pontifical Gregorian University (e.g. Clavius to Possevinus, 10 January 1585 (MS. No. I. 257. Lat. 3) and Clavius to Gabriel Serranus, 21 July 1598 (MS. No. I. 140. Lat. 3) which, through the great courtesy and kindness of Father E. Lamalle, I was able to examine on microfilm at the Jesuit Archives; and finally, a letter from Nicholas Reymers Ursus to Clavius from Prague, 25 March 1594 (MS. No. II. 227. Lat. 6) of which Father Lamalle has kindly provided me with a copy. In addition to the Clavius correspondence, I should also mention DR Bib. Naz. di Roma 1543 (No. 201. 39. I. 26) which was clearly owned by Clavius, as I shall show elsewhere.

This brings us to the second question: Why did Tycho own at least two volumes of the same work? There are, of course, many possible reasons. For one thing, he would have been curious, naturally, to know of any emendations or changes from one edition to another. Furthermore, it is possible, as Horský hypothesizes, that Brahe's commentary in DR Prague was designed for later publication with a future edition of Copernicus' work—although there appears to be no explicit evidence on this point. A third conceivable motive to which I should like to call attention, highlights new conditions which facilitated the exchange of information and thereby encouraged the sharing of interpretations in the scientific community.

The fifteenth-century revolution in printing from a "scribal culture" to a "typographical culture"[8] had affected not only the quantity, fixity and standardization of data,[9] but also facilitated elucidation of the text. The existence of a common text in the hands of widely-dispersed astronomers now increased the probability that the same passage could be discussed more easily without fear of aberrations introduced by a sleepy copyist. One interesting indication of the private communication of textual commentaries is provided by two copies of *De revolutionibus* owned by the Scottish astronomer and friend of Tycho Brahe, Duncan Liddel (1561-1617).[10] The first edition is virtually devoid of textual annotations but it contains several additional folios where Liddel has constructed several working paper models of various planets. By comparison, the 1566 edition has additional interfoliated leaves on which Liddel has copied the manuscript of Copernicus' *Commentariolus,* a work which was not published until the nineteenth century. The text itself is heavily annotated and there is at least one folio on which he has copied down information found in the identical location in Tycho's DR Vatican. One wonders, indeed, whether Liddel had access to Reinhold's manuscripts

[8]Cf. Elizabeth Eisenstein, "The Advent of Printing and the Problem of the Renaissance," *Past and Present,* 1969, *45*: 19-89; "Some Conjectures about the Impact of Printing on Western Society and Thought: A Preliminary Report," *The Journal of Modern History,* 1968, *40*: 1-56.

[9]*Ibid.*

[10]See Figure 8. These copies are today located in the Aberdeen University Library. I intend to discuss them more fully in another place.

as well. This might possibly explain how it was that Reinhold's copy of *De revolutionibus* arrived in Scotland.[11]

Tycho, like his friend Liddel, also adds further working space to both of his copies. DR Prague contains twenty-one blank folios in front of the text and seventy-one folios after it.[12] By comparison, DR Vatican has only thirty additional folios at the back, but these are filled with important diagrams which, among other things, show Tycho systematically working through the Copernican models based upon the text. With the exception of Chapter 10, Book I, DR Prague receives more frequent annotations than its companion copy, but both give considerable attention to the remaining Books. Several folios in the two copies bear in common the same glosses (e.g. iij^r, iiij^r, 67^r, 91^v). For example, fol. iij^r has the common gloss, "De his vide Plura in Narratione Rhetici folio 202," which could only refer to the *Narratio Prima* edition of 1566.

Another interesting set of common glosses links the two Brahe copies with the *De revolutionibus* owned by Erasmus Reinhold. On his trip to Saalfeld in 1575, Tycho evidently took the opportunity to copy out or paraphrase a considerable quantity of glosses from Reinhold's copy into one or both of his own volumes.[13] Particularly striking is the motto on the title page of DR Crawford—"Axioma Astronomicum..."—which is also inscribed on the frontispiece of DR Vatican by the unknown sixteenth-century annotator of DR Wolfenbüttel (1543),[14] as well as DR Liège. This transmission of glosses suggests that lines of communication among sixteenth-century astronomers, often carried on by correspondence, could be and were supplemented by personal visits and that the accumulation and integration of new information was facilitated by the printing of *De revolutionibus* as a common text.

Our final problem is the dating of the notes. It is an issue, however, which cannot be resolved fully apart from an analysis of the internal evidence itself. My contention is that the great bulk of the notes in *both*

[11]Reinhold's copy is today in the Crawford Library of the Royal Astronomical Observatory, Edinburgh—another discovery of Professor Gingerich, who describes it in "Erasmus Reinhold and the Prutenic Tables," *op. cit.*, pp. 56-58.

[12]Horský, *op. cit.*, p. 15.

[13]These concern figures for the obliquity of the ecliptic, equant-less models and the mechanisms of libration among other topics.

[14]Also fols 142^{r-v} and 143^r.

copies fall into the decades of the 1570's. Horský suggests 1573 as the earliest date for the beginning of the commentary in DR Prague because the latest book referred to is John Dee's *Parallacticae Commentationis Praxeos Nucleus quidam.* But this fact alone is too weak to bear the weight of his conclusion: "This means that the commentary was written after 1573 or at the earliest in that year."[15] The reference to Dee merely shows that that particular note, and all those derivative from it, could not have been written before the date in question. We must allow, then, that other notes could have been written at earlier times. The probability of Horský's dating is improved somewhat, however, by the fact that Tycho does not appear to use *De revolutionibus* in his first published work, *De Stella Nova* (1573). The earliest clear indication does not emerge until his unpublished Copenhagen lectures of 1574-75.[16] DR Vatican possesses fewer references to modern works and these few are common glosses with DR Prague (e.g. the reference to Petrus Nonius on fol. 45^V). The following note, however, indicates that the volume was owned at least as early as 1572: A° 1572 D. 8 Novemb: Eclipsabat ♃ 21 Stella ♊ Longi: ♃ fuit secundum supp: 21° 7′ ♈.[17] The young Tycho may well have been studying *De revolutionibus* before 1572, but if he was, he did not begin to publicize his efforts, so far as we know, until 1574.

For our purposes, however, a terminal date is more important since we want to know how, why and when his thinking changed. Tycho himself claimed that he had discovered his new theory in 1583.[18] There are no indications that I have found in DR Prague of the genesis of the geoheliocentric system. We now have dramatic evidence from the manuscript notes bound into the back of DR Vatican, that the first step toward his final system was initially formulated, on paper at least, on February 17, 1578.[19] Since the diagram of the near-Tychonic system comes at the very end of the manuscript notes, it is probably safe to assume that most of the annotations end in that year as well.

[15]Horský, *op. cit.*, p. 14.

[16]*TBOO, op. cit.*, I. p. 149.

[17]DR Vatican, fol 57^V.

[18]Horský, *op. cit.*, p. 15, note 6. *TBOO,* IV, p. 156.

[19]See Figure 12.

COMMENTARY: FICTIONALISM AND REALISM IN 16TH CENTURY ASTRONOMY

Peter K. Machamer

Ohio State University

Robert Westman has written an exciting and important paper. In laying out how three mid-16th Century astronomers differed in their responses to the previous Wittenberg legacy, Westman has made important progress toward our understanding of the development of post-Copernican astronomy. His outline for this development stresses the shared intellectual legacy of Praetorius, Maestlin, and Tycho Brahe. All three men had Wittenberg connections though, in the case of Maestlin, the connection was not close. All three men began their work on the assumption that Copernican astronomy was to be considered as a set of predictive planetary models and not as a cosmological system. From this starting point, there was, in Westman's words, a "growing consciousness of Copernicus' serious intent to propound a realist astronomy."

Much of Westman's work, in this paper and others, invokes, relies upon, and discusses the traditional dichotomy between mathematical and physical astronomy. This ancient distinction has come to be identified with a philosophical distinction between two supposedly antagonistic and mutually exclusive ways of considering scientific theories. In this modern guise the distinction also has been used as a historiographic tool for categorizing and analyzing various positions in astronomy. The modern jargon for this Duhemian dichotomy is realism vs. instrumentalism or fictionalism.

One should fight against dogmas which plague current work. The present battle is against the Duhemian dichotomy. The fight is not to be taken into quarters where the astronomers themselves refer to hypotheses as calculating devices as contrasted with hypotheses taken as having existential or ontological import. There is no doubt that many people for many reasons have espoused such views. Where the dichotomy breaks down is not in what people say about the method they are using, but, in fact, in how they use it.

In the 16th Century cases I shall consider, the inapplicability of the classical distinction is clear. Scholars of 16th century astronomy have been uncritical in their adherence to this distinction as an explanatory construct. "Only a fiction" or "merely mathematical" is used currently, and indeed, was used in the 16th century writings as a phrase of pardon; likewise "is realistic" or "is physical" served as a predicate to characterize the opposition or, at least, to call into question the failures of alternative schemes.

Such generality runs before my discussion. As Westman nicely brings out, prior to the late 16th century there did not exist a situation in which alternative cosmological schemes presented viable choice options for scientists and philosophers. It is not that the possibility of such choices was missing in any logical or abstract sense—the alternatives were there. In fact, a choice situation did not exist. Was this because, as Westman says in passing, that the mid-16th century astronomers theologians and natural philosophers did not see the technical models and parameters in *De Revolutionibus* "as a new Weltanschaung"? Such talk is not helpful or explanatory and serves only to restate the problem. Why is it that the early thinkers perceived Copernicus' work as only calculational and not as realistic? Westman says that this was the Protestant Wittenberg interpretation. Following the lead of Osiander, Reinhold and Peucer they treated the Copernican theory in a non-realistic way. Why did the Wittenberg group leave out of account all concern with the symmetry and systematic character of the Copernican planetary scheme? Such concern would in no way, by itself, commit them to realistic or untoward consequences.

In a moment I shall argue that they were not just fictionalists despite their methodological pronouncements to the contrary. Likewise it is my contention, though I shall not argue it here, that in a thorough-going sense it would be impossible to treat a set of hypotheses merely as instrumental, phenomena-organizing devices.

One can begin to examine the question as to whether or not a given writer takes an instrumental or realistic line by looking at his writings.

This cannot be done by looking solely at the methodological bits. One must also, and primarily, look at the arguments given for and against various positions. In the 16th century, one important form of argument is that which involves attributing an absurdity to an opponent's position. Roughly characterized, this seems to involve showing that the opponent's theory or point of view leads to a conclusion which contradicts (or at least, seems incompatible with) an accepted basis, i.e. a substantive or philosophical claim accepted as true by the proponent (and his friends). If one studies such accepted bases and attempts to see how they function in arguments and what their characteristics are, one will be in a position to comment upon the actual nature of the astronomical writings.

Before turning to the writings themselves, it is worth noting that no one has given an account of the difference between pure mathematical statements and statements concerning the quantitative properties of physical objects. More specifically, in order to avoid a tiff with mathematicians, and worries over the "empirical" basis of mathematics, no one has given a clue as to the difference between pure mathematical astronomy and physical astronomy. This is not to say that one cannot separate something properly called mathematical as distinct from something called physical. Rather, it is that in applied mathematics it is difficult to say where the application leaves off and the mathematics begins.

In order to give substance to these general points let me begin by briefly looking at Copernicus himself and then touch those figures with which Westman deals. It is commonplace to argue, as Maestlin did, that Osiander's preface does not reflect the proper intentions of Copernicus. Copernicus, so most say, was not only arguing hypothetically or mathematically that his system was viable, he was claiming physical truth for his system. Copernicus held that, in point of physical fact, the Sun was at, or near, the center of the universe. The attribution of this realistic sun-centered cosmological view to Copernicus can be supported by reference to the physical arguments in Book I of *De revolutionibus.* But, as Curtis Wilson argued in his presentation, the chief reason behind Copernicus' adoption of the heliostatic system was his insistence on using only uniformly moving circles in the modeling of the universe and the consequent rejection of Ptolemaic equants. Is this reason, the necessity of having true circular motion about a given center, to be classified as physical or mathematical? This is an important question, for circular motion and the rejection of equants assume pride of place in Westman's account of the later developments.

In Book I, chapter 4, Copernicus gave his reasons for requiring true circular motion and they are, seemingly, quite physical in character. He spoke of motor virtues, the planets being moved by spheres, and the like. But even this section of *De revolutionibus* seems not to be composed of purely realistic arguments. Copernicus was quite willing to talk about compounding uniform circular motions about a center. Such compounding completely undercuts the possibility of giving anything like an Aristotelian argument for the necessity of circular motion. The other reasons brought out in I.4 are aesthetic, Christian considerations concerning how things must be in this, the best of all systems. Thus, given only the mathematical/physical alternatives, the basic premiss of uniform circular motion must be a mathematical truth for Copernicus. It was so called by Tycho, when he praised Copernicus for his good mathematics while condemning him for his absurd physics.

The premiss of circular motion might be called a mathematical truth in that it functioned as a constraint upon the construction of mathematical planetary models. But the models were not taken to be purely mathematical. They were designed to accord with observations of physical facts, despite their geometrical character and despite Copernicus' occasional "adjustments" of the facts.[1] When developed, these models determined certain physical properties and relations of the objects observed, e.g. the arrangement and relative distances of the planets, including the sun's position.

Even so, things are not complex enough. Without further arguments, the relations between physical objects which can be read off from the combined planetary models cannot be taken as purely physical. Without some premiss, such as Maestlin and Kepler were later to use, e.g. that such harmonies must betoken reality in all senses (physical and mathematical), there is insufficient ground for a physical interpretation of all the relations determined by the mathematical models. It was the treatment of such physical-mathematical relations and considerations

[1]Cf. Noel Swerdlow's paper in this volume.

of their interpretations which were lacking in most of the mid-16th century astronomical discussions. Alternatively, one might consider the 16th century struggle as the sorting out of what the mathematical-physical relations must be. Part of the interest in Renaissance astronomy lies in the fact that there is not a set of basic physical arguments and premisses accepted by all, nor is there a theory of the relation between mathematics and the world which is detailed enough to be useful.

In Westman's paper this situation is clearly brought out in his discussion of Praetorius' work. Praetorius, as quoted in Westman, claimed that everyone approves the calculations of Copernicus, while abhoring his hypotheses concerning the earth's multiple motion. Praetorius's goal was to rework the Copernican mathematical constructions retaining only the Ptolemaic physical suppositions. What most impressed Praetorius was the Copernican rejection of the equant and its replacement by models employing only eccentrics, epicycles and deferents. The use of an equant was seen as an absurd hypothesis because it contradicted the Axioma Astronomicum.

Westman also points out that Praetorius was worried about the physical ordering of the planets and their maximum, minimum and mean distances.[2] Presumably, such worries went beyond what would be required for mathematically saving the appearances. Officially, Praetorius held that astronomy, by definition, is only concerned with hypotheses which save the phenomena. On this ground he criticized Kepler for doing physics in his *Mysterium Cosmographicum*. What is the difference between Kepler's speculations concerning the regular solids and Praetorius' concern with cosmological ordering and the dimensions of the planetary system? The difference seems to be that Kepler introduced *mathematical* models, i.e. the inscribed regular solids, to show the harmony of the arrangement of the planets. This is what Praetorius seems to object to as physics.

Praetorius' own positive position exhibited the same confusion. If pure astronomy is meant only to save the phenomena, and if "the astronomer is free to devise or imagine circles, epicycles and similar

[2]Tycho Brahe will worry about this also, basing at least one of his arguments against Copernicus on the physical "fact" that he found the parallax of Mars to be greater than that of the Sun.

devises although they might not exist in nature," then why and on what grounds is the equant to be rejected? As noted, the arguments against the equant and for uniform circular motion about a given center are always physically (or, better, ontologically) based. They are based on either the nature of the fifth element, the character of the divine, or, more physically, the fact that circular motion about a center involves no change of place (locomotion) with respect to that center. The absurdity of equants must have its basis in such non-mathematical claims.

In the early stages of Tycho Brahe's thinking, which Westman makes public for the first time,* there was this same internal tension. Brahe called the uniform circularity axiom a mathematical axiom. He said its contravention in Ptolemaic astronomy entails absurdities because the motions of the heavenly bodies on their epicycles and eccentrics would be disposed in an irregular manner with respect to the centers of these circles. The basis which allowed Brahe to claim that the equant is absurd was his conception of natural motion. As Westman quoted, Tycho supported the principal axiom of astronomy by arguing metaphysically that it is the nature of a sphere to be turned in a circle. This is the motion which realizes its form.

Westman comments on this section of Tycho's notes by saying that such physical grounding of the astronomical axiom could hardly fail to win the support of those 16th century astronomers eager to reject the equant. But on the classical account either the astronomers were mathematicians and fictionalists and would be untouched by Tycho's physical claims, or they were not pure mathematical astronomers and if they thought at all about the problem would have realized that there was *no other way* than physically (or, metaphysically) to argue for the axiom. Most probably, these astronomers like Tycho Brahe were beginning to be aware of the problems inherent in the attempt to treat astronomy purely mathematically.

As soon as an astronomer begins to have recourse to rigorous argument and begins to make explicit his own accepted basis for rejecting as absurd his rival's position, he must begin to realize that astronomy and

*Editor's note. Professor Gingerich's identification of DR Vatican as a Tycho Brahe copy was first announced at a meeting of the International Astronomical Union in Warsaw, September 1973. Cf. the Appendix to Westman's paper for further discussion of this newly discovered evidence.

natural philosophy are quite interconnected. the forming of such arguments must make him more sensitive to consistency problems. specifically, he must become aware, as Tycho put it, of the problems of finding a system which accords with both mathematical and physical principles and which would not incur the censure of theologians. When one gets as far as Tycho did in arguing against Copernicus using explicitly physical and metaphysical basis, the pressure naturally would be to develop a coherent system which could not be argued against on such grounds.

Tycho made use of such physical or metaphysical bases as the lack of efficiency in having such a large space between Saturn and the fixed stars, the heavy earth's being unfit to move, and the incredibility of the earth's having a triple motion. These were unlike the observationally based arguments which he gave, e.g. concerning the parallax of Mars or the systematic inability of Copernicus' theory to account for the lack of retrogression in the orbits of comets. In both kinds of cases, the elucidation of such bases in argument contexts served to bring them up explicitly for examination. In a climate, such as the late 16th century, unlike the earlier Renaissance, concern for truth, valid argument and system building had a place.

The making explicit of the accepted bases led not only to their reconsideration but also to asking about their connections with other such bases. Once the physical nature of movement had to be made explicit, it became questioned and, derivatively, questions concerning the things which move, e.g., the crystalline spheres, received closer examination. Early in the 16th century there seems to have been a rather vague, unarticulated agreement concerning the truths of natural philosophy, but no real concern or examination as regards the reasons behind them. Later, as the various elements or bases become clearer and articulated in their natures, grounds and relations come under scrutiny.

This move toward unity comes out in a different form at the end of Westman's paper. He cites Maestlin's annotation on the preface to *De revolutionibus,* where Maestlin said Copernicus is to be treated as an astronomer and not as a natural philosopher. Westman glosses this correctly as a realistic position. What Maestlin seems to have meant is that reality can only be grasped or described appropriately by mathematics.

Mathematics and natural philosophy are seen as unified. This was a new sort of realism.

To develop this line further would take too long; suffice it that this claim by Maestlin ties with what I said before concerning the renewed interest in system building and the developing concern for argumentation and objective truth in the late Renaissance. Specifically in astronomy, I think it has to do with the revival of systematic neo-Platonic views of cosmology and epistemology. In such a system, as found in Kepler's *Epitome*, physical force (virtus), mathematical description and harmonic consistency are all related. No longer could one invoke the astronomical/physical distinction to rebut an antagonist. By the early 17th century the distinction seems to have lost its force.

In all I have said I do not think I disagree with anything Westman has written. His paper excellently brings out aspects of the post-Copernican period which were before unknown. I have only tried to raise a few problems which lay in the way of my understanding of that period; how is it that the emergence of cosmological alternatives comes to the fore. One line I have suggested is that the conception of astronomy as a pure calculating device began to supersede the Renaissance (and Reformation) traits of argumentation as moral persuasion, one finds general attacks and worries about the physical and cosmological elements in the order of nature. In all this there is an increasing demand for a unified theory in which astronomy is part of the theory of nature as a whole. It is only in Kepler that these strands—the physical, mathematical and metaphysical—were brought together consciously, using the concept of *virtus* as central.

X

WHY DID COPERNICUS' RESEARCH PROGRAM SUPERSEDE PTOLEMY'S?

Imre Lakatos and Elie Zahar
London School of Economics

Introduction

1. *Empiricist Accounts of the Copernican Revolution.*
2. *Simplicism.*
3. *Polanyiite and Feyerabendian Accounts of the Copernican Revolution.*
4. *The Copernican Revolution in the Light of the Methodology of Scientific Research Programs.*
5. *The Copernican Revolution in the Light of Zahar's New Version of the Methodology of Scientific Research Programs.*

INTRODUCTION

I first should like to offer an apology for imposing a philosophical talk upon you on the occasion of the quincentenary of Copernicus' birth. My excuse is that a few years ago I suggested a specific method for using history of science as an arbiter of some authority when it comes to debates in philosophy of science and I thought that the Copernican revolution might in particular serve as an important test case between some contemporary philosophies of science.

I am afraid that first I have to explain—very roughly—what philosophical issues I have in mind and how historiographical criticism may help in deciding some of them.

The central problem of philosophy of science is the problem of normative appraisal of scientific theories; and, in particular, the problem of stating *universal* conditions under which a theory is scientific. This latter limiting case of the *appraisal problem* is known in philosophy as the *demarcation problem* and it was dramatized by the Vienna Circle and especially by Karl Popper who wanted to show that some *allegedly* scientific theories, like Marxism and Freudism, are pseudoscientific and hence that they are no better than, say, astrology. The problem is not an unimportant one and much is still to be done towards its solution. To mention a minor example, the Velikovsky affair revealed that scientists cannot readily articulate standards which are understandable to the layman (or, as my friend Paul Feyerabend reminds me, to themselves), and in the light of which one can defend as rational the rejection of a theory which *claims* to constitute a revolutionary scientific achievement.

This problem of appraisal is completely different from the problem of why and how new theories emerge. Appraisal of change is a normative problem and thus a matter for philosophy; explanation of change (of actual acceptance and rejection of theories) is a psychological problem. I take this Kantian demarcation between the "logic of appraisal" and the "psychology of discovery" for granted. Attempts to blur it have only yielded empty rhetoric.[1]

The generalized demarcation problem is closely linked with the problem of the rationality of science. Its solution ought to give us guidance as

---

*This talk was first given at the Quincentenary Symposium on Copernicus of the British Society for the History of Science, on 5th January 1973. The paper is the result of joint efforts by the co-authors; but it is narrated in the first person by Imre Lakatos. Previous versions were criticised by Paul Feyerabend and John Worrall.

[1]This draft is concerned only with the normative aspect of the problem indicated in the title of the paper. It does not attempt to go into the socio-psychological study of the Copernican Revolution.

to when the acceptance of a scientific theory is rational or irrational. There is still no agreed universal criterion on the basis of which we can say whether the rejection of the Copernican theory by the Church in 1616 was rational or not, or whether or not the rejection of Mendelian genetics by the Soviet Communist Party in 1949 was rational. (Of course, we, hopefully, all agree that both the *banning* of *De revolutionibus* and the *murder* of Mendelians were deplorable.) Or to mention a contemporary example, whether or not the present rejection by so-called American liberals of the application of genetics to intelligence by Jensen and others is rational, is a highly controversial question.[2] (We may nevertheless, agree that even if it were decided that a theory ought to be rejected, this decision should not carry with it physical threats to its tenacious protagonists; and that "...[nothing] be condemned without understanding it, without learning it, without even hearing it."[3])

1. *Empiricist Accounts of the "Copernican Revolution"*

Let me first define the term "Copernican Revolution." Even in the descriptive sense, this term has been ambiguously applied. It is frequently interpreted as the acceptance by the "general public" of the belief that the Sun, and not the Earth is the center of our planetary system. But neither Copernicus nor Newton held this belief.[4] Anyway, *changes* from one popular belief to another fall outside the province of the history of *science* proper. Let us, for the time being, forget about beliefs and states of mind and consider only *statements* and their objective (in Frege's and Popper's sense, "third-world"[5]) contents. In particular, let us regard the Copernican Revolution as the hypothesis that it is the Earth that is moving around the Sun *rather* than *vice versa,* or,

[2]According to Urbach (Urbach [1974]) it is irrational. But whether Urbach is right or wrong, the decision of Stanford University not to allow Nobel prize winner Shockley to lecture on race and intelligence is as shocking as the decision of Leeds University to refuse him his honorary doctorate in engineering because Lord Boyle and Jerry Ravetz (a brilliant Copernican scholar!) found that he held a theory which was contrary to so-called "liberal" doctrine.

[3]Galileo [1615].

[4]Cp. *e.g.* Price [1959], pp. 204-5.

[5]Cp. *e.g.* Popper [1972], especially Chapters 3 and 4.

more precisely, that the fixed frame of reference for planetary motion is the fixed stars and not the Earth. This interpretation is held mostly by those who hold that isolated hypotheses are the proper units of appraisal (rather than research programmes or "paradigms").[6] Let us take different versions of this approach in turn, and show how each version fails.

I first discuss the views of those people who attribute the superiority of the Copernican hypothesis to *straightforward empirical considerations.* These "positivists" are either inductivists or probabilists or falsificationists.

According to the *strict inductivists* one theory is better than another if it was deduced from the facts while its rival was not (otherwise the two theories are both mere speculations and rank equal). But even the most committed inductivist has been wary of applying this criterion to the Copernican Revolution. One can hardly claim that Copernicus deduced his heliocentrism from the facts. Indeed, now it is acknowledged that both Ptolemy's and Copernicus' theories were inconsistent with known observational results.[7] Yet many distinguished scholars, like Kepler, claimed that Copernicus derived his results "from the phenomena, from effects, from the consequences, like a blind man who secures his steps by means of a stick..."[8]

Strict inductivism was taken seriously and criticized by many people from Bellarmine to Whewell and was finally demolished by Duhem and Popper,[9] although some scientists and some philosophers of science, like Born, Achinstein and Dorling, still believe in the possibility of deduc-

---

[6]Cp. *below,* sections *3, 4,* and *5.*

[7]Let me quote on this point an authoritative source: "Ptolemy's theory was not very accurate. The positions for Mars, for example, were sometimes wrong by nearly 5°. But...the planetary positions predicted by Copernicus...were nearly as bad..." (Gingerich [1972]). This error was known to Kepler and he complained about it in the preface to his *Rudolphine Tables.* It was even known to Adam Smith as it transpires from his [1773]. Gingerich also reminds us that "in Tycho's observation books, we can see occasional examples where the older scheme based on the *Alfonsine Tables* yielded better predictions than could be obtained from the Copernican *Prutenic Tables.*" (Gingerich [1973]; cp. especially his footnote 6 in the same paper.)

[8]Kepler [1604]. Jeans describes the idea of the moving Earth as Copernicus' "theorem" [1948], p. 359 and claims that "Copernicus had proved his case" (*ibid.,* p. 133).

[9]Cp. Lakatos [1968*a*] and [1971*b*].

tion or valid induction of theories from (selected?) facts.[10] But the downfall of Cartesian and, in general, psychologistic logic and the rise of Bolzano-Tarski logic sealed the fate of "deduction from the phenomena." *If scientific revolution lies in the discovery of new facts and in valid generalizations from them, then there was no Copernican [Scientific] Revolution.*

Let us turn then to the *probabilistic inductivists.* Can *they* explain why Copernicus' theory of celestial motions was better than Ptolemy's? According to probabilistic inductivists one theory is better than another if it has a higher probability relative to the total available evidence at the time. I know of several (unpublished) efforts to calculate the probabilities of the two theories, given the data available in the 16th century, and show that Copernicus' was the more probable. All these efforts failed. I understand that Jon Dorling is now trying to elaborate a new Bayesian theory of the Copernican Revolution. He will not succeed. *If scientific revolution lies in proposing a theory which is much more probable given the available evidence than its predecessor, then there was no Copernican [Scientific] Revolution.*

Falsificationist philosophy of science can give two independent grounds on which the superiority of Copernicus' theory of celestial motions might rest.[11] According to one version, Ptolemy's theory was irrefutable (that is, pseudoscientific) and Copernicus' theory refutable (that is, scientific). If this were true, we really should have a case for identifying the Copernican revolution with the Great Scientific Revolution: it constitutes the switch from irrefutable speculation to refutable science. In this interpretation Ptolemaic heuristic was inherently *ad hoc*: it could accommodate *any* new fact by increasing the incoherent mess of epicycles and equants. Copernican theory, on the other hand, is interpreted as empirically refutable (at least "in principle"). This is a somewhat dubious reconstruction of history: Copernican theory might well use any number of epicyclets with no difficulty. The myth that the Ptolemaic theory included an indefinite number of epicycles which could be manipulated to fit any planetary observations, is anyway a myth invented after the discovery of Fourier series. But, as Gingerich

[10]Cp. Born [1949], pp. 129-34; Achinstein [1970] and Dorling [1971].

[11]For a third, cp. *below*, p. 364.

recently discovered, this parallel between epicycles-on-epicycles and Fourier analysis was not seen either by Ptolemy or by his followers. Indeed, the recomputation of the Alfonsine Tables by Gingerich shows that for actual computations Alfonso's Jewish astronomers used only a single-epicycle theory.

Another version of falsificationism claims that both theories were for a long time equally refutable. They were mutually incompatible rivals, both unrefuted; *finally,* however, some later crucial experiment refuted Ptolemy while corroborating Copernicus. As Popper put it: "Ptolemy's system was not refuted when Copernicus produced his. . . It is in these cases that crucial experiments become decisively important."[12] But Ptolemy's system (any given version of it) was commonly known to be refuted and anomaly-ridden long before Copernicus. Popper cooks up his history to fit his naive falsificationism. (Of course, he might *now* [in 1974] distinguish between mere anomalies which do not refute, and crucial experiments which do. But this general *ad hoc* manoeuvre which he produced in response to my criticisms[13] will not help him to specify in general terms the alleged "crucial experiment."[14]) As we have seen, the alleged superiority of Reinhold's Prutenic tables over the Alfonsine ones could not provide the crucial test. But what about the phases of Venus discovered by Galileo in 1616? Could *this* have formed the crucial test which showed Copernicus' superiority? I think that this might be a quite reasonable answer if not for the ocean of anomalies in which both rivals were equally engulfed. The phases of Venus may have established the superiority of Copernicus' theory over Ptolemy's, and if they did, would make the Catholic decision to ban Copernicus' work in the very moment of its victory all the more horrifying. But if we apply the falsificationist criterion to the question of when Copernicus' theory superseded not only Ptolemy's but also Tycho Brahe's (which was very well known in 1616), then falsificationism has only an absurd reply: that it did so *only in 1838.*[15] The discovery of stellar parallax by Bessel was the

[12]Popper [1963], p. 246. Popper, ignoring Tycho, thinks that the phases of Venus decided the issue for Copernicus.

[13]Cp. my Postscript to my [1971*a*] and my [1974], footnote 49.

[14]Indeed, once a Popperian "potential falsifier" can be seen either serious or unserious according to the great scientist's authority, Popper's whole philosophy of science collapses.

[15]*Not* in 1723, when there occurred a "crucial experiment" on the aberration of light.

crucial experiment between the two. But surely we cannot uphold the view that the abandonment of geocentric astronomy by the whole scientific community could only be defended *rationally* after 1838. This approach requires strong—and implausible—socio-psychological premises in order to explain the rash switch away from Ptolemy. Indeed, the late discovery of stellar parallax had very little effect. The discovery was made a few years *after* Copernicus' work had been removed from the *Index* on the grounds that Copernicus' theory had already been proved to be true.[16] Johnson surely must be wrong when he writes: "The fact that should be emphasized and re-emphasized is that there were no means whereby the validity of the Copernican planetary system could be verified by observation until instruments were developed, nearly three centuries later, capable of measuring the parallax of the nearest fixed star. For that length of time the truth or falsity of the Copernican hypothesis had to remain an open question in science."[17]

Something must be wrong with the falsificationist account. This is a typical example of how history of science can undermine a philosophy of science—too much of the actual history of science was irrational if scientific rationality is falsificationist rationality.[18] *If a scientific revolution lies in the refutation of a major theory and in its replacement by an unrefuted rival, the Copernican Revolution took place* (*at best*) *in 1838.*

2. *Simplicism*

According to conventionalism, theories are accepted by convention. Indeed we can, given sufficient ingenuity, force the facts into *any*

[16]This is very reminiscent of the story of the role of the determination of the speed of light in media optically denser than air in the optical revolution. Prior to Fresnel's work it was agreed both by the corpuscular and the wave theorists that the discovery of the speed of light in, say water would be the decisive factor in the debate. But when Foucault's and Fizeau's results in the 1850's eventually came out in favour of the wave theory, they had little effect—the issue had already been decided. (Cp. Worrall [1976b]).

[17]Johnson [1959], p. 220. Johnson's mistake is made even worse by conflating verification and truth. Watkins too seems to have held, in his otherwise excellent criticism of Kuhn, that the rivalry between the Copernicans and their adversaries was decided by the crucial experiment of 1838. (Watkins [1970], p. 36.)

[18]For the outlines of a general theory of how history of science can be a test of its philosophical "rational reconstructions" cp. my [1971*a*] and [1971*b*].

conceptual framework. This Bergsonian position is logically impeccable,[19] but it leads to cultural relativism (a position assumed both by Bergson and Feyerabend) *unless* a criterion for when one theory is better than another (even though the two theories may be observationally equivalent) is added to it. Most conventionalists try to avoid relativism by adopting some form of *simplicism*. I use this rather ugly term for methodologies according to which one cannot decide between theories on empirical grounds: a theory is better than another if it is simpler, more "coherent," more "economical" than its rival.[20]

The first man to claim that the chief merit of Copernicus' achievement was to produce a simpler, and *therefore* better system than Ptolemy's, was, of course, Copernicus himself. If his theory at the time had been observationally equivalent (if restricted to celestial kinematics) to Ptolemy's, this would have been understandable.[21] He was followed by Rheticus and Osiander; and Brahe too judged there was something in the claim. The superior simplicity of Copernicus' theory of celestial "orbs" became an unchallenged *fact* in the history of science from Galileo to Duhem: all that Bellarmine questioned was the *further* inference from impressive simplicity to Truth. Adam Smith, for example, in his beautiful *History of Astronomy,* argued for the superiority of the Copernican hypothesis on the basis of its superlative "beauty of simplicity."[22] He disclaimed the inductivist idea that the Copernican tables were more accurate than their Ptolemaic predecessors and that therefore, Copernican theory was superior. According to Adam Smith the new, accurate observations were equally compatible with Ptolemy's system. The advantage of the Copernican system lay in the "superior degree of coherence, which it bestowed upon the celestial

[19]Cp. Lakatos [1970], pp. 104-106 and p. 188.

[20]Cp. Lakatos [1970], p. 105.

[21]This "observational equivalence" is actually a great simplicist myth; cp. *below*, p. 364. It should be, however, remembered that Copernicus thought that this greater simplicity will also provide, *eo ipso,* better astronomical tables, that is, it will lead to saving *more* phenomena. Thus he did not believe in the "observational equivalence" of his theory with Ptolemy's.

[22]Smith [1773], p. 72.

appearances, the simplicity and uniformity which it introduced into the real directions and velocities of the Planets."[23]

But the superior simplicity of Copernican theory was just as much of a myth as its superior accuracy. The myth of superior simplicity was dispelled by the careful and professional work of modern historians. They reminded us that while Copernican theory solves certain problems in a simpler way than does the Ptolemaic one, the price of the simplifications is unexpected complications in the solution of other problems.[24] The Copernican system is certainly simpler since it dispenses with equants and with some eccentrics; but each equant and eccentric removed has to be replaced by new epicycles and epicyclets. The system is simpler in so far as it leaves the eighth sphere of fixed stars immobile and removes its two Ptolemaic motions; but Copernicus has to pay for the immobile eighth sphere sphere by transferring its irregular Ptolemaic movements to the already corrupt earth which Copernicus sets spinning with a rather complicated wobble; he also has to put the centre of the universe, not at the Sun, as he originally intended, but at an empty point fairly near to it.

I think it is fair to say that the "simplicity-balance" between Ptolemy's and Copernicus' system is roughly even. This is reflected in de Solla Price's remark that Copernicus' system was "more complicated but more economical";[25] and also in Pannekoek's view that "the new world structure, notwithstanding its simplicity in broad outline, was still extremely complicated in the details."[26] According to Kuhn, Copernicus' account of the *qualitative* aspect of the major problems of planetary motion (*e.g.* the retrograde motion) is much neater, much "more economical," than Ptolemy's, "but this apparent economy . . . is [only] a propaganda victory . . . [and in fact] is largely an illusion."[27] When it comes to details, "[Copernicus'] full system was little if any less cumbersome than Ptolemy's had been." As he succinctly puts it: Copernicus

[23]*Ibid.*, p. 75.

[24]Cp. *e.g.* Kuhn [1957] and Ravetz [1966*a*].

[25]Price [1959], p. 216. According to Price, Copernicus "*increased* the complexity of the (Ptolemaic) system without increasing the accuracy" (my italics).

[26]Pannekoek [1961], p. 193.

[27]Kuhn [1957], p. 169.

introduced a "great, and yet strangely small" change.[28] While the Copernican theory has more "aesthetic harmony," gives a more "natural" account of the *basic* features of the heavens, has "fewer *ad hoc* assumptions," it is in the end "a failure. . . neither more accurate nor significantly simpler than its Ptolemaic predecessors."[29] According to Ravetz, the "irregularly moving stellar sphere" in Ptolemy's system brought with it a "fundamental measure of time [as] a motion along an irregularly moving orbit." In Ravetz's judgment this is *"strictly incoherent,"* but, if this irregularity in the motion of the stars is transferred to the motion of the Earth, as it is in Copernicus' system, we get a "*coherent*" astronomy.[30] But if so, coherence seems to be in the eye of the beholder. Simplicity seems to be relative to one's subjective taste.[31] *If dramatic increase in simplicity of observationally equivalent theories is the hallmark of scientific revolution, the Copernican Revolution cannot be regarded as one* (even if some people like Kepler thought that its superiority was due to the beautiful harmony which it introduced[32]).

Let us now return to Popperian falsificationism. Popper lays great stress on crucial experiments and, in this respect, he is, on my terms, an empiricist. Man proposes and Nature disposes. But at the same time he proposes a new brand of simplicism: he claims that even *before* Nature disposes we should already regard a theory as better than its rival if it has more falsifiable content, more potential falsifiers.[33] Since Popper offered his 1934 falsifiability criterion as an explication of "simplicity,"[34] his *Logic of Scientific Discovery* should be regarded as a new, original brand of simplicity. In this sense, then, especially in its

[28]Kuhn [1957], p. 133.

[29]*Ibid.*, p. 174.

[30]Ravetz [1966*b*].

[31]The most beautiful argument for this statement is on pp. xvi-xvii of Santillana [1953]. One glance suffices to demonstrate the point.

[32]For why Kepler *thought* he preferred Copernicus to Ptolemy and to Brahe *cp.* Westman [1972]. Why he did prefer it is more difficult to say.

[33]He strengthened his empiricism in his "third requirement" (I called it "acceptability$_2$"; *cp.* Lakatos [1968*a*], pp. 379 ff.)

[34]Popper [1953], Chapter VII.

realist interpretaton,[35] the Copernican theory may have been better than Ptolemy's already in 1543, even had they been *observationally* equivalent at the time.

But the two theories were not observationally equivalent. Simplicists usually take it *too easily* for granted that the rival theories which they appraise are either logically or in some other strict sense equivalent so that the claim that only simplicity, and not facts, can decide should sound more plausible. The conventionalist idea that Ptolemy's and Copernicus' theories are *bound to be in some strong sense equivalent* is common currency among "simplicists": after all, they accept conventionalism but want to find a way out of its relativist implications. The idea has been propounded by Dreyer, the Halls, Price, Kuhn and others.[36] Hanson is right in saying, in his criticism of their views, that "in no ordinary sense of 'simplicity' is the Copernican theory simpler than the Ptolemaic"; but he still preserves their "Line of Sight Equivalence."[37]

3. *Polanyiite and Feyerabendian Accounts of the Copernican Revolution*

All the philosophies so far discussed are based on universal demarcation criteria. According to them *all* major changes in science can be explained using the *same* single criterion of scientific merit. But none of these philosophies has been able to give a clear and acceptable account of any rational grounds on which geocentric theories were inferior to Copernicus' *De revolutionibus*. The failure of "demarcationists" to solve this problem (and other similar problems) has led to a situation in

[35]Cp. Feyerabend [1964]; an excellent paper from his almost-Popperian period. Agassi holds that Copernicus' theory had no *empirical* superiority: indeed, Agassi claims that Copernicus "did not succeed in showing that his system is better than Ptolemy's, let alone in refuting him" (Agassi [1962], p. 5).

[36]For a criticism of Dreyer's, the Halls's, Price's, Kuhn's overstatements, cp. Hanson [1973], pp. 200-220. That he himself overemphasizes simplicity ("systemacity") transpires from his arguments and from absurd statements like: "(Copernicus), like Newton after him, and Aristotle before, revealed no new data, nor did he seek any." (*Ibid.*, p. 87).

[37]Hanson [1973], p. 233 and 212. Ironically, on p. 233, Hanson absentmindedly exchanged in his manuscript "Ptolemaic" and "Copernican" and the editor of the posthumous work did not notice or correct the slip of pen.

which some, if not most, scientists and quite a few philosophers of science *deny* that there can ever be any valid universal demarcation criterion or system of appraisals for judging scientific theories. The most influential contemporary protagonist of this view is Polanyi, according to whom the search for a universal rationality criterion is utopian. There can be only a *case law,* no *statute law* for deciding what is scientific and pseudoscientific, what is a better and what is a worse theory. It is the jury of scientists which decides in each separate case and as long as scientific autonomy—and *eo ipso* the independence of this jury—is upheld, nothing will go very wrong. If Polanyi is right, the Royal Society's refusal to sponsor philosophy of science is quite reasonable: ignorant *philosophers* of science should not be allowed to *judge* scientific theories, that is the scientist's own business. The Royal Society is of course, willing to finance *historians* of science who *describe* their activities as constituting triumphant progress.[38]

In the Polanyiite view, in each individual case of rivalry between two scientific theories, one has to leave it to the inarticulable *Fingerspitzengefühl* (Holton's favorite expression) of the great scientists to decide which theory is better. The great scientists are the ones who have "tacit knowledge" of the way things will go. Polanyi writes about the "foreknowledge the Copernicans must have meant to affirm when they passionately maintained, against heavy pressure, during one hundred and forty years before Newton proved the point, that the heliocentric theory was not merely a convenient way of computing the paths of planets, but was really true."[39] But of course, this *"foreknowledge"*—unlike a simple conjecture—cannot be articulated and made available to the layman-outsider. Toulmin seems to have a similar view of the Copernican Revolution.[40] So does Kuhn. Kuhn claims that,

---

[38]The Royal Society gives financial support to history of science, but none to philosophy of science.

[39]Polanyi [1966], p. 23. Also cp. his [1958], *passim.*

[40]I take it that the following passage bears this claim out: "If Kepler and Galileo preferred Copernicus' new heliostatic system, their reasons for doing so were far more specific, varied, and sophisticated than are hinted at by such vague terms as 'simplicity' and 'convenience': especially at the outset, indeed the Copernican theory was by many tests substantially less simple or convenient than, the traditional Ptolemaic analysis. When we consider the conceptual changes between successive physical theories, therefore, the

> . . . to astronomers the initial choice between Copernicus' system and Ptolemy's could only be a matter of taste, and matters of taste are the most difficult of all to define or debate. Yet as the Copernican Revolution itself indicates, matters of taste are not negligible. *The ear equipped to discern geometric harmony* could detect a new neatness and coherence in the sun-centered astronomy of Copernicus, and if that neatness and coherence had not been recognized, there might have been no Revolution.[41]

According to a *later* account of Kuhn's,[42] Ptolemaic astronomy was by 1543 in a state of "paradigm-crisis" which is the inevitable prelude to any scientific "revolution", *i.e.* mass-conversion: "The state of Ptolemaic astronomy was a recognized scandal before Copernicus proposed a basic change in astronomical theory, and the preface in which Copernicus described his reasons for innovation provides a classic description of the crisis state."[43] But how many apart from Copernicus felt this communal "crisis"? After all there was not much of a "scientific community" in Copernicus' time. And if Kuhn thinks that his full analysis of scientific revolutions applies to the Copernican case, why did so few scholars join the Copernican "bandwagon" before Kepler and Galileo?

---

rationality we are concerned with is neither a merely formal matter, like the internal articulation of a mathematical system, nor a merely pragmatic matter, of simple utility or convenience. Rather, we can understand on what foundation it rests, only if we look and see how, in practice, successive theories and sets of concepts are first applied, and later modified within the historical development of the relevant intellectual activity." (Toulmin, [1972], p. 65.)

[41]Kuhn [1957], p. 177; my italics. For a general criticism of this Polanyiite position cp. my [1971*b*], p. 121 and my [1974], p. 372.

[42]Kuhn's position concerning the Copernican Revolution changed radically from the essentially internalist simplicism of his [1957] to his radically sociologistic [1962] and [1963].

[43]Kuhn [1963], p. 367. For Kuhn a "crisis" *must* precede a "revolution" exactly as for a naive falsificationist a refutation *must* precede a new conjecture. No surprise that Kuhn writes that there is "unequivocal historical evidence" that "the state of Ptolemaic astronomy was a scandal before Copernicus's announcement" Kuhn [1962], pp.67-8. Gingerich [1973] showed that Kuhn conjures up a scandal where there was none. (Of course, a progressive "research program" [in my sense] need not be preceded by the degeneration of its rival.)

In Kuhn's judgment there is no *explicit* criterion on the basis of which Copernicus' system can be judged superior to Ptolemy's. But the scientific elite with an inarticulable and esoteric "ear for geometric harmony" or crisis-sensitive psyche could tell which theory was better. It seems, however, that once it comes to details, Kuhn's account is no more trouble-free than the accounts of the demarcationists. He has to invent a socio-intellectual "crisis" in the scientific elite working in the Ptolemaic paradigm in the sixteenth century and then a sudden switch to Copernicanism. *If these are necessary conditions for a scientific revolution, then the Copernican Revolution was not a scientific Revolution.*

For Feyerabend, the failure both of demarcationists and elitists is only to be expected. For him, our brilliant leading cultural relativist, the Ptolemaic system was just one system of belief, the Copernican system another. The Ptolemaists did their thing and the Copernicans did theirs and at the end the Copernicans scored a propaganda victory. To quote Westman's summary of his position: "We are given two theories, the Copernican and the Ptolemaic, both of which provide reliable predictions, but where the former contradicts the accepted laws and facts of the contemporary terrestrial physics. Belief in the success of the new theory cannot be based upon methodological assumptions for no such principles can ever certify the correctness of a theory at its inception; nor, at the start, does there exist any new factual support. Therefore, the acceptance of the Copernican theory becomes a matter of metaphysical belief."[44] According to Feyerabend *nothing more can be said.* Feyerabend's account is much more difficult to rebut than anybody else's. Indeed, we may in the end have to admit that Copernicus' and Kepler's and Galileo's adoption of the heliocentric theory and its victory is not rationally explicable, that it was largely a matter of taste, a *Gestalt-switch,* or a propaganda victory. But even if this *did* turn out to be the case we need not allow ourselves to be steamrollered by Feyerabend into *general* cultural relativism or by Kuhn into *general* elitism. Fresnel's wave theory of light, for example, was by 1830 clearly better than Newton's corpuscular theory on explicit objective criteria, but

[44]Westman [1972], p. 234. In his [1972], Feyerabend slips into a Polanyiite view: he thinks that Copernicans achieved a victory of *Reason* with the help of their "*Lebendigheit des Geistes.*"

Fresnel's first adoption of the old wave idea was clearly a question of taste.[45] If it were irrational to *work on* a theory whose superiority was not yet established then almost all of the history of science would indeed be rationally inexplicable. But, as it happens, the Copernican Revolution can be explained as rational on the basis of the methodology of scientific research programs.

4. *The Copernican Revolution in the Light of the Methodology of Scientific Research Programs*

The methodology of scientific research programs is a new demarcationist methodology (*i.e.* a *universal* definition of progress) which I have been advocating now for some years and which, it seems to me, improves on previous demarcationist methodologies and at the same time escapes at least some of the criticisms which elitists and relativists have levelled against inductivism, falsificationism and the rest.

Let me first explain the rough central features of this methodology.[46]

First of all my unit of appraisal is not an isolated hypothesis (or a conjunction of hypotheses): a research program is rather a special kind of "problem shift."[47] It consists of a developing series of theories. Moreover, this developing series has a structure. It has a tenacious *hard core,* like the three laws of motion and the law of gravitation in Newton's research program, and it has a *heuristic,* which includes a set of problem-solving techniques. (This, in Newton's case, consisted of the program's mathematical apparatus, involving the differential calculus, the theory of convergence, differential and integral equations.) Finally, a research program has a vast belt of auxiliary hypotheses on the basis of which we establish initial conditions. The protective belt of the Newtonian program included geometrical optics, Newton's theory of atmospheric refractions, and so on. I call this belt a *protective belt* because it protects the hard core from refutations: anomalies are not taken as refutations of the hard core but of some hypothesis in the pro-

[45]Cp. Worrall [1976b].

[46]For my use of the technical term "methodology" cp. my [1971*a*], p. 92 and my [1971*b*] footnotes 5 and 13.

[47]Cp. Lakatos [1968*a*], p. 386, Lakatos [1968*b*] and Lakatos [1970], p. 118 ff.

tective belt. Partly under empirical pressure (but partly *planned* according to its heuristic) the protective belt is constantly modified, increased, complicated, while the hard core remains intact.

Having specified that the unit of mature science is a research program, I now lay down rules for appraising programs. A research program is either progressive or degenerating. It is *theoretically progressive* if each modification leads to new unexpected predictions and it is *empirically progressive* if at least some of these novel predictions are corroborated. It is always easy for a scientist to deal with a *given* anomaly by making suitable adjustments to his program (*e.g.* by adding a new epicycle). Such manoeuvres are *ad hoc,* and the program is *degenerating,* unless they not only explain the given facts they were intended to explain but also predict some new fact as well. The supreme example of a progressive program is Newton's. It successfully anticipated novel facts like the return of Halley's comet, the existence and the course of Neptune and the bulge of the Earth.

A research program never solves all its anomalies. "Refutations" always abound. What matters is a few dramatic signs of empirical progress. This methodology also contains a notion of *heuristic progress:* the successive modifications of the protective belt must be in the spirit of the heuristic. Scientists rightly dislike artificial *ad hoc* devices for countering anomalies.

One research program *supersedes* another if it has excess truth content over its rival, in the sense that it predicts progressively all that its rival truly predicts and some more besides.[48]

Before we apply this new and perhaps a bit too elaborate philosophical framework[49] to appraising the rival theories, or, rather, rival *programs,* of Ptolemy and Copernicus, one important remark has to be made.

Any two rival research programs can be made observationally equivalent by producing observationally equivalent falsifiable versions of the two with the help of suitable *ad hoc* auxiliary hypotheses. But such equi-

[48]For an interesting discussion of "superseding" versus "Incommensurability" cp. Feyerabend [1974].

[49]For more careful formulations the reader has to be referred to my [1968*b*), [1970], [1971*a*] and [1971*b*]. Also cp. my [1974].

valence is uninteresting. Two rival research programs are only equivalent if they are identical. Otherwise the two different heuristics proceed at different speeds. Even if two rival programs explain the same range of evidence, the same evidence will give more support to the one than to the other depending on whether the evidence was, as it were, "produced" by the theory or explained in an *ad hoc* way. The weight of evidence is not merely a function of a falsifiable hypothesis and the evidence; it is also function of temporal and heuristic factors.[50] The starting point of the methodology of scientific research programs is the normative problem posed by "revolutionary conventionalism."[51] But if revolutionary conventionalism is correct, observational equivalence can always be produced between two rival theories. Simplicism concluded that empirical evidence loses its weight: only the degree of simplicity counts. Popper's falsifiability and Lakatos' and Zahar's degree of progressiveness discard the ambiguity and pitfalls of degrees of coherence and rehabilitate, in radically new ways, a "positivistic" respect for facts.

The descriptive aspect of the methodology of scientific research programs is clearly superior to the descriptive aspect of the methodologies previously discussed. Both Ptolemy and Copernicus worked on *research programs*: they did not simply test conjectures or try to harmonize a vast conjunction of observational results, nor did they *commit* themselves to any community based "paradigms." I shall offer a description of the two research programs (this, I take it, will be fairly uncontroversial) and I shall also offer an appraisal of their respective progress and degeneration.

Both *programs* branched off from the Pythagorean-Platonic program whose basic principle was that since heavenly bodies are perfect, all astronomical phenomena should be saved by a combination of as few uniform circular motions (or uniform spherical rotations about an axis) as possible. This principle remained the cornerstone of the heuristic of both programs. This proto-program contained no directives as to where the center of the universe lies. The heuristic in this case was primary,

[50]Zahar's achievement lies primarily in producing an improved notion of "weight of evidence," cp. *below*, pp. 376-377.

[51]Cp. Lakatos [1970], p. 104.

the "hard-core" secondary.[52] Some people, like Pythagoras, believed the center was a fireball invisible from the inhabited regions of the earth; others, like some Platonists, that it was the sun; and still others, like Eudoxos, that it was the Earth itself. The geocentric hypothesis "hardened" into a real hard core assumption only with the development of an elaborate Aristotelian terrestrial physics, with its natural and violent motion and its separation of the terrestrial or sublunar chemistry of four elements from the pure and eternal celestial *quinta essentia*.

The first rudimentary geocentric theory of the heavens consisted of concentric orbs around the earth, one for the stars and one for each other celestial body. But this was known to be a false "ideal model" and, as Eudoxus already realised, even if the rudimentary scheme worked for the stars, it definitely did not work for the planets. As is well known, Eudoxus devised a system of rotating spheres in order to account for planetary motion. He introduced twenty-six such spheres in order to explain—or rather to save—the stations and retrogressions of the planets. The model predicted no novel facts and it failed to solve some serious anomalies like the varying degrees of brightness of the planets. After this system of rotating spheres was abandoned, every single move in the geostatic program ran counter to the Platonic heuristic. The eccentric displaced the earth from the center of the circle; the Appollonian and Hipparchan epicycles meant that the real path of the planets about the earth was not circular; and finally the Ptolemaic equants entailed that even the motion of the epicycle's empty center was not simultaneously uniform and circular—it was uniform but not circular. when viewed from the equant point, it was circular but not uniform when viewed from the center of the deferent: uniform circularity was replaced by quasi-uniform, quasi-circularity.

The use of the equant was tantamount to the abandonment of the Platonic heuristic. No wonder then that at an early stage of this development astronomers like Heracleides and Aristarchus started to experiment with partially or completely heliocentric systems. Each move in

[52]The demarcation between "hard core" and "heuristics" is frequently a matter of convention as can be seen from the arguments proposed by Popper and Watkins concerning the intertranslatability of what they called "metaphysics" and "heuristic" respectively. (Cp. especially Watkins [1958].)

the geocentric program had dealt with certain anomalies but had done so in an *ad hoc* way. No novel predictions were produced, anomalies still abounded and certainly each move had deviated from the original Platonic heuristic.[53]

Copernicus recognised the heuristic degeneration of the Platonic program at the hands of Ptolemy and his successors. He assumed that the periodicity of planetary motion was connected with—and, indeed, exhausted by—combinations of uniform circular motions.[54] Copernicus levelled three charges of *ad hocness* against Ptolemy.

(*a*) The introduction of equants violated the heuristic of Ptolemy's own program. It was heuristically *ad hoc* (*ad hoc*$_3$[55]). This objection occurs in the third paragraph of the *Commentariolus*. In the second paragraph Copernicus mentions Callippos' and Eudoxus' vain effort to save the phenomena by a system of concentric spheres.

(*b*) Because of the difference between the solar and the sidereal years, Ptolemy gave two distinct motions to the stellar sphere: the daily rotation and a rotation about the axis of the ecliptic. This was already a major defect of the Ptolemaic system, since the stars, being the most perfect bodies, ought to have a single uniform motion.

In his *Commentariolus,* Copernicus pointed out that the sidereal year provides a more accurate unit of time than the solar year. According to Ravetz,[56] Copernicus must have started from erroneous data and concluded that the difference between the solar and the sidereal years varies irregularly; the stellar sphere must therefore rotate irregularly about the axis of the ecliptic. Thus the Sun moves non-uniformly about the earth. This is yet another violation of the Platonic heuristic, and constitutes further heuristic degeneration.[57]

---

[53]Kuhn claims that "there were no good reasons for taking Aristarchus seriously" (Kuhn [1962], p. 76). But it is clear that there were—the geocentric program had already heuristically degenerated.

[54]In view of what we know about Fourier expansions of periodic functions, this is a remarkable mathematical conjecture. Cp. *e.g.* Kamlah [1971].

[55]Cp. Lakatos [1970], p. 175, footnote 3 and p. 176, footnote 1.

[56]Ravetz [1966*a*]. But cp. Gingerich's remark in his [1973], footnote 19.

[57]This "incoherence," in Ravetz's view, suggested to Copernicus that the stars rather than the Earth determine the primary frame of reference for physics. Of course from the point of view of our *present* problem, it does not matter at all what actually triggered off

(*c*) Despite all these violations of the Platonic heuristic, the geostatic program remained empirically *ad hoc,* that is, it always lagged behind the facts.

Copernicus did not create a completely new program; he revived the Aristarchan version of the Platonic program. The hard core of this program is the proposition that the stars provide the primary frame of reference for physics. Copernicus did not invent a new heuristic but attempted to restore and rejuvenate the Platonic one.[58]

Did Copernicus succeed in creating a more truly Platonic theory than Ptolemy? He did. According to the Platonic heuristic, the stars, being the most perfect bodies, ought ideally to have the most perfect motion, namely a single uniform rotation about an axis. Note that uniform circular motion is perfect because it can be assimilated to a state of rest: all points of the circle being equivalent, uniform circular motion is indistinguishable from rest or absence of change. We have seen that, in Copernicus' time, Ptolemaic astronomers imparted to the stellar sphere (at least) two distinct motions: a daily rotation and a rotation about the axis of the ecliptic. Also, due to erroneous data, they made this second rotation irregular.

Copernicus, however, fixed the stars, thus making them genuinely immutable. It is true that he had to transfer their motion to the Earth; but in his system the Earth is a planet and planets are anyway less perfect than stars, if only on account of their multiple epicyclic motions. (These multiple epicyclic motions were accepted both by Ptolemaists and by Copernicans.)

Copernicus got rid of the equant and produced a system which, despite the elimination of the equant, contained only about as many circles as the Ptolemaic system.[59]

Copernicus' imagination. We are not now concerned with the psychological causes of Copernicus' achievement, but with its appraisal.

[58]It was Kepler who framed the heuristic of the "new" astronomy, namely the principle that the motion of the planets ought to be explained in terms of heliocentric forces.

[59]This mutual replaceability was already known by Islamic astronomers like Ibn-ash-Shatir. As Neugebauer pointed out (cp. Neugebauer [1958] and [1968]), Copernicus used a few equants but since these equants can be replaced by secondary epicycles, they are irrelevant. Copernicus considered uniform circular motions as the only permissible motions in astronomy; this need not prevent him from using equants as computational devices.

In addition to its heuristic superiority over the *Almagest,* Copernican astronomy was no worse off in saving the phenomena than Ptolemaic astronomy. Indeed, Copernicus' lunar theory was a definite empirical advance over Ptolemy's. Using the Earth as an equant point, Ptolemy had succeeded in describing the angular motion of the moon; however, the moon would have had, at certain points of its path, twice its (observable) apparent diameter. Copernicus not only dispensed with equants, but also, through replacing equants by epicycles, he happened to improve on the fit between theory and observation.[60]

Copernicus' program was certainly theoretically progressive. It anticipated novel facts never observed before. It anticipated the phases of Venus. It also predicted stellar parallax, though this was very much a qualitative prediction, because Copernicus had no idea of the size of the planetary system. It was *not,* as Neugebauer put it, "a step in the wrong direction" from Ptolemy.[61]

But the phases of Venus prediction was not corroborated until 1616. Thus the methodology of scientific research programs agrees with falsificationism to the extent that Copernicus' system was not *fully* progressive until Galileo, or even until Newton, when its hard core was incorporated in the completely different Newtonian research program which was immensely progressive. The Copernican system may have constituted heuristic progress within the Platonic tradition, it may have

[60]In Neugebauer's view, this empirical success might have led Copernicus to believe that the elimination of the equant, apart from restoring the Platonic heuristic to its original purity, would also improve the predictive power of the new theory. But the Copernican system remained anomaly-ridden even in its most highly developed version. One of the most important anomalies in the Copernican program were the comets whose motion could not be explained in terms of circular motions. This was one of Tycho's most important arguments against Copernicus, and it was one that Galileo found difficult to counter.

[61]Neugebauer [1968], p. 103. He claims "Modern historians, making ample use of the advantage of hindsight, stress the revolutionary significance of the heliocentric system and the simplifications it had introduced. Had it not been for Tycho Brahe and Kepler, the Copernican system would have contributed to the perpetuation of the Ptolemaic system in a slightly modified form, but *more pleasing to philosophical minds.*" *Which* philosophical minds? One wonders how a man of Neugebauer's stature can end a paper on such a casually inaccurate note. But, alas, even the most professional historians who are *in principle* against the philosophy of science, end up with *philosophically motivated blunders.*

been theoretically progressive but it had no novel *facts* to its credit until 1616.[62] *It seems then that the Copernican Revolution only became a fully fledged scientific revolution in 1616, when it was almost immediately abandoned for the new dynamics-oriented physics.*

From the point of view of the methodology of scientific research programs the Copernican program was not further developed but rather abandoned by Kepler, Galileo and Newton. This is a direct consequence of the shift of emphasis from "hard core" hypotheses to heuristic.[63]

This, rather unwelcome, conclusion seems to be inevitable so long as we regard the prediction of only temporally novel facts as the criterion of progress. However, Zahar was led by considerations quite independent of the history of the Copernican Revolution to propose a new criterion of scientific progress—a criterion which is a very important amendment to that provided by the methodology of scientific research programs.[64]

5. *The Copernican Revolution in the Light of Zahar's New Version of the Methodology of Scientific Research Programs*

I originally defined a prediction as "novel," "stunning," or "dramatic" if it was inconsistent with previous expectations, unchallenged background knowledge and, in particular, if the predicted fact was forbidden by the rival program. The best novel facts were such which might never have been observed if not for the theory which anticipated it. My favorite examples of such predictions which were corroborated (and hence dramatically supported the theory on the basis of which they were made) were the return of Halley's comet, the discovery of Neptune, the Einsteinian bending of light rays, the Davisson-Germer

[62]According to Kuhn, the phases of Venus for the heliostatic system was "not proof, but . . . propaganda" [1957], p. 224. Of course, it was not proof but it was, in the light of most empirical appraisal, including that of the methodology of scientific research programs, an *objective* sign of progress. Kuhn seems to agree two pages later: "Though the telescope *argued much*, it proved nothing." (*Op. cit.*, p. 226.)

[63]It is then wrong to say that "The Copernican system of the world had developed into Newton's theory of gravity." (Popper [1963], p. 98.)

[64]Cp. Zahar [1973].

experiment.[65] But on this view, the Copernican program became *empirically* progressive only in 1616! If this is so, one can well understand why its early protagonists, in want of corroborated excess content, emphasized so much its superior "simplicity."

Interestingly, Elie Zahar's modified methodology of scientific research programs gives a very different picture. Zahar's modification lies primarily in his new conception of "novel fact." In his view the explanation of Mercury's perihelion gave crucial empirical support, "dramatic corroboration," to Einstein's theory, even though, as a low-level empirical proposition, it had been known for almost a hundred years.[66] This was no novel fact in my original sense, yet it was "dramatic." But in what sense "dramatic"? "Dramatic" in the sense that in Einstein's original design Mercury's anomalous perihelion played no role whatsoever. Its exact solution, was, as it were, an unexpected present from Schwarzschild, a result which was an unintended by-product of Einstein's program. The same holds for the role of the Balmer formula in Bohr's program. Bohr's original problem was not to discover the secrets of the hydrogen spectrum but to solve the problem of the stability of the nuclear atom; therefore, Balmer's formula gave "dramatic" evidential support for Bohr's theory even though it was, temporally speaking, no novel fact.

Let us then now look at the situation in 1543 and see if Copernicus' program had *immediate* support from facts which were novel in Zahar's sense.

Copernicus' proto-hypothesis was that the planets move uniformly on concentric circles enclosing the Sun; the moon moves on an epicycle centered on the Earth.[67] Zahar's claim is that several important facts concerning planetary motions are straightforward consequences of the original Copernican assumptions and that, although these facts were previously known, they lend much more support to Copernicus than to Ptolemy within whose system they were dealt with only in an *ad hoc* manner, by parameter adjustment.

---

[65]Later I wanted to turn old empirical observations like the Balmer formula into novel facts with respect to Bohr's program: cp. my [1970], p. 156. But Zahar solved the problem in a superior way.

[66]Cp. Zahar [1973].

[67]Cp. The figure drawn by Copernicus on p. 10 of his *De Revolutionibus*.

From the basic Copernican model and the assumption that inferior planets have a shorter period while superior planets have a longer period than that of the Earth,[68] the following facts can be predicted *prior to any observation*:

(*i*) *Planets have stations and retrogressions.*

Let us remember that Eudoxus' 26 concentric orbs were already doctored to account for the carefully observed stations and retrogressions. In Copernicus' program stations and retrogressions are simple logical consequences of the rough model. Moreover, in Copernicus' program this explain the previously puzzling and unresolved variations in the brightness of planets.

(*ii*) *The periods of the superior planets, as seen from the Earth are not constant.*

For Ptolemy this observational premiss is very difficult to explain; for Copernicus it is a theoretical triviality.

(*iii*) *If an astronomer takes the Earth as the origin of his fixed frame, he will ascribe to each planet a complex motion one of whose components is the motion of the Sun.*

This is an immediate consequence of the Copernican model: a change of origin leads to the addition of the Sun's apparent motion to the motion of every other mobile.

For Ptolemy this is a cosmic accident which one has to accept *after* a careful study of the facts. Thus Copernicus *explains* what for Ptolemy is a fortuitous result in the same way Einstein explains the equality of inertial and gravitational masses, which was an accident in Newtonian theory.[69]

(*iv*) *The elongation of the inferior planets is bounded and the (calculated) periods of the planets strictly increase with their (calculable) distances from the Sun.*

In order to account for the fact that Venus' elongation from the Sun is bounded, Ptolemy resorted to the arbitrary assumption that the Earth, the Sun and the center of Venus' epicycle remain collinear. It follows by Zahar's criterion of empirical support that Venus' bounded elongation lends little or no support to the Ptolemaic system.

[68]In the first chapter of *De Revolutionibus* Copernicus explains that this assumption is part of accepted background knowledge, common both to Ptolemy and Copernicus.

[69]Zahar [1973], pp. 226-7.

Copernicus for his part needs no *ad hoc* assumptions. His theory implies that a planet is inferior if and only if its elongation is bounded. Hence Venus is an inferior planet. Similarly Mars is a superior planet because its elongation is unbounded. *This hypothesis is independently testable* as follows. Let P denote any planet—superior or inferior—and let T_P be the period of P, T_E the period of the earth (*i.e.* one year), and t_P the time interval between two successive retrogressions of P. Then a simple calculation shows that, since retrogression occurs when the planet passes the Earth, the following relations between T_P, T_E and t_P hold good:

$$\frac{1}{T_P} - \frac{1}{T_E} = \frac{1}{t_P} \qquad \text{if } P \text{ is an inferior planet; and}$$

$$\frac{1}{T_E} - \frac{1}{T_P} = \frac{1}{t_P} \qquad \text{if } P \text{ is a superior planet.}$$

Note that T_P is measurable and that T_E is known and equal to one year. Thus the above equations enable us to calculate T_P.

In the case of a superior planet it follows from the second equation that $1/T_E > 1/t_P$; *i.e.* $T_E < t_P$. Hence we can predict that if a planet's elongation is unbounded, then the interval between two successive retrograde motions of the planet is greater than one year. This is a novel—though well known—fact predicted, and therefore "explained," by the Copernican program. It gives support to Copernicus' program without giving support to Ptolemy's. Neugebauer has a point in claiming that "the main contribution of Copernicus to astronomy [was] the determination of the absolute dimensions of our planetary system."[70]

Once he has obtained the periods of the planets Copernicus goes on to calculate their distances from the Sun. One such method of calculation is described by Kuhn.[71] The period of a planet strictly increases with its distance from the Sun, *i.e.* from the origin of the Copernican frame of

[70]Neugebauer [1968].

[71]Kuhn [1957], p. 176.

reference. This is consistent with accepted background knowledge. The Ptolemaic program, as such, has no place for planetary distances but only for the angular motions of the planets. *Hence the determination of planetary distances represents excess content of Copernicus' theory over Ptolemy's.*

Ptolemaic astronomy too may be made to yield planetary distances by laying down arbitrarily that

$$\frac{r}{R} = \frac{\text{radius of epicycle}}{\text{radius of deferent}} = \text{distance of an inferior planet from the Sun (the Earth's distance being taken as unit)}$$

$$\frac{R}{r} = \text{distance of a superior planet.}$$[72]

One can then use these equations to calculate the mean distances of the planets from the Earth. But these equations are grafted in an *ad hoc* way onto the Ptolemaic program. And it is found, that, although Mercury, Venus and the Sun have approximately the same period, their distance from the Ptolemaic origin, *i.e.* from the Earth, differ widely; and this contradicted the background hypothesis commonly held at the time that the period grows with the distance from the fixed center to which the motion is referred.

A historical thought-experiment may illuminate the corroborating strength of these facts. Let us imagine that in 1520—or before—all we knew about the heavens was that the Sun and the planets move periodically relatively to the earth; but our records, because of, say, the cloudy Polish skies, were so scanty that stations and retrogressions have never been experimentally ascertained. Because of his Sun-worship and his belief in the Platonic heuristic, astronomer *X* proposes the basic Copernican model. Astronomer *Y* who adheres to the Platonic heuristic but also to Aristotelian dynamics puts forward the corresponding geocentric model: the Sun and the planets move uniformly on circles

[72]Neugebauer [1968]. One may use also the Aristotelian "doctrine of plenitude" to arrive at distances; but this doctrine is again heuristically *ad hoc,* besides being both false *and,* within Ptolemy's program, unfalsifiable.

centered around the Earth. If so, then *X*'s theory would have been dramatically confirmed by observations carried out later on the coasts of the Mediterranean. The same observations would have refuted *Y*'s hypothesis and compelled *Y* to resort to a series of *ad hoc* manouevres (assuming that *Y* was not so disheartened as to instantly abandon his program).

Zahar's account thus explains Copernicus' achievement as constituting genuine progress compared with Ptolemy. The Copernican Revolution became a great *scientific revolution not because it changed the* European *Weltanschauung*, not—as Paul Feyerabend would have it—*because* it became also a revolutionary change in man's vision of his place in the Universe, but simply because it was *scientifically* superior. It also shows that there were good objective reasons for Kepler and Galileo to adopt the heliostatic assumption, for already Copernicus' (and indeed, Aristarchus') *rough model* had excess predictive power over its Ptolemaic rival.[73]

Why was then Copernicus not content with his *Commentariolus*? Why did he work for decades to complete his system before having it printed? Because he was not content with mere progress for his program, but wanted actually to *supersede* Ptolemy's; *i.e.* instead of merely predicting "novel" facts which Ptolemy's system had not "predicted," he wanted to explain *all* the true consequences of the Ptolemaic theory. This is why he had to write *De revolutionibus*. But it turned out that apart from his initial successes, Copernicus could save all the Ptolemaic phenomena only in an *ad hoc* and, in its dynamical aspects, very unsatisfactory, way.[74] So Kepler and Galileo took off from the *Commentariolus* rather than from *De revolutionibus*. They took off from the point where the steam ran out of the Copernican program. Because of the initial success of the rough model and the degeneration of the full

[73]Note that this statement does not say whether and why Kepler and Galileo actually became "Copernicans."

[74]Zahar's concept of heuristic progress can, of course, be regarded as an *objective* (and "positivist") explication of "simplicity" without running into the inconsistencies of naive simplicists like the ones discussed above, in Section *2*.

program Kepler discarded the old heuristic and introduced a revolutionary new one, based on the idea of heliocentric dynamics.[75]

Let me end with a trivial consequence of this exposition, which I hope at least *some* of you will find outrageous. Our account is a narrowly internalist one. No place in this account for the Renaissance spirit so dear to Kuhn's heart; for the turmoil of Reformation and Counter-reformation, no impact of the Churchman; no sign of any effect from the alleged or real rise of capitalism in the 16th century; no motivation from the needs of navigation so much cherished by Bernal. The whole development is narrowly internal; its progressive part could have taken place at any time, given a Copernican genius, between Aristotle and Ptolemy or in any year, say, after the 1175 translation of the *Almagest* into Latin, or for that matter, by an Arab astronomer in the 9th century. External history *in this case* is not only secondary; it is nearly redundant.[76] Of course, the system of patronage of astronomy through Church sinecures played a role; but studying it will contribute nothing to our understanding of the Copernican scientific revolution.

[75]This pattern is not unique: after all, Bohr's old quantum theory was abandoned soon after it was accepted and de Broglie's new quantum theory took off from his first crude model rather than from Sommerfield's and others' sophisticated calculations.

[76]Of course, our analysis implies that there is a very important, but purely "external," question to be answered in socio-psychological terms: why did the Copernican Revolution take place when it did, rather than at any other time since Ptolemy? But the answer, if indeed it is possible to give one, will not affect the *appraisal* that we have attempted here. This is a good example of how internal (methodological) history can define what are the important external problems, and why it is therefore of primary importance!

REFERENCES

Achinstein, P. [1970]: "Inference to Scientific Laws," in R. Stuewer (*ed.*): *Historical and Philosophical Perspectives in Science, Minnesota Studies in the Philosophy of Science, 5,* pp. 87-111.

Agassi, J. [1962]: *Towards an Historiography of Science.*

Born, N. [1949]: *Natural Philosophy of Cause and Chance.*

Dorling, J. [1971]: "Einstein's Introduction of Photons: Argument by Analogy or Deduction from the Phenomena?", *The British Journal for the Philosophy of Science, 22:* 1-8.

Feyerabend, P.K. [1964]: "Realism and Instrumentalism: Comments on the Logic of Factual Support," In N. Bunge (*ed.*): *The Critical Approach,* pp. 280-308.

Feyerabend, P.K. [1972]: "Von der beschränkten Gültigheit methodologischer Regeln," in Bubner, Cramer and Wiehl (*eds.*): *Dialog als Methods,* pp. 124-71.

Feyerabend, P.K. [1974]: *Against Method.*

Galileo [1615]: Letter to the Grand Duchess.

Gingerich, O. [1973]: "The Copernican Celebration," *Science Year,* pp. 266-267.

Gingerich, O. [1975]: " 'Crisis' versus Aesthetic in the Copernican Revolution," in A. Beer (*ed.*): *Vistas in Astronomy, 17.*

Hanson, N.R. [1973]: *Constellations and Conjectures.*

Jeans, J. [1948]: *The Growth of Physical Science.*

Johnson, F.R. [1959]: "Commentary on Derek J. de S. Price," in M. Clagett (*ed.*): *Critical Problems in the History of Science,* pp. 219-21.

Kamlah, A. [1971]: "Kepler im Licht der modernen Wissenschaftstheorie," in Lenk (*ed.*): *Neue Aspekte der Wissenschaftstheorie,* pp. 205-20.

Kepler, J. [1604]: *Ad Vitellionem Paralipomena.*

Kuhn, T.S. [1957]: *The Copernican Revolution.*

Kuhn, T.S. [1962]: *The Structure of Scientific Revolutions.*

Kuhn, T.S. [1963]: "The Function of Dogma in Scientific Research" in A.C. Crombie (*ed.*): *Scientific Change,* pp. 347-69.

Kuhn, T.S. [1971]: "Notes on Lakatos," in R.S. Buck and R.C. Cohen (*eds.*): *Boston Studies in the Philosophy of Science, 8*: 137-146.

Lakatos, I. [1968*a*]: "Changes in the Problem of Inductive Logic," in I. Lakatos (*ed.*): *The Problem of Inductive Logic,* pp. 315-417.

Lakatos, I. [1968*b*]: "Criticism and the Methodology of Scientific Research Programmes," *Proceedings of the Aristotelian Society, 69*: 149-86.

Lakatos, I. [1970]: "Falsification and the Methodology of Scientific Research Programmes," in I. Lakatos and A.E. Musgrave (*eds.*): *Criticism and the Growth of Knowledge,* pp. 91-195.

Lakatos, I. [1971*a*]: "History of Science and its Rational Reconstructions," in R.C. Buck and R.S. Cohen (*eds.*): *Boston Studies in the Philosophy of Science, 8*: 91-136.

Lakatos, I. [1971*b*]: "Popper zum Abgrenzungs-und Induktionsproblem," in H. Lenk (*ed.*): *Neue Aspekte der Wissenschaftstheorie,* pp. 75-110. It appeared in English as "Popper on Demarcation and Induction," in P. Schilpp (*ed.*): *The Philosophy of Karl Popper,* 1974.

Lakatos, I. [1974]: "The Role of Crucial Experiments in Science," in *Studies in the History and Philosophy of Science, 4*: 357-373.

Neugebauer, O. [1958]: *The Exact Sciences in Antiquity.*

Neugebauer, O. [1968]: "On the Planetary Theory of Copernicus," *Vistas in Astronomy, 10*: 89-103.

Pannekoek [1961]: *A History of Astronomy.*

Polanyi, M. [1958]: *Personal Knowledge.*

Polanyi, M. [1966]: *The Tacit Dimension.*

Popper, K.R. [1935]: *Logik der Forschung.*

Popper, K.R. [1963]: *Conjectures and Refutations.*

Popper, K.R. [1972]: *Objective Knowledge.*

Price, D.J. de S. [1959]: "Contra-Copernicus: a Critical Re-estimation of the Mathematical Planetary Theory of Ptolemy, Copernicus, and Kepler," in M. Clagett (*ed.*): *Critical Problems in the History of Science,* 1959, pp. 197-218.

Ravetz, J. [1966*a*]: *Astronomy and Cosmology in the Achievement of Nicolaus Copernicus.*

Ravetz, J. [1966*b*]: "The Origins of the Copernican Revolution," *Scientific American, 215*: 88-98.

Santillana, G. de [1953]: "Historical Introduction" to Galileo Galilei: *Dialogue on the Great World Systems.*

Smith, A.: "The Principles which Lead and Direct Philosophical Inquiries Illustrated by the History of Astronomy," in Adam Smith: *Essays on Philosophical Subjects,* edited by Dugald Stewart, 1799. (Adam Smith wrote this essay some time before 1773, when he mentioned it in a letter to David Hume.)

Toulmin, S. [1972]: *Human Understanding, 1.*

Urbach, P. [1974]: "Progress and Degeneration in the 'IQ Debate,' " *The British Journal for the Philosophy of Science, 25,* March.

Watkins, J.W.N. [1958]: "Confirmable and Influential Metaphysics," *Mind, 67*: 344-365.

Watkins, J.W.N. [1970]: "Against Normal Science," in I. Lakatos and A.E. Musgrave, (*eds.*): *Criticism and the Growth of Knowledge,* pp. 25-37.

Westman, R.S. [1972]: "Kepler's Theory of Hypothesis and the 'Realist Dilemma,' " *Studies in History and Philosophy of Science, 3*: 233-64.

Worrall, J. [1976*a*]: "Thomas Young and the 'Refutation' of Newtonian Optics," in C. Howson (*ed.*): *Research Programmes in Physics.* Cambridge University Press.

Worrall, J. [1976*b*]: "The Victory of Wave Optics," forthcoming.

Zahar, E.G. [1973]: "Why Did Einstein's Research Programme Supersede Lorentz's?," *The British Journal for the Philosophy of Science, 24*: 95-123 and 223-262.

COMMENTARY

Stephen E. Toulmin
University of Chicago

The general issues raised by Imre Lakatos and Elie Zahar in this present paper are of great interest and significance for philosophy of science, as well as for our historical understanding of Copernicus and his successors in planetary astronomy. In this comment, I propose to take up two of these general issues. One of them has to do with the extraordinary influence—amounting at times to domination—that the example of planetary astronomy has exerted on the philosophy of science, and on the methodology of other sciences, during the centuries since Newton. The other has to do with the "agonizing reappraisal" which I detect in Lakatos' own philosophical account of science as a result of his collaboration with Zahar.

In discussing this second topic, I am tempted to follow Imre Lakatos' own example. A few years ago he published what purported to be an historical analysis of Karl Popper's changing views about the philosophy of science. He divided Popper's career into three phases and labelled these, with solemn scholarly seriousness, "Popper_1," "Popper_2" and "Popper_3." But, on closer inspection, these labels proved to be a trifle disingenuous. As the attentive reader soon discovered, the position Imre called "Popper_3" was not one that Sir Karl had ever, in real life, brought himself to adopt. Rather, it turned out to be one that Imre Lakatos thought that Popper *ought* (if he understood his own intellectual development) to adopt: indeed, it was in fact *Lakatos' own view* or, to speak more exactly, his own *current* view—what we may conveniently label "Lakatos_2." That being so (I say) it is tempting to treat the historical development of Imre's philosophy of science in the same way; and, in that case, his present paper on Copernicus would have to be construed as marking the transition from "Lakatos_2" to "Lakatos_3." Yet to do this would again be a bit disingenuous; for, in fact, "Lakatos_3"—the position towards which Imre will, in my view, be irreversibly driven as he follows out the consequences of Elie Zahar's emendations—is *my own* current view!

To say this is not just to display egotistical bad manners. Instead, my aim in this comment is to draw attention to one fundamental philosophical point, which has always previously divided the two of us, but about which Imre is here at last showing signs of changing his mind. In many respects, our two positions have always had more in common than either of us was at first ready to admit. (What I call "intellectual strategies" he happens to call "research programs," what he calls "progressive" I happen to call "fruitful," and so on; but at most points our accounts of scientific change run closely parallel.) The one fundamental difference between us—and it is a point that deeply divides contemporary philosophers of science—is the following. Imre Lakatos, like some others who have come to the subject from the philosophy of mathematics, has always firmly insisted on analysing the intellectual content of any science in a predictivist manner, as legitimately representable only by a formal and systematic network of propositions. By contrast, those of us who have come, rather, from physics and history tend to reject "predictive success" as a supreme (or even, as a major) index of scientific merit, and regard the intellectual content of any science as representable—more generally—by the totality of its explanatory concepts and procedures, as applied to handle typical problems in its domain of concern. In a few special and restricted cases (we would happily concede) these explanatory procedures may *include* the construction of coherent systems of general propositions interlinked by formal relations. But this is not *in general* the case; and, even where it is, formal systematization is merely one instrument of scientific explanation and understanding—one among others, and one of limited and conditional applicability—certainly not its *essence*.

Thus, Lakatos' central claim, that the essential achievement of a good new theory is that it "predict (or anticipate) novel facts" is open to attack for being either over-simplified, or ambiguous, or both. The actual business in which we find working scientists engaged in practice—the business of developing and deploying new explanatory concepts and procedures, so as to broaden and enrich our intellectual command over the relevant domain of scientific understanding—may include some bits of authentic "predicting" from time to time. But it can be boiled down to yield a purely predictivist essence, only at the price of evaporating away all the distinctive variety and complexity of the natural sciences: the variety and complexity that mark the differences, both between different phases in the development of a single science, and between different sciences at any one time. And this variety and complexity are of a sort that is of central importance, even for Imre Lakatos' own philosophical purposes in this very paper on Copernicus.

However, the emendations which Zahar has apparently induced Lakatos to accept in his "methodology of scientific research programs" do now put him in a position to make a great leap forward. For (1) they at last demonstrate just how sloppy and misleading the philosophical use of terms like "predict," "prediction" and "predictive" has become. In addition (2), once this sloppiness is corrected, they compel us to restore to Lakatos' account that very variety and complexity which his "predictivism" has previously concealed. And, this done, we may go on (3) to raise again, in Lakatos' own terms, all those problems that compel us to pay attention to the changing problematics and intellectual strategies (or "research programs") of the different natural sciences *within their own specific intellectual and historical niches.*

At this point, let me stand back from my dispute with Imre, and make a more general point about the influence of astronomy on the philosophy of science. For it is the example of planetary astronomy (I believe) that has encouraged the rise of predictivism, for reasons first pointed out by Henri Poincaré in his discussion of "inertial frames," in *Science and Hypothesis.* The development of Western scientific thought, Poincaré argued, has been heavily influenced by the fact that the sky is not forever covered with clouds, so that men have been able to observe and keep track of the Sun, the Moon and the planets, and have used their motions as the type-example of a "natural phenomenon." This fact has been, at one and the same time, the source both of triumph and of tragedy. On the one hand it made the scientific task of establishing a satisfactory general analysis of kinematics and dynamics—the road which began with Copernicus and ended with Newton—much easier than it would otherwise have been, and that analysis was certainly a scientific triumph. But it has been a source of tragedy, also, because it has encouraged, far beyond the bounds of planetary astronomy, the rise of unrealistic and irrelevant expectations about what *any* science could and should achieve—viz. the capacity to make predictions, in the straightforward, primary sense of "forecasts"—and in this way has saddled other sciences with faulty aims and methodologies.

These wider expectations were unrealistic and irrelevant, because astronomical systems such as the solar system are pretty well unique among all natural objects and systems. They represent perhaps the only empirically indentifiable systems in existence whose individual constituents are effectively shielded from the action of *all but one single* physical force of known magnitude; so that the movements of these constituents can indeed be predicted—that is to say, *forecast*—with almost categorical certainty, for thousands of years in advance. Strictly speaking, of course, we should make all our predictions weakly conditional, even in the planetary case: e.g. "Venus will rise at 5:27 p.m. P.S.T. on January 27th, in the year 3,000, *provided that* in the meantime no unknown heavenly body cataclysmically disturbs the solar system from outside, *provided that* men do not meanwhile decide to rocket-propel the Earth into a different orbit, etc., etc." Still, the impression created by Newton's initial predictive successes has been hard to overcome, and scientists in other fields have manfully but misguidedly set themselves to do for their own subjects what Newton—in their view—did for dynamics and astronomy: namely, develop techniques for generating categorical forecasts about systems of other kinds. If Plato successfully sold men the example of geometry as illustrating the ideal *structure* for human knowledge, the example of Newtonian planetary theory was thus seen as illustrating its proper *function*, and the damaging effects of this astronomical model have remained with us to the present day.

In practice, physical scientists themselves quickly and quietly gave up the ambition to make categorical forecasts of naturally-occurring events in general. They soon came to realize that (to use Kant's vivid imagery) Nature will never testify to us spontaneously, but only if we put her on the rack and extort answers to our questions. By the time they reached this level of sophistication, however, the idea that "predicting" is the central task of all science was so firmly embedded that they merely went on talking about experimental measurements and observations conducted in wholly artificial situations in the same terminology—as concerned with much more conditional "predictions." In this way, they obliterated the distinction between categorical forecasts—e.g. "The Moon will be totally eclipsed at Los Angeles at 8:34 p.m. P.S.T. on such-and-such a day"—and highly conditional guesses at the likely outcomes of elaborately contrived experiments. And this in turn helped to perpetuate the original confusion, by which the intellectual task of building up a well-founded account of the general processes going on in the world around us was treated, for philosophical purposes, as though it were a superior form of *divination*.

The extraordinary persistence of this predictivist mode of thought is apparent if we return to the text of Imre Lakatos' paper. In summarizing his original "methodology of scientific research programs," he writes:

> A research program is either progressive or degenerating. It is *theoretically progressive* if each modification leads to new unexpected predictions and it is *empirically progressive* if at least some of these novel predictions are corroborated. . . . The supreme example of a progressive program is Newton's. It successfully anticipated novel facts like the return of Halley's comet, the existence and the course of Pluto and the bulge of the earth. . . . One research program *supersedes* another if it has excess truth content over its rival, in the sense that it predicts all that its rival truly predicts and some more besides.

Two things in this passage need to be commented on. On the one hand: if we were to take this account at its most strict and literal face-value, Lakatos' "methodology of scientific research programs" would then be applicable only to research programs in planetary astronomy—or, more exactly, to research programs on the computative techniques of forecasting used in the preparation of astronomical tables—since only there can one properly give "predicting" the crucial status Lakatos allots to it. On the other hand, if we are to follow Imre in generalizing his account, and extending it to other branches of natural science, we shall land ourselves in precisely the confusions that I was talking about a moment ago. For notice Lakatos' own examples. To predict the return of Halley's comet would have been to make a straightforward astronomical *forecast*

—though Edmund Halley himself had of course done this by a simple extrapolation, without benefit of Newton's dynamics, some years before the *Principia* was written—to infer "the phases of Venus" or "the existence and the course of Pluto" was to propose a previously unforeseeable *discovery;* while to relate the flattening of the Earth at its poles to the centrifugal effects of its rotation involved using Newton's general theory to render a hitherto-mysterious geodesic observation newly *intelligible.* And a philosophy of science that debars itself from distinguishing between categorical forecasts, conditional discoveries and novel explanatory interpretations is (I submit) incoherent.

But that is all in the past—a defect in the original "methodology of scientific research programs," i.e., "Lakatos$_2$," that is now only history. What pleases me about Imre's present paper, by contrast, is his readiness to move away from that position to a new one: one that implicitly concedes the distinctions I have been discussing here, and at any rate begins on the task of facing their consequences. For, as Elie Zahar argues—and Lakatos goes along with him—the phrase "predicting novel facts" is misleading even in the case of Copernican astronomy. Almost without exception, the "facts" that supported Copernicus' position were not "novel." Rather, his intellectual strategy had the virtue of connecting together, in an intelligible way, a whole string of features of the planetary system that had either been left unaccounted for, or had at best been handled separately and one at a time, by Ptolemy and his successors. Copernicus thus took phenomena that were already well-attested, connected them together by novel relations, and so made them jointly intelligible as they had never been before. In a word, the central aim of Copernicus' whole scientific enterprise was not to *predict* "novel" facts which Ptolemy's system had not "predicted." (Notice the significant quotation-marks in this phrase.) On the contrary, it was to *explain,* by a single consistent treatment, all the phenomena that Ptolemy had treated separately, and more besides.

That's enough for now. The ambiguities inherent in any analysis of scientific merit in terms of such notions as "prediction" and "truth-content" only have the effect of submerging the fundamental problems about scientific change; and, once we replace the word "prediction" by "explanation," those problems are free to surface again. If Zahar is finally dislodging Lakatos from his commitment to a predictivist account of the function of science, he is accordingly to be commended for plucking Imre from the burning. As to the question, what kinds of things are, or have been, accepted or acceptable as "explanations" in one natural science or another, at one stage of its development or another, this *clearly* lands us back up to the neck in all that variety and complexity which single-minded predictivists would so much like to avoid. In particular, it lands us in a position at which we are compelled to reappraise the particular processes and problematics of conceptual change in science against their specific historical backgrounds.

To state the position Lakatos has hitherto rejected, but will now have to take more seriously: the intellectual claims to be satisfied by any theoretical innovation in science which is to justify itself here and now (or there and then) are too plural and specific to be distilled into any single, timeless and universal index of "merit." Rather, we might say, the fundamental history of any particular science is the history of men's experience in exploring the applicability and consequences of different intellectual strategies in that particular field of enquiry. And, if we ask ourselves the question, what features have indeed been legitimately *perceived as* "merits" in this science or that, at this time or that, in relation to some particular strategy, problematic or research program, only the most starry-eyed among us will imagine that this question can be given a single, simple and universal answer.

I realize that Imre will find even this conclusion an unwelcome one. The idea that the problematics of conceptual change in natural science involve such inescapable subtleties and conflicts will probably offend his mathematical soul. But to those of us who long ago rejected the geometrical model of a science as a formal network of propositions, and who came to see the problems of conceptual choice in science as arising out of complex and varied historico-intellectual contexts, there need not be the same sense of loss. For there is no reason to expect that terms like "simple" and "coherent," "fact" and "observation," "predict" and "explain," will be any easier to apply unambiguously to the actual problems of science *in all contexts,* than (say) terms like "right" and "responsibility," "consequence" and "public good," "excuse" and "justify," are to apply unambiguously to the actual problems of law and practical life *in all contexts.* On the contrary: acknowledging this complexity in the problematics of natural science is merely the beginning of wisdom. That is where the real work has to start. So, if Elie Zahar has finally induced Imre Lakatos to concede the difficulties built in to the terms "predict," "predictive" and "prediction," and to begin taking the concept of "explanation" seriously in its own right, I for one can happily say to him, "Welcome to the Real Problems!"

Index